NONLINEAR SYSTEM DYNAMICS

W. Richard Kolk
and
Robert A. Lerman

VNR VAN NOSTRAND REINHOLD
New York

Softcover reprint of the hardcover 1st edition 1992

Library of Congress Catalog Card Number 90-24187

Manufactured in the United States of America

Published by Van Nostrand Reinhold
115 Fifth Avenue
New York, New York 10003

Chapman and Hall
2-6 Boundary Row
London, SE1 8HN, England

Thomas Nelson Australia
102 Dodds Street
South Melbourne 3205
Victoria, Australia

Nelson Canada
1120 Birchmount Road
Scarborough, Ontario M1K 5G4, Canada

16 15 14 13 12 11 10 9 8 7 6 5 4 3 2 1

ISBN-13: 978-1-4684-6496-2 e-ISBN-13: 978-1-4684-6494-8
DOI: 10.1007/978-1-4684-6494-8

Library of Congress Cataloging-in-Publication Data

Kolk, W. Richard.
Nonlinear system dynamics/W. Richard Kolk and Robert A. Lerman.
p. cm.
Includes bibliographical references and index.

ISBN-13: 978-1-4684-6496-2

1. Nonlinear control theory. I. Lerman, Robert A. II. Title.
QA402.35.K65 1989
629.8'312—dc20 90-24187
CIP

To Janet . . . sine qua nil.

Dick

To Ellen

Our life together continues to be a series of delightful experiences.

Bob

To Joanne, Kenneth, Beth, Carla, and Andrea

The more you know, the better prepared you are for life's adventures.

Dad

Contents

Preface

Engineers, scientists, and applied mathematicians are habitually curious about behavior of physical systems. More often than not they will model the system and then analyze the model, hoping to expose the system's dynamic secrets. Traditionally, linear methods have been the norm and nonlinear effects were only added peripherally. This bias for linear techniques arises from the consummate beauty and order in linear subspaces and the elegance of linear independence is too compelling to be denied. And the bias has been, in the past, fortified by the dearth of nonlinear procedures, rendering the study of nonlinear dynamics untidy. But now a new attractiveness is being conferred on that nondescript patchwork, and the virtue of the hidden surprises is gaining deserved respect.

With a wide variety of individual techniques available, the student and the engineer as well as the scientist and researcher, are faced with an almost overwhelming task of which to use to help achieve an understanding sufficient to reach a satisfying result. If linear analysis predicts system behavior sufficiently close to reality, that is delightful. In the more likely case where nonlinear analysis is required, we believe this text fills an important void.

We have tried to compile and bring some order to a large amount of information and techniques, that although well known, is scattered. We have also extended this knowledge base with new material not previously published. Included in these new procedures and techniques are Chapter 2 (finding analytic solutions to nonlinear differential equations), Chapter 7 (investigations into the stability of digital recursion and understanding the events of numerical analysis which lead to higher-order furcations, chaos, etc.), and Chapter 9 (a tutorial which shows how to use the desk top personal computer to analyze systems).

Finding a closed form solution to nonlinear differential equations has long been a study in frustration. In Chapter 2 we extend the class of equations with known solutions by use of a variety of transformations. Relationships between the known solutions to linear and nonlinear differential equations and those of derivable classes of other nonlinear differential equations are developed yielding a means of solving these new equations.

There is no guarantee that the appropriate transformation exists (it is, after all, nonlinear), so the analyst often turns to digital simulation, in the hope that it will adequately approximate the solution. A tutorial is supplied in Chapter 9 which enables the modeler to simulate his system on a spread-sheet-equipped personal computer. It presents methods of approximating ordinary differential equations with finite difference equations which are then easily recursed to provide solutions which are rapidly plotted.

A topic of increasing interest which naturally evolves from recursions is chaos. In Chapter 7 we have presented some of its aspects, notably the stability of recursions starting with Jacobi's iteration of linear algebraic equations, then the nonlinear Mechanic's rule for extracting square roots, followed by the furcations in the logistics map, and finally, recursions of Lin's method for extracting quadratic factors from polynomials, showing convergence, divergence, furcations and essential chaos.

Just as there is a little order in the nonlinear world, the order of our presentation is not sacrosanct. Having taught this material repeatedly in courses involving Nonlinear Controls or System Dynamics, we are comfortable with this order, but leave it to the reader to select his own. Only an introductory knowledge of differential equations and linear controls is presumed, and, the material can be used in either undergraduate or graduate curricula.

The completion of the manuscript required more sacrifice and pain than we'd ever imagined at its initiation. We could not have finished without the unflinching support of our wives, Janet Kolk and Ellen Lerman, the encouragement and helpful criticism of Dr. Dan DeBra of Stanford University, and Professor George Murphy of Lowell University, nor the unselfish assistance of Mrs. Lee Brown in preparing the manuscript. We thank them sincerely, one and all.

W. RICHARD KOLK
Portland, Connecticut

ROBERT A. LERMAN
West Hartford, Connecticut

Introduction
Linearity—A Shangri-la

I.1 WHY LINEARITY?

Linear theory is a formidable modeling tool which is unmatched by its nonlinear counterpart. While, as engineers, we are fully aware of Mother Nature's ubiquitous nonlinearity, we are nevertheless lured by the power of eigenvalues to describe a system's dynamics. Their compact explanation of an aircraft's Dutch Roll or the attenuation in an operational amplifier is satisfying and conclusive. That they are unique to the linear world and do not exist in Mother Nature's Kingdom is something we tend to overlook, making us susceptible to the illusion that her Kingdom is truly linear. As Arnold Tustin[1] expresses it . . .

> . . . closed sequence systems are delightfully simple to understand and—even more important—very easy to handle in exact mathematical terms. Because of this, most introductory accounts of control systems either brazenly or furtively assume that all such systems are linear . . ."

Linearity invades much of our daily life. The change in our pocket is the same no matter how the coins are counted. If an aspirin is good, then two are better where we vaguely equate better to "twice-as-good." There is, of course, a loose contradiction that the whole bottle can be harmful, but this is not considered a rational option. Real and complex numbers add like change, as do vectors. The world abounds in linear things, but occasionally we encounter unusual quantities like nonplanar Eulerian angles which "add" differently, according to their order. The Gas Law is an awkward relation between pressure, density, and temperature, and Maxwell's equations are even more challenging. These are, however, a distinct minority, or so it would seem, because by and large our formal engineering training follows the safe path of linear analysis.

---

[1] *Automatic Control*, A Scientific American Book, Simon and Schuster, 1955.

1.2 FROM WHENCE LINEARITY

The linear problem has evolved as dominant in both algebra and differential equations because the proposition of linearity bears a myriad of analytic methods that, proof upon proof, has produced a vast body of analysis.

Chronologically, algebra developed before differential equations, yet it is interesting that linear algebra, per se, is a relatively recent development, while linear ordinary differential equations have been studied for 300 years. Linear algebraic equations were not intentionally avoided prior to 1950; on the contrary, most of what then rapidly developed as linear algebra was well-known hundreds of years before, but it resided in that amorphous aggregate known simply as algebra. World War II fostered an interest in mini-max problems, Von Neuman developed game theory, and computers made linear pogramming practical. By then Heaviside's operational calçulus had been clothed in the respectability of LaPlace Transforms, which reduced differential equations to polynomials. These factors were recognized by the mathematics community, which responded by sifting out the linear part of what was already available and created linear algebra.

1.3 WHAT IS LINEARITY?

Starting with common addition, a commutative binary rule of combination, any two quantities can be "added." Next, consider a special process, P, which can operate on either or both the quantities x and y. P can be a function, an operation, a transformation, or a mapping, but what makes it special is its capability to distribute over addition,

$$P(x + y) = P(x) + P(y)$$

This defines a linear process, and any process failing the equality is by default nonlinear. It follows that for a, a scalar, and scalar multiplication defined through the continuum of real numbers by successive additions,

$$P(ax) = aP(x)$$

An interesting observation is the provable fact that among all of the real continuous[2] algebraic functions, $f: R \to R$, imposing linearity on f; that is,

$$f(x + y) = f(x) + f(y) \qquad \text{for } x, y \in R$$

the only thing f can be is scalar multiplication, so,

$$f(x) = kx \quad \text{for } k \in R$$

Thus, the only continuous algebraic function permitted within the orderliness of linearity is a straight line through the origin. The dogma of linearity eschews quadratics, cubics, polynomials in general, and even straight lines with a dc-component. This abstract and perhaps stark approach to linearity points up its rather confining aspect by conceding that a single unique function from among an infinity of applicants is the sole survivor.

A more pleasing approach is usually followed, in which some as yet unknown function of specific variables is hypothesized. With little justification, the function is assumed to have a Taylor series expansion about an operating point (often selected to be the origin), displacements from which are argued to be sufficiently small to truncate their products and powers. The result is identical to the formal imposition of linearity, producing a linear combination of the displacements, which represents the total derivative with partials evaluated at the operating point.

An example is the aerodynamic force on a body which is influenced by many variables, such as air density ρ, speed v, temperature T, viscosity ϑ, Mach number M, and Reynolds number Re, which is expressed as,

$$\text{Force} = f(\rho, v, T, \vartheta, M, Re)$$

The procedure expresses the increment of force in terms of variable increments.

A linear combination of perturbations

$$\begin{aligned}\Delta\text{Force} = &(\partial f/\partial\rho)\Delta\rho + (\partial f/\partial v)\Delta v + (\partial f/\partial T)\Delta T + (\partial f/\partial\vartheta)\Delta\vartheta \\ &+ (\partial f/\partial M)\Delta M + (\partial f/\partial Re)\Delta Re + \cdots + [\text{Higher order} \\ &\text{terms (HOTs), i.e., products and powers of perturbations}]\end{aligned}$$

where the partials are evaluated as scalars at an operating point, a direct consequence of discarding HOTs.

I.4 THINGS LINEAR OR NOT LINEAR

Typically the linear process, P, is one of the familiar operations, normally occurring in well-behaved engineering problems, a few of which are listed in Table I.1.

TABLE I-1 Some Linear Processes

Scalar multiplication	$a(x + y) = ax + ay$
Differentiation	$d(x + y) = d(x) + d(y)$
Integration	$\int (x + y)\, dt = \int x\, dt + \int y\, dt$
Real (imaginary) part	$Re(x + y) = Re(x) + Re(y)$
Fourier (Laplace) transforms	$F(x + y) = F(x) + F(y)$

[2]Here "continuous" means differentiable and Taylor series-expandable, thus avoiding the undifferentiable functions of Bolzano and Weierstrasse, which have not yet appeared in engineering.

There are more nonlinear than linear processes but their abundance goes unnoticed simply because they are not as easily manipulated. A few are listed in Table I-2.

TABLE I-2 Some Nonlinear Processes

Circular (elliptic) functions	$\text{Sin}\,(x + y) \neq \text{Sin}\,(x) + \text{Sin}\,(y)$
Sign of	$\text{SGN}\,(x + y) \neq \text{SGN}\,(x) + \text{SGN}\,(y)$
Logarithm (exponential)	$\ln\,(x + y) \neq \ln\,(x) + \ln\,(y)$
Powers (roots)	$(x + y)^\alpha \neq (x)^\alpha + (y)^\alpha$
Magnitude (phase)	$\lvert x + y\rvert \neq \lvert x\rvert + \lvert y\rvert$

I.5 OUR SHANGRI-LA

One need not look far for evidence of Mother Nature's nonlinearity. Some of the greatest minds in the physical sciences have ratified that premise with the Gas Law, both Newton's second law and his Law of Gravitation, and Bernoulli's flow equation to name a few. That each expresses a form of nonlinearity is simply Her disregard of conformity.

But what if She were linear? Would it be helpful? As things stand, if our system doesn't work we can blame it on nonlinearity. Were it linear, however, we could analyze it, understand its function, and avoid frustration. And engineering would cease to exist because our Hiltonian Systems would all have known outputs, and malfunctions just couldn't occur.

In the following chapters, we investigate the nonlinear world to introduce a range of techniques which provide comfort, useful tools, and additional insights.

Chapter 1
Four Interesting Equations

1.0 THE EQUATIONS

Four differential equations, (i)–(iv), are presented to show that the consummate uniformity of linear equations is not shared by nonlinear equations

$$\dot{x} + x = 0 \tag{i}$$

$$\dot{x} + |x| = 0 \tag{ii}$$

$$|\dot{x}| + x = 0 \tag{iii}$$

$$|\dot{x}| + |x| = 0 \tag{iv}$$

which is illustrated by addressing three essential features of the solutions.

1. *The Method of Solution:* The solution to (i), while easily found today, was obscure to its discoverer who knew only algebra and differentiation. Take away the number e, integration, the logarithmic and exponential functions, and the method is no longer straightforward. And, whatever method is employed, its existence is implied.
2. *Existence of a Solution:* Whether a solution exists is (or should be) of fundamental importance in engineering. $1 + 2 = 0$ is not an equation but a contradiction, unless the addition is modulo 3. Likewise a differential "equation" can be a disguised but unforgiving contradiction as in (iv). Linearity bestows the guarantee that a solution exists—a fact that is often taken for granted, and inappropriately extended to nonlinear equations.
3. *Uniqueness of a Solution:* If one solution is found, can there be others? Or, is this the only solution? If so, the solution is unique as in (i) and (ii). Should there be more than one, as in (iii), the solutions are not unique. And if there is no solution, the question of uniqueness is moot as in (iv)

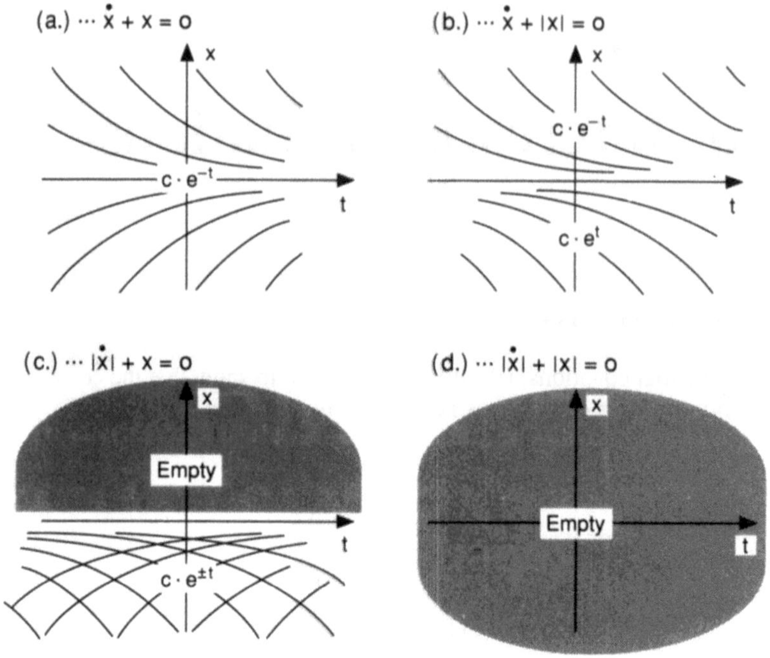

Figure 1-1. Solutions to four equations.

1.1 THE SOLUTIONS—A PREVIEW

The solutions to (i)–(iv) are shown in Fig. 1.1. The unifying relations are the real exponentials, $e^{\pm t}$, one or both of which satisfy the equation if it can be satisfied.

1.2 SOLVING EQUATION (i): $\dot{x} + x = 0$

Four methods of solution are presented, each with its own interpretation. It is helpful to slightly generalize the equation by defining a scaled time, T.

$$T = t/\lambda$$

so $dT = \lambda dT$, and thus $x' + \lambda x = 0$, where $x' = dx/dT$

1.2.1 Method 1—Separating Variables

Ordinary differentials, dx and dT, follow the rules of common algebra, yielding

$$dx/x = -\lambda dt$$

From elementary calculus, the left side integrates to the logarithm, the inverse of which is the exponential.

Remark: We have, here, presumed knowledge of both the logarithm and exponential, as well as integration. Were we ignorant of these (as was the first person to formulate the equation), it would be necessary to develop them from only those facts expressed by the equation.

The equation's *solution* is:

$$\ln x = -\lambda T + c_0$$

where $c_0 \equiv \ln x_0$, or

$$x = e^{-\lambda T + c_0} = ce^{-\lambda T}$$

where $c \equiv x_0 = e^{c_0}$.

This is plotted for $\lambda = +1$ [Figure 1.1(a)], where the entire solution plane x vs. t or x vs. T is included in the set of all solutions. There are no regions omitted, and so we confidently assert that the solution exists everywhere.

But is there another solution which the method neglected to expose? The general question of existence and the related question of uniqueness are discussed in Section 1.7, where it is established that this solution passes the sufficiency tests for existence and uniqueness throughout the entire x–T plane. A property of all linear ordinary differential equations with constant coefficients is the existence and uniqueness of solutions, making them a special target of analysis.

1.2.2 Method 2—Identity Under Differentiation

If in the generalized form of equation (i), we choose $\lambda = -1$, then $x' = x$, which asks for a function that reproduces itself under differentiation. Such a property is displayed by the exponential function, e^t.

It is noteworthy that this function can be designated as the operation which inverts the logarithm. That it is a shorthand notation for an infinite series is unnecessary to the solution of the equation. The important point is to recognize that differentiation is the exponential function's identity element.

1.2.3 Method 3—Recursive Approximation

Following Polya's prompt, we guess the answer. Thereafter the guess is refined using the equation to produce a better guess. Hopefully, successive iterations

TABLE 1-1 Solving $x' \lambda \cdot x$ Recursively

Step	Generally	Specifically
1. Guess solution, x_1	$x_1 = \text{guess}$	$x_1 = a = \text{scalar}$
2. Substitute	$x' = -\lambda x_1$	$x' = -\lambda a$
3. Integral, $\int x'\,dT = x$	$x = \int dx + c_1$ $= \int x_1 dT + c_1$	$x = -\int \lambda a dT + c_1$ $= -\lambda a T + c_1$
4. Force $x(t_0) = x_1(t_0)$ to determine the integration constant	$x(t_0) = x_1(t_0)$ determines c_1	$-\lambda a T_0 + c_1 = a$. Let $t_0 = 0 \Rightarrow c_1 = a$, so $x = a(a - \lambda T)$
5. Treat x as a new guess x_2	$x = x_2$; so, $x_2 = \int x_1 dT + c_1$	$x_2 = a(1 - \lambda T)$
6. Repeat 3, 4, and 5 recursively	$x_3 = \int x_2 dT + c_2$ $x_n = \int x_{n-1}\, dT + c_{n-1}$	$x_3 = a(1 - \lambda T + \frac{\lambda^2}{2}T^2)$ $x_m = a \sum_{m=0}^{n-1} (-\lambda T)^m/m!$
7. Assume convergence,* as $n \to \infty$	$\lim_{n\to\infty} x_n = \text{solution} = x,$	$x = a \sum_{k=0}^{\infty} (-\lambda T)^k/k!$

*Uniform convergence is due to Picard.

converge to the exact solution, although there is no guarantee. The approach is similar to Jacobi's recursion for approximating roots to linear algebraic equations as is discussed below, and is expanded in more detail in Chapter 7.

There are seven steps in the recursive approximation (Table 1-1), and the method is not restricted to first-order equations. Whether it converges toward a solution is never clear, a priori, and isn't necessarily easy to establish.

Observation: Term-by-term differentiation (allowed as it is a linear process) shows we have a solution

$$x' = a \sum_{k=1}^{\infty} - \lambda k \frac{(-\lambda T)^{k-1}}{k!} = -a \sum_{k=0}^{\infty} \frac{(-kT)^k}{k!} = -a\lambda x$$

Denote series as

$$e^{-\lambda T} \equiv \sum_{k=0}^{\infty} \frac{(-\lambda T)^k}{k!}$$

And note $e^{-\lambda t}$ has the necessary property of reproducing itself under differentiation.

1.2.4 Method 4—Graphical Approximation

Note that x and x' are linearly related since this is the assertion of our equation, $x' = -\lambda x$. Thus, plot x' vs. x as in Figure 1-2, generally called the phase portrait. This method is discussed in more detail in Chapter 5.

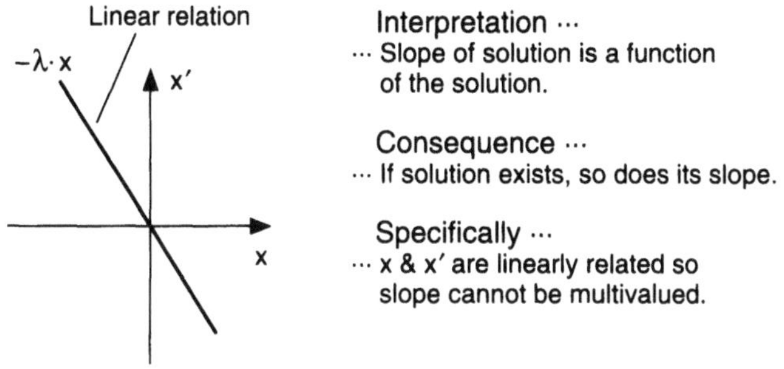

Figure 1-2. Phase portrait, $x' = \lambda x$.

The solution is generated by fairing a curve through the indicated slope at each point as shown in Figure 1-3.

Isoclines are loci in the solution plane, x vs. T, of constant slope. In general, for, $\dot{x} = f(x)$, let m_i = slope of ith locus, so $m_1 = f(x_1)$ and $x_1 = f^{-1}(m_1)$. In our case, $x = m_1$, so all horizontal lines are isoclines.

"Fairing" the smooth curve is an interpolation procedure corresponding to "solving the equation." It is therefore a form of graphical integration and can be reduced to a systematic process amenable to a computer program.

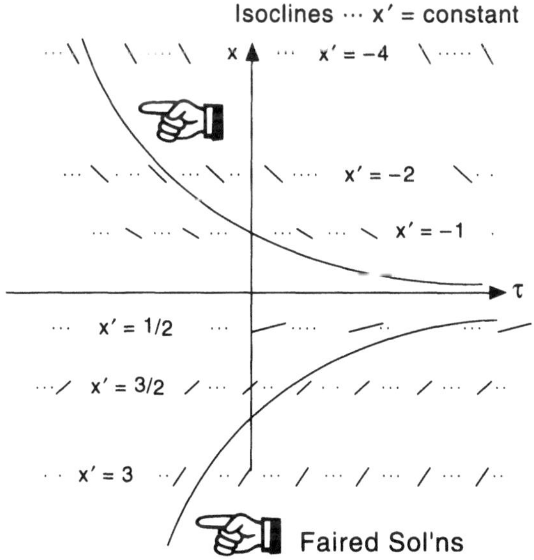

Figure 1-3. Solution faired through isoclines.

1.3 SOLVING EQUATION (ii): $\dot{x} + |x| = 0$

Inspection reveals that this equation can be interpreted as two equations: the first being

$$\dot{x} = -x, \qquad x \geq 0, \qquad \text{and no equation for } x < 0$$

and the second,

$$\dot{x} = x, \qquad x \leq 0, \qquad \text{and no equation for } x > 0$$

At $x = 0$, the origin is the admissible equation's solution. Thus, each equation has, in the x–t plane,

- A region of solution;
- A region of no solution (i.e., the "equation" is really a contradiction as it has no meaning in terms of the physical system under study).

The regions of existence and nonexistence do not overlap, except at the origin (Figure 1-4), so there is a solution everywhere (just as there is also a contradiction).

Clearly, the trivial solution $x = 0$ is also a solution, since if $x = 0$ everywhere, $dx/dt = 0$ everywhere as well.

Conjecture: There may be an equation whose solution plane contains either two (or more) solutions or no solution whatsoever. This is true of Clairaut's equation, where the demarcation is itself a solution not contained in the general solution, leading to the notion of a "singular" solution. Here $x = 0$ qualifies as a singular solution—that is, one which bounds the region of existence.

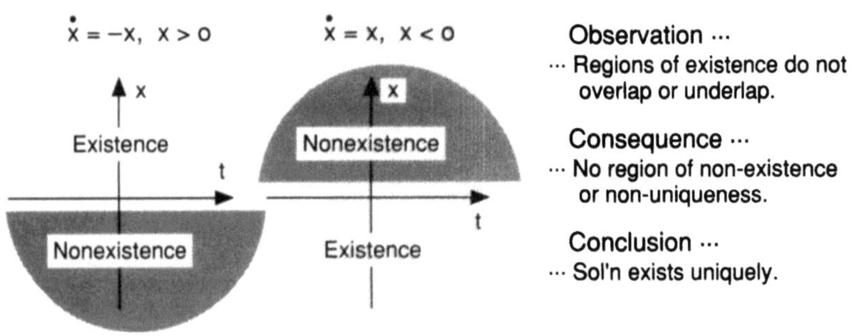

Figure 1-4. Solution space for $\dot{x} + |x| = 0$.

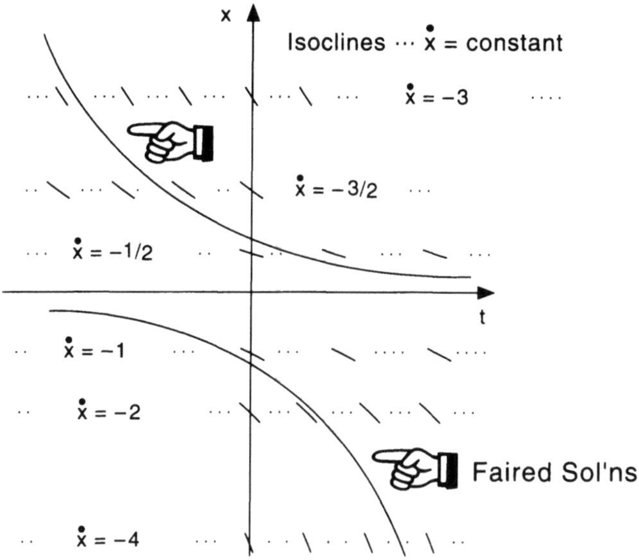

Figure 1-5. Solution faired through isoclines.

Solutions to $\dot{x} + |x| = 0$ are $x = ce^{\pm t}$, $x > 0$, $x < 0$ [Figure 1-1(b)], which can be approximated by constructing isoclines and fairing curves as illustrated in Figure 1-5.

The phase portrait relating x and $\dot{x}$ yields information relating to the existence and uniqueness of solutions. Figure 1-6 presents the phase portraits of (i) and (ii), where in both equations the slope $\dot{x}$ (or x') is defined for all x.

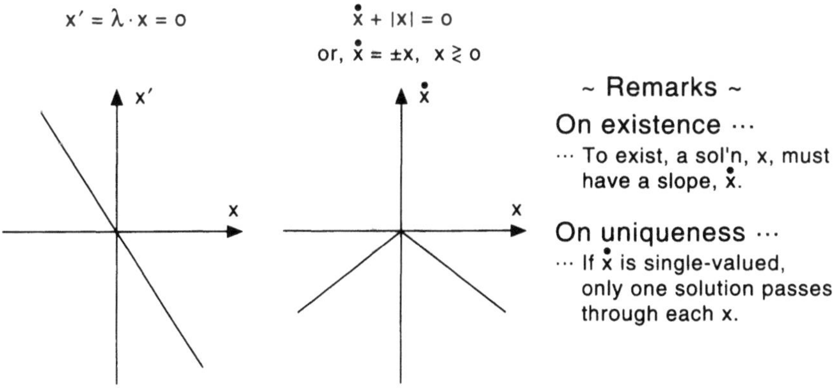

Figure 1-6. Phase portraits.

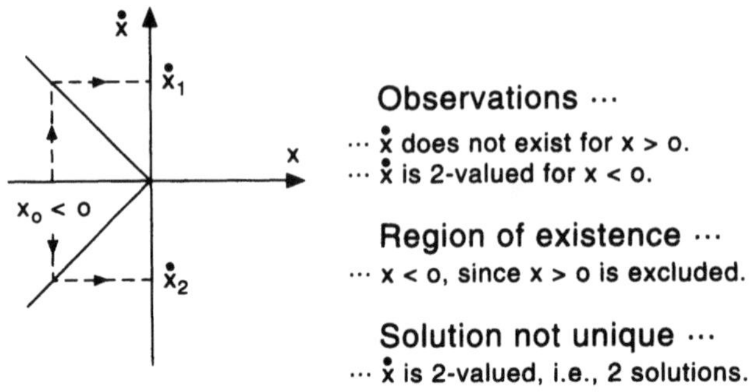

Figure 1-7. Dual solution slopes for $|\dot{x}| + x = 0$.

Thus, the solutions exist for all x, and they are unique, which is evidenced since their slopes are single-valued.

1.4 SOLVING EQUATION (iii): $|\dot{x}| + x = 0$

Here $\dot{x}$ is defined only for $x \leq 0$, since the "equation" is a contradiction for $x > 0$ (has no solution). Just as in equation (ii), this can be considered as two equations and a contradiction. We have $\dot{x} = \pm x$, $x \leq 0$, with solution $x = ce^{\pm t}$; and $\dot{x}$ = undefined for $x > 0$, with no solution.

The solution exists only in the lower half of the solution plane, where every point has two slopes and hence two solutions, as shown in Figure 1-1(c).

The phase portrait, $\dot{x}$ vs. x (Fig. 1-7) illustrates the region of existence ($x \leq 0$) and the nonuniqueness of the solution due to the two-valued nature of $\dot{x}$. We note that at a typical point on a solution trajectory, the tangent to the curve has two possibly admissible values. In the region $x \geq 0$, no slope exists nor does a solution.

1.5 SOLVING EQUATION (iv): $|\dot{x}| + |x| = 0$

Except for the trivial case of $x = 0$, the "equation" asks for the sum of two positives to be zero—a contradiction; so the equation has no solution as shown in Figure 1-1(d). The phase portrait, $\dot{x}$ vs. x (Figure 1-8) also indicates that the derivative does not exist for any value of x.

1.6 A VARIATION ON EQUATION (iv): $|\dot{x}| = |x|$

In this case, there are everywhere two equations, $\dot{x} = \pm x$ for all x. The phase portrait indicates two slopes at each point, generating two solutions everywhere, as shown in Figure 1-9.

Observation ···
··· Equation asks the sum of two positives to be zero.

Conclusion ···
··· Slope (direction) of sol'n not defined.

Consequence ···
··· There is no solution (except x = o).

Consequence ···
··· Question of uniqueness is moot.

Figure 1-8. Solution space for $|\dot{x}| + |x| = 0$.

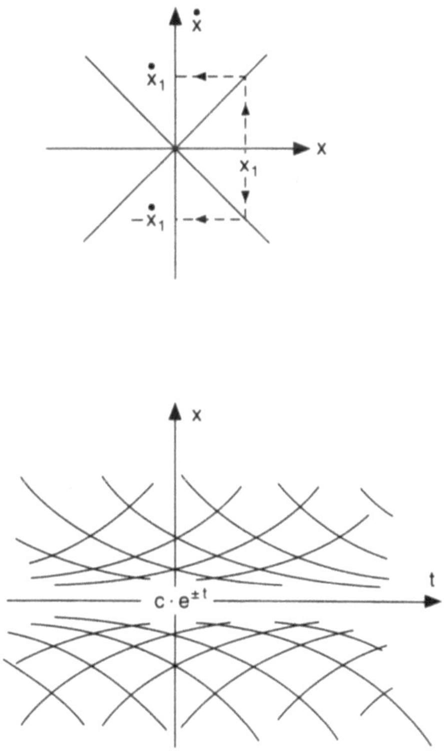

Observation ···
··· Slope is 2-valued for all x.

Conclusion ···
··· 2 directions of sol'n at each x.

Consequence ···
··· There are two solutions.

And so ···
··· Solution is not unique.

Figure 1-9. Multiple solutions to $|\dot{x}| = |x|$.

1.7 EXISTENCE/UNIQUENESS OF SOLUTIONS

1.7.1 Observations on the Four Equations

The existence and uniqueness of the solutions, $x(t)$, are seemingly related to the existence and uniqueness of its derivative, $\dot{x}(t)$, summarized in Figure 1-10, where $x(t)$ only exists if $\dot{x}(t)$ exists. Furthermore, if $\dot{x}(t)$ is anywhere multivalued, $x(t)$ is correspondingly multivalued.

1.7.2 Conjectures on Existence and Uniqueness

If $x(t)$ is to exist continuously, then $\dot{x}(t)$ must also exist to provide one slope (direction) from which $x(t)$ can proceed. If $x(t)$ is to exist uniquely, then $\dot{x}(t)$ must be single-valued, allowing only one (unique) direction for $x(t)$ to proceed.

1.7.3 The Initial-Value Problem Implies Existence and Uniqueness

To question the existence and uniqueness of solutions to $\dot{x} = f(x, t)$, is the same as solving the associated initial-value problem

$$\dot{x} = f(x, t), \qquad x(t_0) = x_0$$

1.7.4 Existence Theorem (Sufficient, Not Necessary)

The initial-value problem has at least one solution, $x(t)$, in the interval $|t - t_0| < \alpha$ if (not iff) $|f(x, t)| \leq k$ and f is continuous in some rectangular region, R, such that,

$$R\colon \underbrace{|t - t_0| < a, \qquad |x - x_0| < b,}_{\text{This much defines a rectangular region } R_1 \subset R.} \qquad \text{and } \alpha = \min(a, b/k)$$

The region, R, is shown in Figure 1-11.

A double wedge formed by slopes, $\pm k$, is shown in Figure 1-12. Since k is a bound on $|f|$, it is the bound on the slope, $|\dot{x}|$. Hence, within R_1 there is a double wedge where the solution $x(t)$ is confined (Figure 1-12).

1.7.5 The Interval of Existence: $|t - t_0| < \alpha$

Superimposing the double wedge of Figure 1-12 on R_1 (Figure 1-13) gives the two possibilities of R which define an interval, α, in which the solution exists.

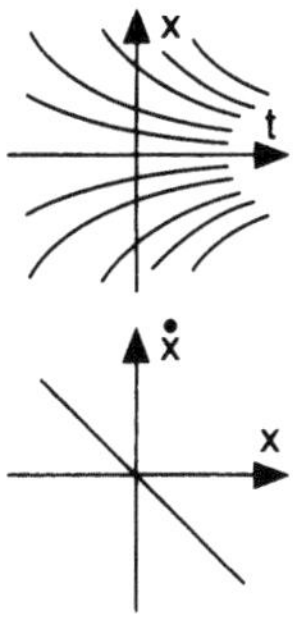

$\dot{x} + x = 0$

Sol'n covers entire x-t plane.
i.e., exists elsewhere.

$\dot{x}$ exists for all x.

$\dot{x}$ is single-valued for all x.
i.e., just one sol'n ··· unique.

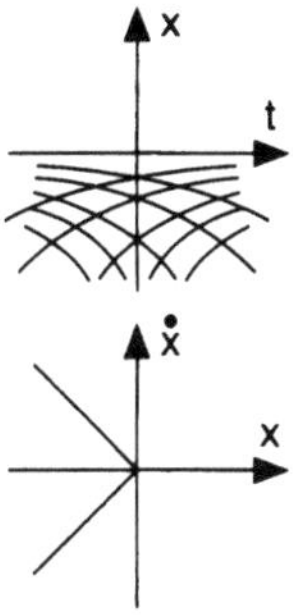

$|\dot{x}| + x = 0$

Sol'n covers only lower-half plane.
i.e., no sol'n in UHP.

$\dot{x}$ doesn't exist for $x > 0$.
i.e., "sol'n" has no direction (slope).

$\dot{x}$ is 2-valued for $x < 0$.
i.e., 2 sol'ns in BHP.

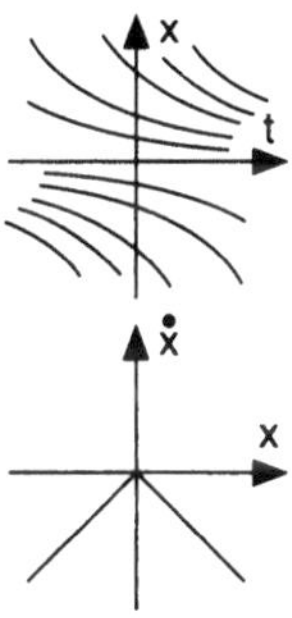

$\dot{x} + |x| = 0$

Sol'n covers entire x-t plane.
i.e., exists everywhere.

$\dot{x}$ exists for all x.

$\dot{x}$ is single-valued for all x.
i.e., just one sol'n ··· unique.

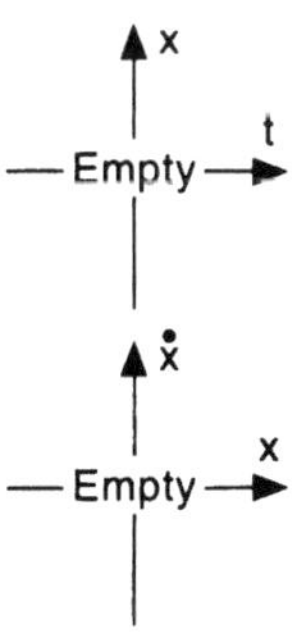

$|\dot{x}| + |x| = 0$

Sol'n plane is empty.
i.e., no sol'n anywhere.

Phase plane is empty.
i.e., no slope from which to start.

"Eq'n" is a contradiction.

Figure 1-10. Summary: solutions and phase portraits of the four equations.

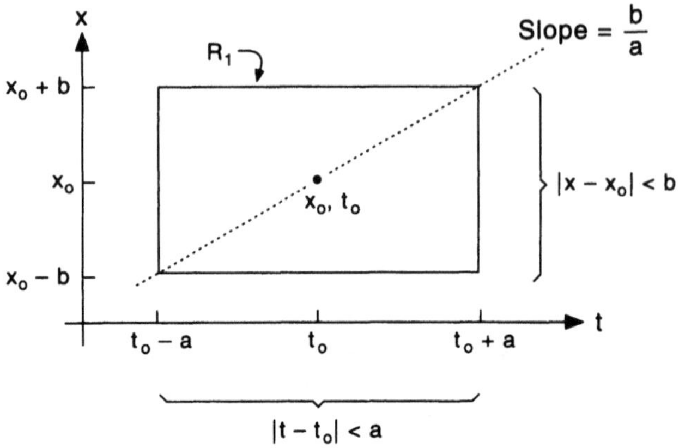

Figure 1-11. The rectangle of possible solutions.

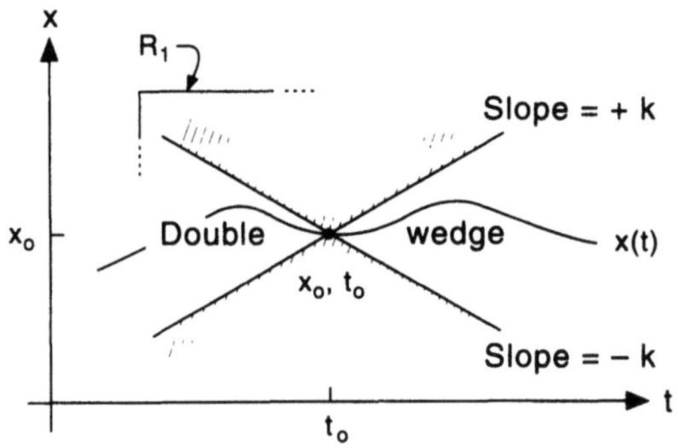

Figure 1-12. Region of validity for solutions.

1.7.6 Uniqueness Theorem (Sufficient, Not Necessary)

The initial-value problem has only one solution, $x(t)$, in the interval $|t - t_0| < \alpha$ if (not iff):

$$|f| \equiv |f(x, t)| \leq k \tag{i}$$

$$|f_x| \equiv \left|\frac{\partial f}{\partial x}\right| \leq M \tag{ii}$$

where f and f_x must be continuous and single-valued, in R, a region all of whose points have an interval of existence, α.

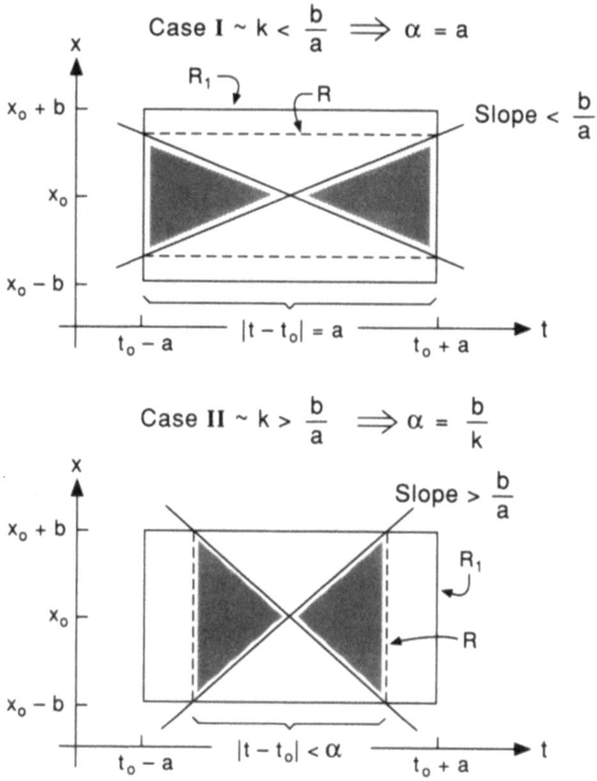

Figure 1-13. The interval of existence: $\alpha = \min(a, b/k)$.

1.7.7 Summarizing the Theorems

For the initial-value problem, $\dot{x} = f(x, t)$, $x(t_0) = x_0$.

Inside a region, R, all of whose points have an interval of existence α:

Existence if (not iff)

$$f = \begin{cases} \text{real and discontinuous} \\ \\ \text{Bounded} \end{cases}$$

Uniqueness if (not iff)

$$f_x = \begin{cases} \text{Real and continuous} \\ \text{Bounded} \\ \text{Single-valued} \end{cases}$$

1.7.8 Implications of Boundedness on f and f_x

For f to be bounded, the slope of the solution can never get to infinity, and thus:

1. x is sufficiently bounded;
2. x is continuous in the sense that it has neither finite nor infinite discontinuities; and
3. x cannot "doubleback" on itself as in a loop. (This is not essential to existence.)

For f_x to be bounded, the slope of the phase portrait is bounded, and cannot "loop back" on itself, preventing f from being multivalued. This guarantees only one direction (slope) for a solution to proceed from every value of $x(t)$, hence one (unique) solution.

1.7.9 Finding the Unique Solution to the Initial-Value Problem (Picard Iterates)

Convert the initial-value problem to an implicit equation—that is,

$$\dot{x} = f(x, t), \qquad x(t_0) = x_0$$

is integrated, so,

$$\int_{t_c}^{t_1} \dot{x}\, dt = \int_{t_0}^{t_1} f(x, t)\, dt \Rightarrow x(t) = x_0 + \int_{t_0}^{t_1} f(x, t)\, dt$$

Form a recursion formula:

1. Guess $x_1(t)$, consistent with x_0.
2. Substitute $f(x, t) \equiv f_1$, evaluated at $x = x_1$.
3. Integrate to find a "new" x, say $x_2 = x_0 + \int_{t_0}^{t_1} f_1\, dt$.
4. Recurse $x_n = x_0 + \int_{t_0}^{t_1} f_{n-1}\, dt$.

Show convergence, $x(t) = \lim_{n \to \infty} x_n(t)$.
Show existence, $x(t)$ satisfies initial value problem.
Show uniqueness—if $Z(t)$ also satisfies initial-value problem, then $Z(t) = x(t)$.

1.7.10 Relaxing the Uniqueness Condition—Lipschitz Condition

The bounding condition, $|f_x| \leq M$, is sufficient, meaning that it is more confining than necessary. From the mean value theorem,

$$f(x_2, t) - f(x_1, t) = (x_2 - x_1) \left.\frac{\partial f}{\partial x}\right|_{x=n}, \text{ such that } x_1 \leq n \leq x_2$$

But with f_x bounded, we have the inequality, $f(x_2, t) - f(x_1, t) \leq (x_2 - x_1)M$, so $|f(x_2, t)| - |f(x_1, t)| \leq |x_2 - x_1| M$ for $(x_1, t), (x_2, t) \in R$.

1.7.11 Interpreting Existence and Uniqueness by an Example (Figure 1-14)

The equation and its initial condition are:

$$\dot{x} = \frac{2}{3\sqrt[3]{t}}, \qquad x(o) = 0$$

For existence:

The region R is defined as all points in the f–t plane except $t = 0$, so $x(t)$ exists for all t except possibly at $t = 0$. However, the initial condition, $x(o) = 0$, guarantees the solution outside R. Thus, the solution exists throughout the x–t plane.

For uniqueness:

$f_x = 0$, which is bounded, continuous, and single-valued in R. Thus, $x = t^{2/3}$ is unique in $\dot{x}$–t except (possibly) at $t = 0$. However, the initial condition, $x(o) = 0$, guarantees only *one* solution outside R.

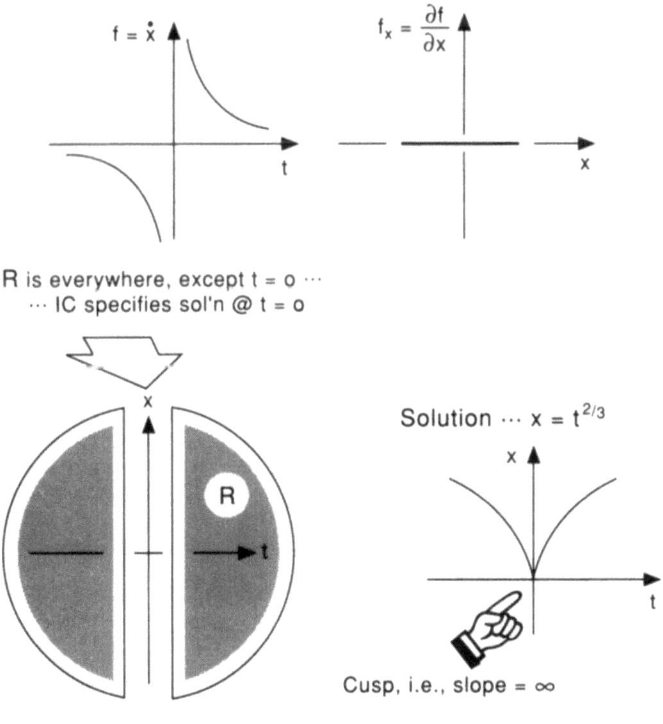

Figure 1-14. Example of existence and uniqueness.

TABLE 1-2 Summary of Existence and Uniqueness for Solutions to the Four Interesting Equations

Equation	Graphs	Existence and uniqueness	Region
$\dot{x} = -x$	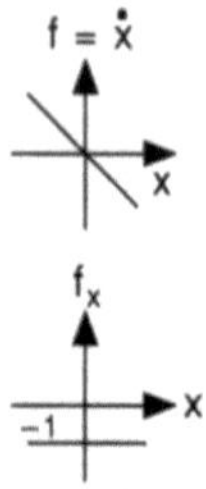	f is bound$^{\underline{d}}$ & continuous ··· ··· for all x (and t implicitly) ∴ R: entire x − t plane. i.e., sol'n exists everywhere. f is bound$^{\underline{d}}$ & contn$^{\underline{us}}$ ··· for all x (and t implicitly) i.e., sol'n is unique everywhere.	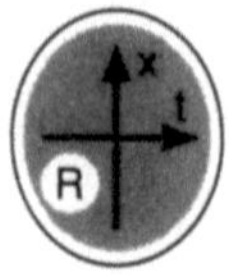
$\dot{x} = -\lvert x \rvert$	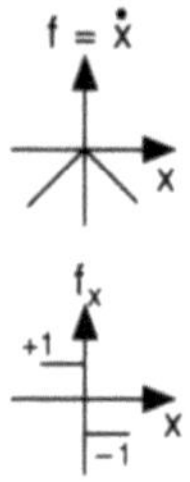	f is discontinuous at x = o ∴ R: x ≠ o, sol'n exists ··· ··· throughout x − t except possibly @ x = o. But, x = o satisfies eqn; so ··· ··· sol'n exists everywhere. f_x is discontinuous at x = o, so ··· ··· solution is unique everywhere except possibly at x = o. But, x = o is the only sol'n @ x = o, so ··· ··· sol'n is everywhere unique.	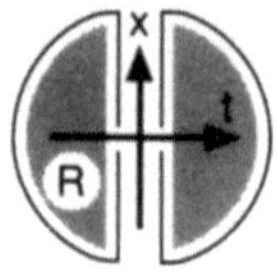
$\lvert \dot{x} \rvert = -x$	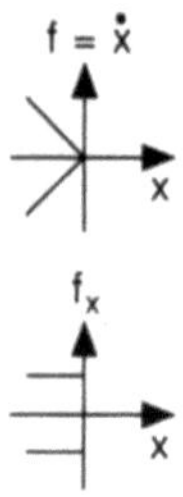	f ··· {bound$^{\underline{d}}$ & contn$^{\underline{us}}$ @ x < o; undefined @ x > o} ∴ R: BHP (x < o) i.e., sol'n exists in BHP only. f_x ··· {bound$^{\underline{d}}$, contn$^{\underline{us}}$, 2-valued @ x < o; undefined @ x > o} i.e., sol'n not unique @ BHP (2 sol'n$^{\underline{s}}$)	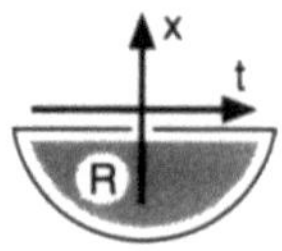
$\lvert \dot{x} \rvert = -\lvert x \rvert$	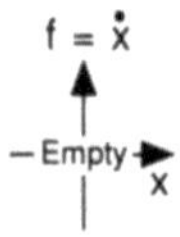	f undefined ... can't solve explicitly for ẋ ∴ R: Nowhere i.e., sol'n nonexistent, and ··· ··· question of uniqueness is moot.	
$\lvert \dot{x} \rvert = \lvert x \rvert$	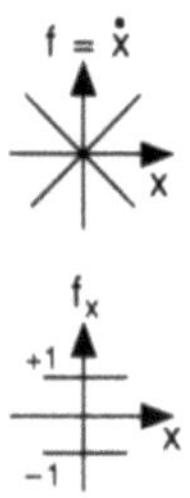	f is bound$^{\underline{d}}$ & contn$^{\underline{us}}$ ··· ··· for all x (and t implicitly) ∴ R: entire x − t plane. i.e., sol'n exists everywhere, ··· including x = o. f is bound$^{\underline{d}}$, contn$^{\underline{us}}$ & 2-valued ··· ··· everywhere. i.e., sol'n is nowhere unique.	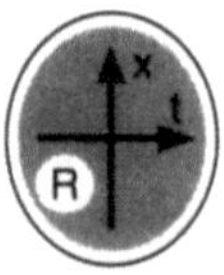

1.7.12 Existence and Uniqueness for Equations of Chapter 1

A summary is shown in Table 1-2.

1.8 PROBLEMS

1. A form of the Riccati equation is

$$\dot{V} = a - bV^2;\ a, b \in R$$

which approximates an aircraft's velocity during takeoff.
(a) Interpret the equation for negative velocity.
(b) Use the method of isoclines to approximate the solution for $a = b = 1$.
(c) Start with an initial guess, $V = 0$; successfully iterate the integration and develop a series solution.
(d) Separate the variables and find the exact solution from $V(o) = 0$.
(e) Find the power series expansion for the exact solution found in part (d). Do you think the series in part (d) converges to the exact solution?

2. The nonlinear, first-order equation (ref. 1) has a region of nonexistence, a region of existence that is not unique, and a boundary separating these regions which is a singular solution.

$$\dot{x}^2 = 1 - x; \qquad x = 1 - \left(\frac{t-c}{2}\right)^2; \quad x_{\text{singular}} = 1$$

(a) Using the method of isoclines, find the graphical approximation to the solutions.
(b) Show graphically that the singular region is the envelope of the solutions.

1.9 REFERENCES AND RELATED LITERATURE

1.9.1 References

1. Van Karman, T., and Biot, M., *Mathematical Methods in Engineering*, McGraw-Hill, New York (1940).
2. Polya, G., *How to Solve It, The World of Mathematics*, pp. 1980–1992, Simon & Schuster, New York (1956).
3. Gerald, Curtis F. and Wheatley, Patrick O., *Applied Numerical Analysis*, Addison-Wesley, 3rd edition, Reading, Mass. (1984).

1.9.2 Related Literature

1. Osgood, W. F., *Advanced Calculus*, Macmillan, New York (1925).
2. Spiegel, M., *Applied Differential Equations*, third edition, Prentice-Hall, Englewood Cliffs, New Jersey (1981).
3. Saaty, Thomas L., *Modern Nonlinear Equations*, Dover, New York (1981).
4. Saaty, Thomas L., and Bram, Joseph, *Nonlinear Mathematics*, Dover, New York (1964).
5. Davis, Harold T., *Introduction to Nonlinear Differential and Integral Equations*, Dover, New York (1962).

Chapter 2
Analytic Solutions to Nonlinear Differential Equations

2.0 INTRODUCTION

Most studies of nonlinear differential equations deal with a variety of techniques of how to best *approximate* the solution to nonlinear differential equations. The resulting solutions are termed "good enough," usually because there are not other techniques to obtain solutions or because practice often shows that the solutions predict performance reasonably well. Such techniques are included in the remaining chapters of this text.

Accordingly, it is reasonable to ask, When can we exactly solve nonlinear differential equations? The answer for first- and second-order linear differential equations is found in uniqueness and existence proofs which were discussed in Chapter 1. Unfortunately, such proofs do not exist in the generalized nonlinear case, although special cases do exist. Therefore, the analyst is left with the problem of how to best approximate the solution or, better yet, determine how the equation can be solved by analytical means. It is this latter approach (i.e., finding analytic solutions to nonlinear differential equations) to which we will devote this chapter. There are many texts which offer specific solutions to classes of equations (see references); in this chapter, we add to those techniques.

To introduce the concept, the following example demonstrates the approach which is the subject of this chapter. We select Bernoulli's equation

$$dy/dx = a(x)y + b(x)y^m, \qquad m = \text{constant} \tag{1}$$

This equation has the generalized nonlinear form

$$dy/dx + a(x)G(y) + b(x)H(y) = 0 \tag{2}$$

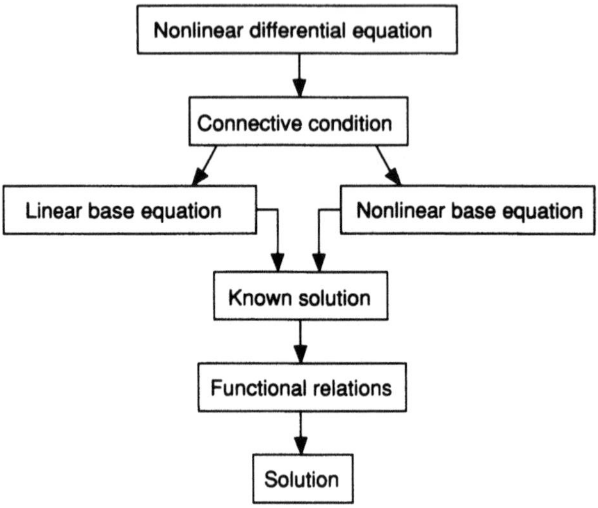

Figure 2-1. The path to a solution.

and we will show the generalized form of the equation can be solved in terms of the solution to the first-order linear differential equation

$$du/dx + a(x)u + b(x) = 0 \tag{3}$$

which itself has the known solution

$$u = u(x) = \exp\left[-\int^x a(t)dt\right]\left[\int^x - b(t)\exp\left\{\int^t a(s)ds\right\}dt + c\right] \tag{4}$$

The methodology presented in this chapter finds the solution to (1) in the form $y = F(u)$, where from (4) we know u. Our task is to define the *functional relation*, F. Equation (3) is termed a *base equation*, and the functional relation can be found if and only if the *connective condition*

$$H(y)\, d/dy[G(y)/H(y)] = 1 \tag{5}$$

is satisfied. These relationships are shown in Figure 2.1.

2.0.1 The Bernoulli Equation

Applying the connective condition (5) to the Bernoulli equation, [equation (1)], and for the purposes of this illustrative example we set $m = 2$. Using $G(y) = -y$ and $H(y) = -y^2$, it is seen that

$$H\, d/dy[G/H] = 1 \tag{6}$$

[Note that if it were initially decided to select $H(y) = y$ and $G(y) = y^2$, or if $G(y) = +y$ and $H(y) = +y^2$, the connective condition would not be satisfied.] Thus, we are seeking the solution of

$$dy/dx + a(x)(-y) + b(x)(-y^2) = 0 \tag{7}$$

Thus, the linear base equation is

$$du/dx + a(x)u + b(x) = 0 \tag{8}$$

and the functional relation (as included in Theorem 1, following) yields

$$y = F(u) = 1/u$$

And we obtain the solution of the Bernoulli equation [using equation (4)] as

$$y = 1/u = \exp\left[\int^x a(t)\,dt\right]\left[\int^x - b(t)\exp\left\{-\int^t a(s)\,ds\right\}dt + c\right]^{-1}$$

2.0.2 Formalizing the Concept of Base Equations

The approach to be followed is one in which we define a class of nonlinear differential equations

$$y' = f_1(y, x, c_0, c_1) \qquad (\ )' = d(\ )/dx \tag{9}$$

$$y'' = f_2(y', y, x, c_0, c_1, c_2) \tag{10}$$

$$y = y(x)$$

$$c_0, c_1, c_2 = \text{constants}$$

and associate them with linear or nonlinear differential equations whose solutions are known (or can be found)

$$u' = g(u, x, a, b) \tag{11}$$

$$u'' = h(u', u, x, c, d, e) \tag{12}$$

$$u = u(x)$$

$$a, b, c, d, e = \text{constants}$$

We term the ordinary differential equation (11) or (12), with which we associate the nonlinear ordinary differential equation (9) or (10), the *base equation*. Although the technique illustrated here is demonstrated for first- and second-order equations, the technique is applicable for nth-order and matrix equations.

The solution to the base equation is coupled to that of the nonlinear equation (9) or (10) by means of a *functional relationship* of the form

$$y = F(u) \tag{13}$$

The question we address is: How do we find the solution to the nonlinear equation (9) or (10) in terms of the solution to the base equation (11) or (12)? That is, what is the relationship designated by the function F of (13)?

By introducing a *connective condition*, we are able to "connect" the terms of the nonlinear equations (9) and (10) to those of the base equation (11) or (12). Then by mathematical (usually algebraic) manipulation, one arrives at the solution function F. As a result, if the connective condition is satisfied, then the solution of the nonlinear equation is expressible in terms of the solution to the base equation in a simple algebraic form. Review Figure 2.1 again to track these relationships.

2.1 FIRST-ORDER BASE EQUATIONS

We begin with the simplest case, that being a first-order linear differential equation of the form (11) or

$$du/dx + a(x)u + b(x) = 0 \tag{14}$$

which has the known (or solvable) solution

$$u = u(x) = \exp\left[-\int^x a(t)dt\right]\left[\int^x - b(t)\exp\left\{\int^t a(s)ds\right\}dt + c\right] \tag{15}$$

We wish to define a class of nonlinear differential equations of the form (9) or

$$dy/dx + a(x)G(y) + b(x)H(y) = 0 \tag{16}$$

whose solution we seek and which is associated with the linear equation (14).

2.1.1 Theorem 1

The solution $y = y(x)$ to equation (16) is

$$y = F(u)$$

where $u = u(x)$ is the solution (15) to the base equation (14) and F is defined by the functional relation (in inverse form)

$$u = G(F)/H(F) \tag{17}$$

or by the separable differential equation

$$dF/du - G(F) = 0 \tag{18}$$

if and only if G and H satisfy the connective condition

$$H(y)\, d/dy[G(y)/H(y)] = 1 \tag{19}$$

To prove this theorem, we assume the solution of the nonlinear differential equation

$$y' + a(x)G(y) + b(x)H(y) = 0$$

is $y = F(u)$, where $u = u(x)$ is the solution of the linear differential equation, the base equation

$$u' + a(x)u + b(x) = 0$$

Differentiating $y = F(u)$ with respect to x and substituting yield

$$y' = -F_u[a(x)u + b(x)]; \qquad F_u = dF/du; \qquad y' = (dF/du)(du/dx)$$

and defining

$$F_u \equiv H(y)$$

and its integral

$$u = \int^y [ds/H(s)]$$

Substituting once again yields

$$y' + a(x)H(y) \int^y [ds/H(s)] + b(x)H(y) = 0$$

which, when rewritten in the form

$$y' + a(x)G(y) + b(x)H(y) = 0$$

is equivalent to the originally assumed equation, provided that the connective condition

$$G(y) = H(y) \int^y [ds/H(s)]$$

is satisfied. This can be rewritten more compactly as

$$H(y)\, d/dy[G/H] = 1$$

and thus we find

$$u = \int^y [ds/H(s)] = G(F)/H(F)$$

and the proof is completed.

2.1.2 Some Comments

It should be recognized that the redefinitions of the Bernoulli equation (which arises in studies of the properties of fluids flowing in pipes) in the form of equation (7) is a specific example of the following concept. That is, if applying the connective condition of Theorem 1 results in a constant k, where $k \neq 1$, then by rewriting equation (16) in the form

$$y' + ka(x)G(y)/k + kb(x)H(y)/k = 0$$

the new functions

$$\begin{aligned} a_1(x) &= ka(x) \\ b_1(x) &= kb(x) \\ G_1(y) &= G(y)/k \\ H_1(y) &= H(y)/k \end{aligned} \tag{20}$$

will satisfy the connective condition. In the case of Bernoulli's equation with the exponent m in the nonlinear term of equation (1), the connective condition is satisfied if $k = m - 1$. Furthermore, this approach will be found to be applicable throughout this chapter.

It is natural to inquire next into which extensions of this technique will prove fruitful. Rather than an exhaustive search for comparative theorems, the theory is extended in a series of natural steps. First, we introduce second-order base equations using second-order linear differential equations (Theorem 2), and show that a wide range of equations are "related" to this class of equations. Then we investigate the use of specific nonlinear differential equations which have known solutions to obtain specific forms of solvable classes of first- and second-order nonlinear differential equations (Theorems 3–5). We then investigate integral forms of the functional relation for first-order equations (Theorem 6), which is extendable to second-order linear and nonlinear base equations; the examination of derivative forms are left to the reader. We return to the first-order linear equation and develop a class of nonlinear matrix equations (Theorem 7). An application to the calculus of variations is shown in Theorem 8; and Theorem 9 (in the problem section) is included for the reader to develop third-order equations using a linear base equation.

It should be clear, therefore, that this technique identifies, expands, and unifies a branch of differential equations to allow the applied mathematician and engineer to analytically solve a wide range of equations.

2.2 SECOND-ORDER BASE EQUATIONS

2.2.1 Theorem 2

The second-order nonlinear differential equation

$$y'' + p(x)y' + q(x)G(y) + r(x)Z(y) = (\dot{Z}/Z)y'^2$$

$$(\ )^{\cdot} = d(\ )/dy \qquad (\ )' = d(\ )/dx \tag{21}$$

has the solution $y = F(u)$, *where* u *satisfies the base equation*

$$u'' + p(x)u' + q(x)u + r(x) = 0 \tag{22}$$

and F *is defined by the functional relation (in inverse form)*

$$u = G(F)/Z(F) \tag{23}$$

or by the separable differential equation

$$dF/du - Z(F) = 0 \tag{24}$$

if and only if G *and* Z *satisfy the connective condition*

$$Z(y)\ d/dy[G(y)/Z(y)] = 1 \tag{25}$$

To prove Theorem 2, we proceed as follows:

We seek the solution to the second-order nonlinear equation

$$y'' + p(x)y' + q(x)G(y) + r(x)Z(y) = (\dot{Z}/Z)y'^2$$

$$(\ )^{\cdot} = d(\ )/dy$$

to be obtained from the solution $u = u(x)$ of the second-order linear base equation

$$u'' + p(x)u' + q(x)u + r(x) = 0$$

in the form $y = F(u)$. Differentiation and substitution yield $y' = F_u u'$ and $d^2y/dx^2 = F_{uu}u'^2 + F_u u''$,

$$y'' + p(x)y' = -F_u[q(x)u + r(x)] + F_{uu}u'^2$$

Defining

$$Z(y) \equiv F_u = dF/du$$

$$u = \int^y [ds/Z(s)]$$

and since $y' = F_u u'$ and $F_{uu} = dZ/du = ZdZ/dy$,

$$y'' + p(x)y' = -q(x)Z(y)\int^y [ds/Z(s)] - r(x)Z(y) + (\dot{Z}/Z)y'^2$$

and by defining the connective condition as

$$G(y) = Z(y) \int^{y} [ds/Z(s)]$$

substitution yields the original equation.

The definition of the connective condition yields

$$Z(y)\, d/dy[G(y)/Z(y)] = 1$$

and

$$u = G(F)/Z(F)$$

which completes the proof.

2.2.2 Equations Solvable by Infinite Series

Linear second-order differential equations can be solved by a variety of techniques. Among these is the use of an infinite series, especially since uniqueness and existence proofs involve the use of series. It is not always possible to use a convergent Taylor series and thus other series have evolved. Among these are the hypergeometric series developed by Karl Frederich Gauss and is associated with Gauss's equation, Legendre's equation (which can be transformed into Gauss's equation), Bessel's equation, and Ricatti's first-order nonlinear equation, which can be transformed into a second-order linear equation.

Furthermore, particular solutions of Laplace's partial differential equation

$$\partial^2 V/\partial x^2 + \partial^2 V/\partial y^2 + \partial^2 V/\partial z^2 = 0$$

which arises as a result of studying potential fields, V, generates second-order linear equations. By transforming this equation into spherical coordinates, Legendre's equation results, and using cylindrical coordinates, we obtain Bessel's equation. Other partial differential equations are associated with particular physical problems. Some are listed below:

The vibrating string problem—

$$\partial^2 y/\partial t^2 = a^2 \partial^2 y/\partial x^2 + F(x, t)$$

The vibrating membrane—

$$\partial^2 z/\partial t^2 = a^2(\partial^2 z/\partial x^2 + \partial^2 z/\partial y^2)$$

The telegraph equation—

$$\partial^2 v/\partial x^2 = A\partial^2 v/\partial t^2 + B\partial v/\partial t + Cv$$

The beam equation

$$\partial^2 y/\partial t^2 + c^2 \partial^2 y/\partial x^2 = 0$$

all of which lead to ordinary differential equations that are solvable by use of Fourier and related series. Each of the equations of this section gives rise to differential equations which are solvable by infinite series—and these series, when different forms of polynomials are assumed, give rise to specific generating equations. Thus, the introduction of boundary value problems lead to considering orthogonal systems which are characterized by Sturm–Liouville formulations. Such formulations are solvable by Legendre, Tschebyscheff, Hermite, and Laguerre polynomials (See Broman, Related Literature).

Gauss's equation,

$$[ax^2 + bx + c]y'' + [dx + e]y' + fy = 0,$$
$$a, b, c, d, e, f = \text{constants}$$
$$b^2 - 4ac \quad \text{nonzero} \tag{26}$$

Legendre's equation,

$$(1 - x^2)y'' - 2xy' + n(n + 1)y = 0, \qquad n = \text{integer} \tag{27}$$

and Bessel's equation,

$$x^2y'' + xy' + (x^2 - n^2)y = 0, \qquad n = \text{constant} \tag{28}$$

are all special forms of the base equation (22). Note that in all three cases in equation (21), if $q(x) = 0$ and $Z(y) = y$, the connective condition (25) defines the function G as $y(k + \ln y)$, k = constant of integration.

Thus, each of the above three equations, as well as those studied by Laguerre (29), Hermite (30), Jacobi (31), and Tschebyscheff (32),

$$xy'' + (c - x)y' - ay = 0, \qquad a, c = \text{constants} \tag{29}$$
$$y' - 2xy' + 2ny = 0 \qquad n = \text{integer} \tag{30}$$
$$x(1 - x)y'' + [a - (1 + b)x]y' + n(a + n)y = 0$$
$$a, b = \text{constants}, n = \text{integer} \tag{31}$$
$$(1 - x^2)y'' - xy' + n^2y = 0, \qquad n = \text{integer} \tag{32}$$

are base equations for the nonlinear differential equation (33)

$$y'' + p(x)y' + r(x)y = y'^2/y \tag{33}$$

and are all special cases of equation (26).

2.2.3 A Polar Coordinate Example

Patel [see equation (10) in ref. 9], formulates the terrestrial brachristochrone problem in polar coordinates (Figure 2-2). This problem seeks the path through

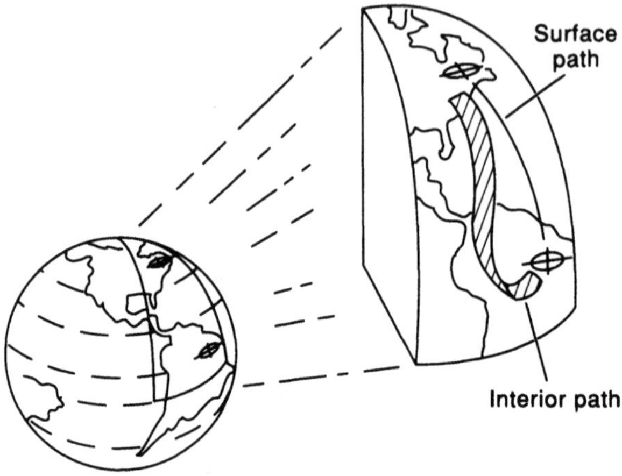

Figure 2-2. Terrestrial brachistochrone.

the earth's surface, which allows a mass to move from point to point in minimum time influenced by an inverse square gravity force without friction.

$$r'' + [rR^2/(r^2 - R^2)] + [(2R^2 - r^2)/r(r^2 - R^2)]r'^2 = 0$$

$$(\ )' = d(\ )/d\theta$$

$$r = \text{point's radial distance from the earth's center}$$

$$R = \text{radius of earth} \tag{34}$$

Applying Theorem 2 and using $G = r'^2/2$ and $Z = \frac{1}{2}$, the functional relation yields $F = u^{1/2}$ and

$$dr/d\theta = F(u) \tag{35}$$

solves the equation.

2.2.4 Lambert's Equation

Consider the nonlinear equation presented by Lambert [ref. 6, equation (9)]

$$y'' + (\omega^2/n)y = [(1 - n)/y]y'^2, \qquad \omega, n = \text{constants} \tag{36}$$

This is a reduced form of the general nonlinear equation (21) of Theorem 2, and an example of a simple second-order system with nonlinear damping. The connective condition (25) cannot be applied nor the equation functions identified until $Z(y)$ and $G(y)$ are defined.

To establish $Z(y)$, the following relation follows from the comparison of the coefficients of y'^2 in (21) and (36).

$$\dot{Z}/Z = dZ/dy(1/Z) = (1 - n)/y \tag{37}$$

which can be integrated to yield

$$Z(y) = cy^{(1-n)}, \qquad c = \text{constant} \tag{38}$$

A first attempt at satisfying the connective condition (25) with $G(y) = y$ yields the constant n rather than unity. However, the equation can be redefined by altering terms to satisfy the connective condition in a manner similar to the discussion associated with equation (20). That is, let $p_1(x) = 0$; $G_1(y) = y/n$; $q_1(x) = \omega^2$; $r_1(x) = 0$; and $Z_1(y) = (c/n)y^{1-n}$, where $Z_1(y)$ also satisfies (37). Then $G_1(y)$ and $Z_1(y)$ satisfy (25).

Hence, the solution is given as $y = F(u)$, where F is found using $F = (cu)^{-n}$ and u is the solution of the linear second-order equation.

$$u'' + \omega^2 u = 0 \tag{39}$$

which is

$$u = x_0 \cos \omega t + (x_0'/\omega) \sin \omega t \tag{40}$$

Therefore, the solution of (36) is

$$y = [c(x_0 \cos \omega t + (x_0'/\omega) \sin \omega t]^{-n} \tag{41}$$

which agrees with the results of ref. 6 for $c = 1$.

If we try to generalize Lambert's equation and establish solutions which are obtainable through Theorem 2, we need only redefine (36) as

$$y'' + (\omega^2/n)y = f(y)y'^2$$

and inquire under which conditions does this technique work? Clearly, the methodology revolves around defining $Z(y)$ and $G(y)$, and so we obtain an equivalent form of (37)

$$dZ/dy(1/Z) = f(y)$$

or

$$Z = c \exp\left[\int^y f(s)ds\right]$$

and applying (25)

$$Z\, d/dy[G/Z] = 1$$

where $G = y/n$, we obtain

$$f(y) = (1 - n)/y$$

which is the same form of the equation as studied by Lambert.

2.2.5 A Comment

Hill's equation is a special form of the second-order linear differential equation [equation (22)], which is the base equation of Theorem 2. If we set $r(x) = 0$, then Hill's equation results. This equation arises in a rather wide variety of applications; and even though it was identified earlier than G. W. Hill's studies of the motion of the moon in 1877, it is attributed to him due to his contributions.

Note that by use of the substitution

$$w = [\exp \tfrac{1}{2}A(x)]u, \qquad p(x) = dA/dx$$

equation (22), with $r(x) = 0$, can be transformed into

$$w'' + Q(x)w = 0$$

where

$$Q = -(\tfrac{1}{2})dp/dx - p^2/4 + q$$

This equation is central to the study of periodic motion and stability theory. For an indepth study of these topics, see ref. 11.

2.3 THE RICATTI EQUATION

The first-order nonlinear equation

$$y' + p(x)y + q(x)y^2 + r(x) = 0$$

was studied by Count Ricatti.

The Ricatti equation arises throughout the physical sciences. For example, the solution of the Ricatti equation provides the optimal gains in the feedback channels of the Kalman–Bucy filter, while it also provides the time evolution of the error covariance matrix. The matrix Ricatti equation generates the optimal time-varying or constant gains in the linear optimal regulator problem (see ref. 12).

Furthermore, note that the Blasius equation, which arises in fluid flow and electronics applications (and also known as the Falkner–Skan equation),

$$y''' + yy'' + k[(1 - y'^2) + c(1 - y')] = 0$$

can be transformed into a Ricatti equation when $k = -1$. In this case, the equation has the form

$$y''' + yy'' + y'^2 = 1 + c(y' - 1)$$

Since this is equivalent to

$$d/dx(y'' + yy') = 1 + c(y' - 1)$$

upon integration twice, while noting

$$d/dx(y^2) = 2yy'$$

there results

$$y' + (1 - c)y^2/2 = (1 - c)x^2/2 + c_1x + c_2$$

which is a form of the generalized Ricatti equation.

It is easily shown that when considering the homogeneous Ricatti equation [i.e., when $r(x) = 0$], Theorem 1 yields the solution. The connective condition is satisfied by rewriting the equation in the form [using equation (20)]

$$y' + (-p)(-y) + (-q)(-y^2) = 0$$

and associating its solution with the first-order linear base equation

$$u' + p(x)u + q(x) = 0$$

through the functional relation $y = F(u)$, where

$$y = 1/u$$

Thus, using equation (4), the solution to the homogeneous Ricatti equation is

$$u = u(x) = \exp\left[\int^x p(t)dt\right]\left[\int^x - q(t)\exp\left\{\int^t p(s)ds\right\}dt + c\right]^{-1}$$

It is natural to inquire if the concepts of this chapter can be applied to yield a means of solving the generalized Ricatti equation. And indeed they do. In fact, by introducing integral functional relations of the form

$$y = \int^u F(s)ds$$

the derivation of the appropriate theorem leads to the use of second-order linear base equations to solve first-order nonlinear equations. This is presented in Section 2.5.1, Theorem 6. Note the similarity to the transformation originally used by Ricatti

$$y = Z'/(qZ)$$

which yields the second-order linear equation

$$Z'' + (p - q'/q)Z' + rZ = 0$$

and which is also comparable to Hill's equation.

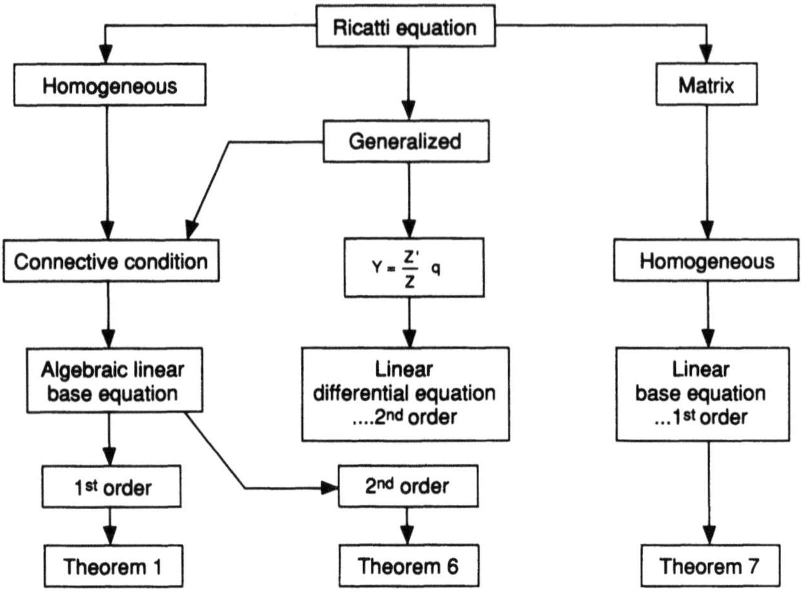

Figure 2-3. Solving the Ricatti equation.

See also Sugai's paper (Related Literature, 4) for the generation of results which occur with the use of

$$y = ru/u'$$

Furthermore, extension of the theory of linear first-order base equations to that of matrix equations is another natural step. Section 2.6 and Theorem 7 provide that extension, which is then applicable to the matrix Ricatti equation. For some further concepts applicable to the Ricatti equation, see Problem 9. The relationships between Ricatti's equation and these various approaches to its solution are shown diagrammatically in Figure 2-3.

2.4 NONLINEAR BASE EQUATIONS

In establishing a class of nonlinear differential equations solvable by these methods, one would logically investigate the generation of solvable nonlinear equations using those nonlinear equations for a base which were generated by use of linear base equations, such as (16) and (21). These new equations would then be solved by means of a double transformation. (This procedure would be equivalent to transforming a general nonlinear equation into a second nonlinear equation, and then performing a second transformation to obtain a solvable linear equation.) However, the newly generated equation will also satisfy the conditions of the linear base theory, and thus no new equations are generated by this double transformation. It is thus necessary to investigate only those base equations which can be solved analytically, and which do not have linear base equations themselves.

2.4.1 First-Order Nonlinear Base Equations

Two first-order nonlinear equations which do not have linear base equations subject to the functional relation $y = F(u)$ are Abel's and Chrystal's. Because of the specialized nature of these equations, only the results of the theory and the associated proofs will be presented.

2.4.1.1 Theorem 3 *Chrystal's equation, the base equation*

$$u'^2 + Axu' + Bu + Cx^2 = 0; \; (\;)' = d(\;)/dx \tag{42}$$

where A, B, and C are constants, can be associated with the general nonlinear equation

$$y'^2 + AxZ(y)y' + BG(y) + Cx^2Z(y)^2 = 0 \tag{43}$$

by means of the variable transformation $y = F(u)$ and the functional relation $u = G(F)/Z(F)^2$ or the separable differential equation $dF/du - Z(F) = 0$, if and only if G and Z satisfy the connective condition

$$\dot{G} - 2\dot{Z}G/Z = Z; \; (\dot{\;}) = d(\;)/dy \tag{44}$$

Proof

Let Chrystal's equation (42) be the base equation and the dependent variable transformation be $y = F(u)$. Differentiating and squaring,

$$y'^2 = F_u^2 u'^2$$

results where F_u is the partial derivative with respect to u. Substituting for u'^2 using (42) yields

$$y'^2 = -F_u^2(Axu' + Bu + Cx^2)$$

Define $Z(y) = F_u$; then,

$$u = \int^y [ds/Z(s)] \tag{45}$$

Substituting

$$y'^2 + Z(y)Axy' + Z(y)^2B\int^y [ds/Z(s)] + Z(y)^2Cx^2 = 0$$

which can be written in the general form as

$$y'^2 + AxZy' + BG(y) + Cx^2Z^2 = 0 \tag{46}$$

where $G(y)$ satisfies the relation

$$G(y) = Z^2 \int^y [ds/Z(s)] \tag{47}$$

or in a differential form as

$$\dot{G} - 2\dot{Z}G/Z = Z; \ (\dot{\ }) = d(\ )/dy \tag{48}$$

Finally, the solution for the transformation F can be found using (47) and the integral for u, (45);

$$u = G(F)/Z(F)^2 \tag{49}$$

which completes the proof.

Thus, the solution of the general nonlinear equation (46) is $y = F(u)$, where the function F satisfies (49); Chrystal's equation (42) is the base equation; and the functions $G(y)$ and $Z(y)$ satisfy either form of the connective condition (47) or (48).

The solution of Chrystal's equation has many forms depending on the magnitude of the constants A, B, C (ref. 7, Section 2.7). For example, if A, B, C satisfy the relation

$$A^2 + AB - 4C = 0$$

a solution of (42) is the parabola

$$4Bu = -ABx^2 \pm (d + Bx)^2 \tag{50}$$

where d is a constant. Then the solution of (46), the general nonlinear equation, is given as $y = F(u)$, where $u(x)$ is given by (50).

2.4.1.2 Theorem 4 *Using Abel's Equation as the base equation,*

$$[u + s(x)]u' + p(x) + q(x)u + r(x)u^2 = 0 \tag{51}$$

the general nonlinear equation can be written as:

$$y'[G(y) + s(x)] + Z(y)[p(x) + q(x)G(y) + r(x)G(y)^2] = 0 \tag{52}$$

which has the solution y in terms of the variable transformation $y = F(u)$ if and only if G, Z satisfy the connective condition

$$Z(y)\ d/dy\ [G(y)] = 1 \tag{53}$$

The functional relation can be written as either

$$u = G(F) \quad or \quad dF/du - Z(F) = 0$$

Proof

If Abel's equation (51) is the nonlinear base equation, and $y = F(u)$ is the variable transformation, by differentiating and substituting for u, with the definition $F_u = Z(y)$, which corresponds to $u = \int^y [dt/Z(t)]$, there results

$$\left[\int^y [dt/Z(t)] + s(x)\right] y'$$
$$= -p(x)Z - q(x)Z \int^y [dt/Z(t)] - r(x)Z\left[\int^y [dt/Z(t)]\right]^2$$

By redefining terms, the general nonlinear equation having Abel's equation as its base equation is

$$[G(y) + s(x)]y' + Z(y)[p(x) + q(x)G(y) + r(x)G^2(y)] = 0$$

where $G(y)$, $Z(y)$ must satisfy the connective condition

$$Z(y)\, d/dy\, [G(y)] = 1$$

The solution is given as $y = F(u)$, where F is found using

$$u = G(F)$$

or $dF/du - Z(F) = 0$, which completes the proof.

2.4.2 Second-Order Nonlinear Base Equations

The study of Duffing's equation

$$x'' + ax' + x + bx^3 = \cos \omega t$$

has received a great deal of attention lately under the heading of Chaos Theory. This equation is solved in terms of elliptic integrals, and the reader is referred to texts such as refs. 7 (Chapters 6 and 7) and 15 for their study.

We are interested here in expanding the class of equations which yield solutions, and thus we assume the solution to the generalized equation

$$u'' = a + bu + cu^2 + du^3$$

and refer to this equation as the elliptic equation.

Duffing introduced damping and a harmonic forcing function to study the equation. We shall state the Theorem and then outline the proof; the application to Chaos Theory, etc., shall be left to the reader.

2.4.2.1 Theorem 5 *The equation*

$$y'' = (\dot{G}/G)y'^2 + aG(y) + bZ(y) + [c/G(y)]Z^2(y) + [d/G^2]Z^3(y) \quad (54)$$

has the general solution $y = F(u)$, *where* u *is a solution of the elliptic equation*

$$u'' = a + bu + cu^2 + du^3 \quad (55)$$

if and only if $G(y)$ *and* $Z(y)$ *satisfy the connective condition*

$$G(y)\ d/dy\ [Z(y)/G(y)] = 1 \quad (56)$$

The functional relation is given as

$$u = Z(F)/G(F) \quad (57)$$

Proof

If u is a solution of the elliptic equation (55) and it is assumed that the solution of the derivable nonlinear equation has the form $y = F(u)$, then the following second-order identity can be written

$$y'' = F_u(a + bu + cu^2 + du^3) + F_{uu}u'^2 \quad (58)$$

Defining the functional relation as $F_u = G(y)$ with its solution in terms of u

$$u = \int^y [ds/G(s)] \quad (59)$$

then equation (58) becomes a function of y and x

$$\begin{aligned} y'' &= aG(y) + bG(y)\left\{\int^y [ds/G(s)]\right\} \\ &+ cG(y)\left\{\int^y [ds/G(y)]\right\}^2 \\ &+ dG(y)\left\{\int^y [ds/G(s)]\right\}^3 + [\dot{G}(y)/G(y)]y'^2 \end{aligned} \quad (60)$$

Rewriting (60), the general second-order nonlinear differential equation having (54) as its base equation is where $Z(y) \equiv G(y) \int^y [ds/G(s)]$

$$y'' = aG(y) + bZ(y) + [c/G(y)]Z^2(y) + [d/G^2(y)]Z^3(y) + [\dot{G}(y)/G(y)]y'^2$$

where G and Z satisfy

$$G(y)d/dy[Z(y)/G(y)] = 1$$

Finally, the function F can be found from the integral of (59) as

$$u = Z(F)/G(F)$$

thereby completing the proof.

Referring to Section 2.4.2, Duffing's equation is given as

$$x'' + ax' + x + dx^3 = \cos \omega t$$

which is related to the elliptic equation

$$u'' = a + bu + cu^2 + du^3$$

and the generalized nonlinear equation of Theorem 5

$$y'' = (\dot{G}/G)y'^2 + aG(y) + bZ(y) + [c/G(y)]Z^2(y) + [d/G^2]Z^3(y)$$
$$\dot{(\ )} = d(\ )/dy$$

Note that the pendulum equation becomes a special case of the elliptic equation when the substitution for sin θ is made in terms of its infinite series, which is then truncated after the second term. For ease of comparison with the material of Chapter 5, we have used $u = \theta$ or $u'' = bu + du^3$. The generalized equation of Theorem 5 then reduces to

$$y'' = (\dot{G}/G)y'^2 + bZ(y) + [d/G^2]Z^3(y)$$

If, for example, $Z = y/n$, $G = (k/n)y^{1-n}$, then the connective condition of Theorem 5 is satisfied and the equation becomes

$$y'' = y'^2/y + k_1 y^{1-n} + k_2 y^{1-3n}$$

Duffing's, the elliptic equation, and therefore the equations of Theorem 5, are not solvable in terms of the normal functions we encounter. The solutions are obtained in terms of elliptic functions and elliptic integrals. We shall not delve too deeply into these matters here, but will refer the reader to Davis' text (ref. 7) and others for the underlying theory. In essence, elliptic integrals have the form

$$\int_0^x [(1 - x^2)(1 - k^2x^2)]^{-1/2}\, dx$$

which are equivalent to

$$\int_0^\phi (1 - k^2 \sin^2 \theta)^{-1/2}\, d\theta$$

with the substitution of $x = \sin \phi$ other variations of the integrand, where different forms of the functions $(1 - ax^2)$ appear in numerators and denominators to create a family of such integrals. A specialized yet standardized notation has developed and, therefore, a simple mathematical jargon must be mastered to work with this material. As a simple example, the notation sn (x, k) refers to the elliptic sine of Jacobi and $K = F(1, k) = F(\pi/2, k)$ is used to represent the complete elliptic integral of the first kind, etc., and are all quite natural shorthand means of simplifying the presentation. Of importance in recognizing the nature of these functions is the dependence on an independent variable x and various constants n, k, etc., which appear in the functions. Classifications are therefore made in terms of how these variables and constants appear in the integrals.

The elliptic functions arose by considering properties of an ellipse and noticing that, by using elliptical coordinates, certain equations could be simplified. The wave equation is the partial differential equation associated with vibrating membranes, etc., which made it natural to consider the relationship of the inner portions of the membrane to the fixed outer circumference. When we use an ellipse, Mathieu's equation results

$$w'' + (a - 2q \cos 2z)w = 0, \qquad (\ )' = d(\ )/dz$$

or, if $t = \cos z$,

$$(1 - t^2)w'' - tw' + (a + 2q - 4qt^2)w = 0$$

Consider the pendulum problem, it is seen that

$$\theta'' + (g/L) \sin \theta = 0$$

has a first integral of

$$\theta'^2 - (2g/L) \cos \theta = c$$

and by separating variables, a second integral in terms of elliptic integrals occurs

$$dt = [c + (2g/L) \cos \theta]^{-1/2}\, d\theta$$

which can be shown to yield

$$\theta = 2 \arcsin [k \text{ sn } (g/L)^{1/2}t, k]$$

where

$k = \sin w/2$

$\theta = w =$ maximum displacement of the pendulum from its equilibrium point.

2.5 DERIVATIVE AND INTEGRAL FUNCTIONAL RELATIONS

We have developed solutions of classes of nonlinear ordinary differential equations using both linear and nonlinear ordinary differential equations as base equations. The solutions of the two classes were related to one another by means of a functional relation of the form $y = F(u)$, provided the connective conditions were satisfied. It is logical to consider investigating the use of derivative and integral forms of F to accomplish the same end.

2.5.1 Integral Functional Relations

Suppose one wishes to investigate the relationship between first- or second-order base equations and their corresponding solvable nonlinear counterparts. By introducing the integral functional relation

$$y = \int^{u} F(s)\,ds$$

we can obtain the following theorem.

2.5.1.1 Theorem 6 *The nonlinear differential equation*

$$y'' + a(y)G(y') + b(y)H(y') = 0;\ (\ )' = d(\ )/dx$$

has the solution

$$y = \int^{u} F(s)\,ds$$

where $u = u(y)$ *is a solution of the linear base equation*

$$du/dy + a(y)u + b(y) = 0$$

if and only if $H(y')$ *and* $G(y')$ *satisfy the connective condition*

$$H(y')\ d/dy'\ [G(y')/H(y')] = y'$$

Furthermore, the functional relation F can be obtained from

$$u = G(F)/H(F)$$

Proof

We note that if

$$u = u(y)$$

and

$$y = y(x)$$

then

$$u = u[y(x)] = u(x)$$

Applying Theorem 1 to the linear base equation

$$\frac{du}{dy} + a(y)u + b(y) = 0$$

yields the nonlinear equation

$$\frac{dp}{dy} + a(y)V(p) + b(y)W(p) = 0$$

where $p = F(u)$ is the functional relation and the connective condition is given by

$$W(p)\; d/dp\; [V(p)/W(p)] = 1$$

Suppose we introduce a new functional relation

$$y = \int^{u} F(s)\,ds$$

from which we obtain

$$y' = \frac{dy}{dx} = F(u) = p$$

$$y'' = \frac{dp}{dy}\frac{dy}{dx} = \frac{dp}{dy}y' = p\frac{dp}{dy}$$

By use of the definitions

$$G(y') = y'V(y')$$

$$H(y') = y'W(y')$$

substitution yields the nonlinear equation

$$y'' + a(y)G(y') + b(y)H(y') = 0$$

and the connective condition

$$H(y')\frac{d}{dy'}\frac{G(y')}{H(y')} = y'$$

which are associated with the base equation

$$du/dy + a(y)u + b(y) = 0$$

and

$$u = V(p)/W(p) = G(y')/H(y') = G(F)/H(F)$$

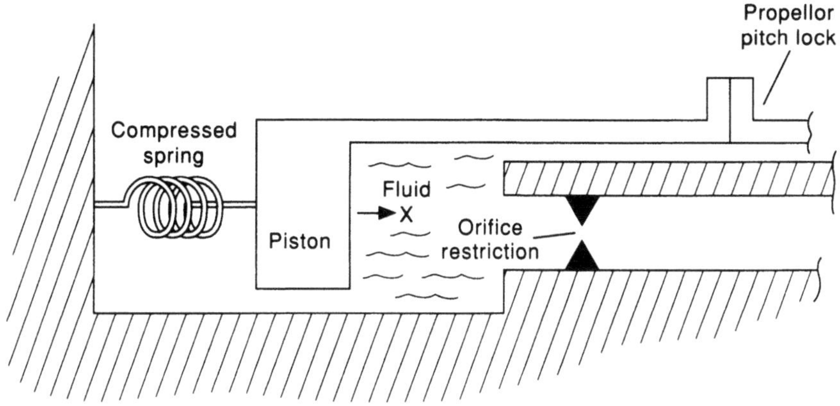

Figure 2-4. Example of a nonlinear system.

It should also be observed that should the connective condition yield a constant multiplier of y', using the technique of equation (20)—that is, by redefining the terms $G_1(y') = G(y')/k$, $H_1(y') = H(y')/k$, and $a_1(y) = ka(y)$ and $b_1(y) = kb(y)$, as with the prior Theorems—the connective condition is also satisfied.

A Nonlinear Hydraulic—Mechanical System

A system to lock a mechanism into a prescribed position is show in Figure 2-4. A piston, constrained by the balancing forces of spring tension and hydraulic fluid pressure controlled through an orifice, engages the locking device dependent upon a sensed signal, which opens or shuts the orifice.

In the unlocked position, hydraulic pressure in the line compresses the spring. When pressure is reduced, the spring forces the piston forward, engaging the locking device.

The piston is assumed to have a mass m, and thus it is acted on by the forces of

Inertia, my''; $\quad (\,)' = d(\,)/dt$

Spring preload, B

Spring force, Cy

and the orifice restriction is assumed to behave nonlinearly in accordance with the square of the velocity or Ay'^2.

Thus, the system equation is written as

$$my'' + Ay'^2 = B - Cy$$

where A, B, C, and m are constants. As the system equation is of the form of that of Theorem 6,

$$y'' + a(y)G(y') + b(y)H(y') = 0$$

by defining $G = y'^2/2$ and $H = 1$, the connective condition is satisfied. Thus the base equation is

$$du/dy + a(y)u + b(y) = 0$$

where

$$a(y) = 2A/m \quad \text{and} \quad b(y) = (Cy - B)/m.$$

Using equation (4), the solution of the base equation is

$$u = c_1 e^{-(2A/m)y} - \frac{C}{2A} y + \frac{B}{2A} + \frac{Cm}{4A^2}$$

where c_1 is a constant of integration.

Note that it is first desirable to define the interim first order nonlinear equation as

$$dp/dy + a(y)V(p) + b(y)W(p) = 0$$

following the proof so that

$$V(p) = V(y') = G(y')/y' = y'/2 = p/2$$

$$W(p) = W(y') = H(y')/y' = 1/y' = 1/p$$

which satisfy the connective condition of Theorem 1

$$W(p) \frac{d}{dp} \frac{V(p)}{W(p)} = 1$$

and thereby we obtain the interim functional relation

$$p = F(u)$$

from which we show

$$u = V(p)/W(p) = V(F)/W(F) = p^2/2 = F^2/2$$

Hence

$$p = F = (2u)^{1/2} = (2)^{1/2} \left[c_1 e^{-(2A/m)y} - \frac{Cy}{2A} + \frac{B}{2A} + \frac{Cm}{4A^2} \right]^{1/2} \tag{60a}$$

satisfies the equation in p

$$\frac{dp}{dy} + \frac{A}{m} p + \left(\frac{Cy}{m} - \frac{B}{m} \right) \frac{1}{p} = 0$$

However, we are interested in the solution to a second-order equation and have a theorem based on $y' = F = p$ where, in our case, the independent variable is time, t. Thus $(\ )' = d(\ )/dt$ and

$$\frac{dy}{dt} = (2u)^{1/2}$$

Substitution from equation (60a) yields

$$\int \left[c_1 e^{-(2A/m)y} - \frac{Cy}{2A} + \frac{B}{2A} + \frac{Cm}{4A^2} \right]^{-1/2} dy = (2)^{1/2} t + c_2$$

Now by reference to G. Petit Bois, *Tables of Indefinite Integrals*, Dover, New York, 1964 one finds on page 58 that

$$\int (ax^2 + bx + c)^{-1/2}\, dx = (a)^{-1/2} \log\, [b + 2ax + 2\{a(ax^2 + bx + c)\}^{1/2}]$$

and also a great many specific formulas which include reduction formulas for specific coefficients.

We will sidestep these issues by use of two considerations. First, if the initial condition is chosen so that $c_1 = 0$, the exponential term disappears and we can use the formula [G. Petit Bois, page 40].

$$\int (ax + b)^{-1/2}\, dx = (2/a)(ax + b)^{1/2}$$

so that

$$(2)^{1/2} t + c_2 = \frac{-4A}{C} \left(-\frac{Cy}{2A} + \frac{B}{2A} + \frac{Cm}{4A^2} \right)^{1/2}$$

where c_2 is a second constant of integration and from which we obtain

$$y = \frac{B}{C} + \frac{m}{2A} - \frac{C}{8A} [(2)^{1/2} t + c_2]^2$$

Alternately, if we expand the exponential into an infinite series and truncate after the second-order term, the first integration formula can be used since

$$c_1 e^{-(2A/m)y} \approx c_1 \left[1 - \frac{2Ay}{m} + \frac{2A^2}{m^2} y^2 \right]$$

and the radicand in the formula becomes

$$\frac{2A^2}{m^2} c_1 y^2 - \left(\frac{C}{2A} + \frac{2Ac_1}{m} \right) y + \frac{B}{2A} + \frac{Cm}{4A^2} + c_1$$

yielding

$$\left\{2c_1\left[\frac{2A^2}{m^2}c_1y^2 - \left(\frac{C}{2A} + \frac{2c_1A}{m}\right)y + \frac{B}{2A} + \frac{Cm}{4A^2} + c_1\right]\right\}^{1/2}$$

$$= \frac{Cm}{4A^2} + c_1 + \frac{mc_2}{2A}e^{(2A/m)(c_1)^{1/2}t} - \frac{2c_1A}{m}y$$

Squaring both sides yields a first order expression in y as a function of t as the quadratic terms cancel.

$$y = \frac{m}{2A} + \frac{Cm^2}{8A^3c_1} + \frac{m^2c_2}{8A^2c_1}e^{(2A/m)(c_1)^{1/2}t}$$

$$- \frac{1}{2c_2}\left(B/A + c_1 - \frac{C^2m^2}{16A^4c_1}\right)e^{-(2A/m)(c_1)^{1/2}t}$$

which approximates the solution as we used a truncated infinite series to represent the radicand for integration purposes.

We note that the expression for $y' = (2u)^{1/2}$ where u is given in terms of y is directly applicable to the phase plane techniques of Chapter 5. Thus an "exact graphical" solution is possible, akin to that obtained for the simple pendulum.

2.6 MATRIX NONLINEAR DIFFERENTIAL EQUATIONS

Just as the opening sections of this chapter extended the class of nonlinear differential equations which can be solved analytically in terms of ordinary differential equations, it is shown in this section that matrix nonlinear differential equations can be solved in the same manner.

We define vectors $\mathbf{y}$ and $\mathbf{u}$ of dimension n as

$$\mathbf{y} = (y_1, y_2, \cdots, y_n); \qquad y_i = y_i(x)$$

or

$$\mathbf{u} = (u_1, u_2, \cdots, u_n); \qquad u_i = u_i(x)$$

and we define matrices

$$\mathbf{A} = A(x); \qquad mxn$$

$$\mathbf{B} = B(x); \qquad mxn$$

where $\mathbf{G} = \mathbf{G}(\mathbf{y})$ and $\mathbf{H} = \mathbf{H}(\mathbf{y})$ are given matrices and their inverses $\mathbf{A}^{-1}$, $\mathbf{B}^{-1}$, $\mathbf{G}^{-1}$, and $\mathbf{H}^{-1}$ all exist.

The differential operators are defined as

$$D_x(\ ) = \frac{d}{dx}(\ ), \qquad \text{and} \qquad D_y(\ ) = \frac{d}{dy}(\ )$$

where we recall from the chain rule $D_x(\ ) = D_y(\ )\mathbf{y}'$. Using these conventions, we state and then prove the following theorem.

2.6.1 Theorem 7

The nonlinear matrix equation

$$\mathbf{y}' + \mathbf{G}(\mathbf{y})\mathbf{A}(x)\mathbf{H}(\mathbf{y}) + \mathbf{G}(\mathbf{y})\mathbf{B}(x) = \mathbf{0} \tag{61}$$

has the solution

$$\mathbf{y} = \mathbf{H}^{-1}(\mathbf{u}) \tag{62}$$

where u is the solution of the linear base equation

$$\mathbf{u}' + \mathbf{A}(x)\mathbf{u} + \mathbf{B}(x) = \mathbf{0} \tag{63}$$

provided the nonlinear matrix functionals **G** *and* **H** *satisfy the connective condition*

$$[\mathbf{G}(\mathbf{y})]^{-1} = D_y[\mathbf{H}(\mathbf{y})] \tag{64}$$

Proof

Beginning with (63), and substituting

$$\mathbf{u} = \mathbf{H}(\mathbf{y}) \tag{65}$$

where

$$\mathbf{u}' = D_x[\mathbf{H}(\mathbf{y})]$$

we get

$$D_x[\mathbf{H}(\mathbf{y})] + \mathbf{A}(x)\mathbf{H}(\mathbf{y}) + \mathbf{B}(x) = 0$$

However, from (65) and the chain rule

$$\mathbf{y} = \mathbf{H}^{-1}(\mathbf{u}), \qquad \text{and} \quad D_x[\mathbf{H}(\mathbf{y})] = D_y[\mathbf{H}(\mathbf{y})]\mathbf{y}'$$

and therefore

$$D_y[\mathbf{H}(\mathbf{y})]\mathbf{y}' + \mathbf{A}(x)\mathbf{H}(\mathbf{y}) + \mathbf{B}(x) = 0$$

and by defining

$$\mathbf{G}(\mathbf{y}) = \{D_y[\mathbf{H}(\mathbf{y})]\}^{-1}$$

there results

$$\mathbf{y}' + \mathbf{G}(\mathbf{y})\mathbf{A}(x)\mathbf{H}(\mathbf{y}) + \mathbf{G}(\mathbf{y})\mathbf{B}(x) = 0$$

provided $\mathbf{G}^{-1}$, $\mathbf{H}^{-1}$, and $\{D_y(\mathbf{H})\}^{-1}$ exist. To explicitly define $\mathbf{H}^{-1}(\mathbf{y})$, and thereby be able to solve for y [see equations (62) and (65)], we must form

$$\mathbf{G}^{-1}(\ ) = D_y\mathbf{H}(\ )$$

and

$$D_y^{-1}\mathbf{G}^{-1}(\ ) = \mathbf{H}(\ )$$

or

$$\mathbf{H}^{-1}(\ ) = [D_y^{-1}\mathbf{G}^{-1}(\ )]^{-1}$$

which completes the proof.

2.6.2 Comparison to the Nonmatrix Equation

The nonlinear matrix equation of Theorem 7

$$\mathbf{y}' + \mathbf{G}(\mathbf{y})\mathbf{A}(x)\mathbf{H}(\mathbf{y}) + \mathbf{G}(\mathbf{y})\mathbf{B}(x) = 0 \tag{66}$$

and the first-order nonlinear differential equation of Theorem 1

$$y' + a(x)f(y) + b(x)g(y) = 0 \tag{67}$$

are obviously similar. In essence, if the vectors $\mathbf{u}(x)$ and $\mathbf{y}(x)$ are of dimension 1, then $n = 1$ or

$$\mathbf{y} = (y_1, y_2, \cdots, y_n) \text{ must be } \mathbf{y} = (y_1) = y$$

and likewise

$$\mathbf{A}(x) = \{a_{ij}(x)\} = a_{11}(x) = a(x)$$

$$\mathbf{B}(x) = \{b_{ij}(x)\} = b_{11}(x) = b(x)$$

further, in the nonmatrix case,

$$b(x)g(y) = g(y)b(x)$$

and therefore

$$\mathbf{G}(\mathbf{y}) = \{g_{ij}(y)\} = \{g_{11}(y)\} = g(y)$$

similarly,

$$\mathbf{H}(\mathbf{y}) = \{h_{ij}(y)\} = \{h_{11}(y)\} = h(y)$$

Thus, to identically satisfy both the matrix equation (66) and the first-order equation (67),

$$\mathbf{G}(\mathbf{y})\mathbf{A}(x)\mathbf{H}(\mathbf{y}) = a(x)f(y)$$

Substituting

$$g(y)a(x)h(y) = a(x)f(y)$$

or

$$f(y) = g(y)h(y)$$

Under these conditions, the matrix equation and the first-order nonlinear differential equation are equivalent.

2.6.3 Comparison to the Ricatti Equation

The homogeneous Ricatti equation is

$$y' + Q(x)y + R(x)y^2 = 0$$

From the prior section, if we set

$$Q(x) = a(x) = \mathbf{A}(x)$$

$$R(x) = b(x) = \mathbf{B}(x)$$

we get the semi-matrix nonlinear equation

$$y' + \mathbf{A}(x)y + \mathbf{B}(x)y^2 = 0$$

or

$$y' + \mathbf{A}(x)y^2 + \mathbf{B}(x)y = 0$$

where to complete the analogy, we need to define the matrices **G** and **H**.

Thus, if $\mathbf{y} = (y_1, \cdots, y_n)$, we need to only satisfy

$$\mathbf{GAH} = ay^2 \qquad \text{and } \mathbf{GB} = by$$

or

$$\mathbf{GAH} = ay \qquad \text{and } \mathbf{GB} = by^2$$

Note that if $\mathbf{A}(x) = 0$ in the matrix equation, $\mathbf{H}(\mathbf{y})$ is arbitrary and can be chosen so that the connective condition is satisfied. Then the solution is related to

$$\mathbf{u}' + \mathbf{B}(x) = 0$$

or

$$\mathbf{u} = \int^x \mathbf{B}(x)\,dt + c$$

2.7 APPLICATION TO THE CALCULUS OF VARIATIONS

The calculus of variations (COV) is the branch of mathematics dealing with minima and maxima of functionals, such as

$$I[y(x)] = \int_{x_0}^{x_1} H(x, y, y')\,dx$$

in an analogous fashion as the ordinary calculus seeks minima and maxima of functions of the form

$$y = f(x)$$

In the calculus, derivatives are used to find these extreme values, whereas the COV uses variations. In the COV, the first necessary condition is expressed in terms of satisfying the Euler–Lagrange equation

$$\frac{d}{dx}(H_{y'}) - H_y = 0 \tag{68}$$

where H is the integrand, the subscripts of equation (68) refer to partial derivatives, and the solution $y = y(x)$ establishes the class of admissible solutions.

The basic difficulty in finding those functions $y = y(x)$ satisfying the Euler–Lagrange equation, which are termed extremals, is that the equation is almost always nonlinear. Hence, it has become necessary to investigate numerical solutions of this equation because of the wide range of problems arising from consideration of the calculus of variations. These problems have developed in such diverse fields as orbital mechanics, high-speed transportation, exterior ballistics, control system design, and process controllers.

2.7.1 Development of the Concept

To illustrate the application of the theory presented in this chapter, consider equation (68) or its expanded version

$$H_{y'y'}y'' + H_{y'y}y' + H_{y'x} - H_y = 0 \tag{69}$$

It is immediately seen that (69) is a second-order equation which, when written in the form of equation (21) of Theorem 2, is given as

$$y'' + [H_{y'y}y' + H_{y'x} - H_y]/H_{y'y'} = 0 \tag{70}$$

where it is assumed that $H_{y'y'}$ is unequal to zero. [$Hy'y' = 0$ reduces (69) to a first-order equation, and Theorem 1 would be considered.]

For ease of comparison, (21) is rewritten here

$$y'' + p(x)y' + q(x)G(y) + r(x)Z(y) = (\dot{Z}/Z)y'^2 \tag{71}$$

The concept we shall now explore concerns two questions:

1. What form should H assume to satisfy the requirements of Theorem 2?
2. Having defined H, what is the definition of the extremals for the functional $I(y(x))$?

By combining (70) and (71), the following equation results

$$(H_{y'y}y' + H_{y'x} - H_y)/H_{y'y'} = p(x)y' + q(x)G(y) + r(x)Z(y) - (\dot{Z}/Z)y'^2 \tag{72}$$

Suppose that $H = H(x, y, y')$ is separable

$$H = H_1(x)H_2(y)H_3(y')$$

substituting into (72) yields the equation

$$\begin{aligned} p(x)y' + q(x)G(y) + r(x)Z(y) - (\dot{Z}/Z)y'^2 &= \frac{1}{H_2}\frac{dH_2}{dy}\left(y'\frac{dH_3}{dy'} - H_3\right)\Big/\frac{d^2H_3}{dy'^2} \\ &+ \left(\frac{1}{H_1}\frac{dH_1}{dx}\frac{dH_3}{dy'}\right)\Big/\frac{d^2H_3}{dy'^2} \end{aligned} \tag{73}$$

To define H_1, H_2, and H_3, we now compare terms, establish the associated differential equations, and solve. One possible set of differential equations is:

$$p(x)y' = \left[\frac{1}{H_1}\frac{dH_1}{dx}\frac{dH_3}{dy'}\right]\Big/\frac{d^2H_3}{dy'^2} \tag{74}$$

and

$$\frac{-2\dot{Z}}{Z}\frac{y'^2}{2} = \frac{1}{H_2}\frac{dH_2}{dy}\left[y'\frac{dH_3}{dy'} - H_3\right]\Big/\frac{d^2H_3}{dy'^2} \tag{75}$$

Considering (74), the definitions $H_1 = H_1(x)$ and $H_3 = H_3(y')$ imply

$$p(x) = 1/H_1\, dH_1/dx$$

or

$$H_1(x) = \exp\left[\int^x p(s)ds\right] \tag{76}$$

and

$$y' = \frac{dH_3}{dy'}\Big/\frac{d^2H_3}{dy'^2} \tag{77}$$

which has the solution

$$H_3(y') = y'^2/2 \tag{78}$$

Returning to (75), the definitions of H_2, H_3, and Z imply

$$\frac{y'^2}{2} = \left(y'\frac{dH_3}{dy'} - H_3\right)\Big/\frac{d^2H_3}{dy'^2} \tag{79}$$

and

$$\frac{-2\dot{Z}}{Z} = \frac{1}{H_2}\frac{dH_2}{dy} \tag{80}$$

Integrating (80) yields

$$H_2(y) = Z(y)^{-2} \tag{81}$$

Hence, we conclude that the answer to our first query is

$$H = H_1 H_2 H_3 = \exp\left[\int^x p(s)ds\right] Z(y)^{-2}(y'^2/2) \tag{82}$$

And, thus, when the definition of terms of (70) are applied to (82), the equivalent of (71) results as

$$y'' + p(x)\exp\left[-\int^x p(s)ds\right] y' = -y'^2/Z(y)^2 \tag{83}$$

which is the Euler–Lagrange equation associated with the function H.

2.7.2 Solving the Euler–Lagrange Equation

Having defined the nonlinear differential equation (82), under the assumption that H is separable, we now seek its solution in terms of the conditions of Theorem 2. Referring once again to (71), we note that the terms equivalent to $q(x)G(y)$ and $r(x)Z(y)$ in (82) are missing. Hence, we conclude that $q = r = 0$, and since $G(y)$ is not explicitly defined, we can define it as we so desire since it has a zero coefficient. Thus the connective condition (25) is satisfied and (82) can be solved by the method of Theorem 2. Specifically, referring to Theorem 2,

$$y = F(u)$$

where $u = u(x)$ is a solution of the linear base equation

$$u'' + p(x)\exp\left[-\int^x p(s)ds\right] u' = 0$$

or

$$u'' + R(x)u' = 0; \qquad R(x) = p(x)\exp\left[-\int^x p(s)ds\right]$$

which has the solution

$$u = c_1 + \int^x c_2 \exp\left[\int^t R(s)ds\right] dt \tag{84}$$

Finally, the functional relation F is found from (23) as

$$u = G(F)/Z(F)$$

where $G(F)$ is defined by the connective condition in terms of Z. If, for example, $Z(y) = y^2$, then the connective condition yields

$$G = cy^2 - y$$

and then $u = c - 1/F$ implies that

$$y = F = 1/(c - u) \tag{85}$$

where u is given in (84). Thus, (85) is the family of extremals satisfying query 2.

The nature of these investigations, then, can be expected to yield useful theorems associated with the variational calculus. It is important to note that the example considered above can be investigated in terms of each of Theorems 1–7, and also in terms of assumed forms for H other than $H = H_1(x)H_2(y)H_3(y')$. For example, the example just discussed can be explored in terms of integral transformations by recognizing that if we redefine (83) using the change of variables $y' = w$, a Riccati equation results which is of central importance in optimization problems.

Thus, the generation of new results as was accomplished in this section, when coupled with practical applications (such as in ref. 12), should be of benefit to the scientific and engineering communities. The results given above are summarized in the following theorem.

2.7.3 Theorem 8

The variational problem in Lagrange form

$$I[y(x)] = \int_{x0}^{xl} H(x, y, y')dx$$

can be solved using Theorem 2 if

$$H = 1/2 \exp\left[\int^{x} p(s)ds\right] \{(y'/Z(y)\}^2$$

where the associated Euler–Lagrange equation is given as equation (83).

Further, the resulting Euler–Lagrange equation and its solution are associated with the Riccati equation and its solution by means of the transformation $y' = w$.

It is reasonable to expect that many more such theorems and attendant applications can be generated using the technique discussed in this chapter. It is left to the readers to explore these possibilities in terms of their needs.

2.8 PROBLEMS

1. Show that if the equation is homogeneous of first degree in y—that is, if

$$f(ay) = a^n f(y); \qquad n = 1$$

multiples of y are also solutions to the nonlinear differential equation of Theorem 1.

2. Show that by using the connective condition of Theorem 2 the coefficient of the nonlinear damping term y'^2 can be

$$(\dot{G} - 1)/G; \qquad (\dot{\ }) = d(\ )/dy$$

3. Apply the results of problem 2 to solve

$$y'' + a(x)y^2 = (2/y)y'^2$$

4. Show that, if in Theorem 2, the nonlinear damping term is missing, that the general nonlinear equation becomes the linear second-order differential equation.

5. Show that the equation

$$\begin{pmatrix} y_1' \\ y_2' \end{pmatrix} + (a\ b)\begin{pmatrix} c & d \\ e & f \end{pmatrix}\begin{pmatrix} g \\ h \end{pmatrix} + (a\ b)\begin{pmatrix} i & k \\ j & l \end{pmatrix} = 0$$

satisfies the matrix equation of Theorem 7, where

$$G(y) = (a\ b) \qquad A(x) = \begin{pmatrix} c & d \\ e & f \end{pmatrix}$$

$$H(y) = \begin{pmatrix} g \\ h \end{pmatrix} \qquad B(x) = \begin{pmatrix} i & k \\ j & l \end{pmatrix}$$

Further, if y_2 = constant, show that $y = (y_1, k)$ also satisfies Theorem 7.

6. (a) Show that Clairaut's equation

$$y = xy' + f(y')$$

can be solved graphically by the substitution $y' = p$, yielding

$$p'[x + f'(p)] = 0$$

(b) Further, show that since by integration

$$f(p) + x^2/2 + C = 0$$

or

$$f(y') + x^2/2 + C = 0$$

that if $f(y')$ satisfies Theorem 1, then a class of analytically solvable Clairaut's equations have the form

$$f[(au + b)F'] + x^2/2 + C = 0$$

(c) Show that if $H = e^{ky}$, that the connective condition yields $G = -1/k$ and that the "nonlinear" Clairaut equation reduces to

$$y' + c_0 e^{ky} + x^2/2 + C = 0$$

(d) Show that if $H = y^k$, k = constant, $G = y/(1 - k)$ satisfies the connective condition.

(e) Why cannot Theorem 6 be used in solving the Clairaut equation?

7. Show that if the nonlinear equation solvable in terms of the elliptic equation $u'' = au + bu^3$ does not include a term explicitly in y'^2 (i.e., if $\dot{G}/G = 0$), Theorem 5, then the equation reduces to the elliptic equation itself.

8. (a) Prove the following theorem for third-order nonlinear equations:

Theorem 9:

The nonlinear differential equation

$$y''' + p(x)y'' + q(x)y' + r(x)Z(y) + s(x)G(y) = [d/dy(\dot{G}/G) - (\dot{G}/G)^2]y'^3 + 3(\dot{G}/G)y'y'' + p(\dot{G}/G)y'^2 \qquad (\dot{\ }) = d(\ )/dy$$

has the solution

$$y = F(u)$$

where $u = u(x)$ is the solution of the base equation

$$u''' + p(x)u'' + q(x)u' + r(x)u + s(x) = 0$$

and the functional relation F can be obtained from

$$u = Z(F)/G(F)$$

if and only if the connection condition

$$G(y)\, d/dy[Z(y)/G(y)] = 1$$

is satisfied.

Hint: From $G = F_u$, form $F_{uu} = G\dot{G}$ and $F_{uuu} = G^2\ddot{G} + \dot{G}^2 G$, and define $Z = uG$.

(b) Show that if G = constant, the nonlinear equation of Theorem 9 becomes linear and is almost identical to the base equation.

(c) Show that if Z = constant = k; $G = c \exp[-y/k]$, c = constant of integration.

9. Ricatti's first-order nonlinear differential equation

$$y' + p(x)y + q(x)y^2 = r(x)$$

is reduced to a second-order linear ordinary differential equation

$$Z'' + (p - q'/q)Z' - qrZ = 0$$

under the transformation

$$y = Z'/qZ$$

(a) Show that if one generalizes the functional relation to

$$y = Z'H(x)F(Z)$$

that this requires (for the transformed equation to be a linear differential equation)

$$H = 1/q \text{ and } F = 1/Z$$

(b) Show that if we associate the base equation

$$u'' + p(x)u' + q(x)u + r(x) = 0$$

with the functional relation of part (a), the second-order nonlinear equation associated with Ricatti's equation becomes

$$Z'' + (P + H'/H)Z' - R/HF = -(F' + HQF^2)Z'^2/F$$

(c) Apply Theorem 2 to the results of (b) and develop the appropriate relationships from the connective condition

$$Z(y)\, d/dy[G(y)/Z(y))] = 1$$

and the inverse of the functional relation

$$u = G(F)/Z(F)$$

10. Using the transformation

$$y' = F(u)$$

where $u = u(x)$ and the linear base equation

$$u' + a(x)u + b(x) = 0; \qquad (\,)' = d(\,)/dx$$

derive the nonlinear equation

$$y'' + a(x)G(y) + b(x)H(y) = 0$$

and the required connective condition.

11. For the system of Figure 2.4, Section 2.5.1,
(a) using the definition

$$G(y') = y'^2 - B/A$$

derive $a(y)$, $b(y)$, and $H(y')$ and show that the connective condition is satisfied. What is the form of the functional relation? Discuss how it modifies the derivation of the solution.

(b) Assume the orifice nonlinearity is described by

$$y'|y'|$$

instead of y'^2. How would you solve the problem in this form?

12. In ref. 14, equations of the form

$$y' = f[(a_1 + b_1x + c_1y)/(a_2 + b_2x + c_2y)]$$

are transformed into separable equations, such as

$$g(y)\,dy = h(g)dx$$

by means of a coordinate transformation

$$x = x_1 + h, \qquad y = y_1 + k$$

and then letting $y = vx_1$. This separation of variables occurs because of the equation's homogeneity.

(a) Develop a theorem that generalizes this concept.

(b) Develop a second theorem for second-order equations.

2.9 REFERENCES AND RELATED LITERATURE

2.9.1 References

1. Pinney, E., The nonlinear differential equation $y'' + p(x)y + cy^{-3} = 0$, *Proc. Amer. Math. Soc.* **1,** 581 (1950).
2. Thomas, J. M., Equations equivalent to a linear differential equation, *Proc. Amer. Math. Soc.* **3,** 899–903 (1952).
3. Herbst, R. T., The equivalence of linear and nonlinear differential equations, *Proc. Amer. Math. Soc.* **7,** 95–97 (1956).
4. Gergen, J. J., and Dressel, F. G., Second-order linear and nonlinear differential equations, *Proc. Amer. Math. Soc.* **16,** 767–773 (1965).
5. deSpautz, J. F., and Lerman, R. A., Equations equivalent to nonlinear differential equations, *Proc. Amer. Math. Soc.* **18,** 441–444 (1967).
6. Lambert, W. G., Solution of nonlinear differential equations by transformation of variable, *Instruments and Control Systems* **39,** 177–178 (1966).
7. Davis, H. T., *Introductions to Nonlinear Differential and Integral Equations*, Dover, New York, (1962).
8. Lerman, R. A., and deSpautz, J. F., Analytical solutions of nonlinear differential equations, *Instruments and Control Systems*, **40,** 99–102 (1967).
9. deSpautz, J., and Lerman, R. A., Nonlinear differential equations and the terrestrial brachistochrone, *AIAA Journal* **7,** 1173–1174 (1969).
10. Cohen, A., *An Elementary Treatise on Differential Equations*, D.C. Heath and Company, New York (1933).

11. Magnus, W., and Winkler, S., *Hill's Equation*, Interscience Publishers, New York (1966).
12. Athens, N., and Levine, W. S., On the numerical solution of the matrix Ricatti differential equation using a Runge–Kutta scheme, Report ESL-R-276, M.I.T. Project DSR 76105, Department of Commerce Contract C-85-65, Massachusetts Institute of Technology (July 1966).
13. Bliss, G. A., *Lectures on the Calculus of Variations*, The University of Chicago Press, Chicago (1946).
14. Ford, L. R., *Differential Equations*, First Edition, McGraw-Hill, New York (1933), pp. 46–49.
15. Arscott, F. M., *Periodic Differential Equations*, Macmillan, New York (1964).

2.9.2 Related Literature

1. Moulton, F. R., *Differential Equations*, Dover, New York (1958).
2. Forsyth, A. R., *Calculus of Variations*, Dover, New York (1960).
3. Akhiezer, N. I., *The Calculus of Variations*, Blaisdell Publishing, New York (1962).
4. Sugai, Iwao, Ricatti's nonlinear differential equation,'' *Amer. Math. Monthly*, **67**, 134–139 (1960).
5. Sneddon, Ian N., *Elements of Partial Differential Equations*, McGraw-Hill, New York (1957).
6. Broman, Arne, *Introduction to Partial Differential Equations from Fourier Series to Boundary-Value Problems*, Dover, New York (1989).
7. Churchill, Ruel V., *Fourier Series and Boundary Value Problems*, McGraw-Hill, New York (1941).
8. Churchill, Ruel V., *Operational Mathematics*, McGraw-Hill, New York (1958).
9. Tenenbaum, Morris, and Pollard, Harry, *Ordinary Differential Equations*, Dover, New York (1985).

Chapter 3
Linearization

3.0 INTRODUCTION

Linearization is commonly used to analyze a nonlinear system for small departures from an operating point. Its stability and transient nature can then be quantified for comparison with the system's requirements. When the perturbations are small enough, their products and powers are arguably negligible, so discarding them reduces the system dynamics to linear equations in the perturbed variables from which the eigenvalues may be found and the motion described. The legacy of small perturbations is just that; their smallness must be assured as otherwise the validity of the approximate linear equations degenerates rapidly, leaving no precise measure of a neighborhood of acceptable solutions. The nature of the nonlinearity plays an important role in determining if there is a neighborhood of validity. Continuous single-valued functions, such as the torque-speed curve of an ac induction motor or the pitching moment curve of an airplane, are typically good applications. Discontinuous functions which are not single-valued, such as gear train backlash or relays, are never the objects of small perturbation analysis. In general, a function expandable in a Taylor series is a candidate for small perturbation analysis, although success requires the truncation error after two terms to be acceptably small. Since each succeeding term in the series contains the corresponding derivative, those functions having discontinuities will have extremely large values in their higher derivatives and, accordingly, their truncation errors cannot be ignored

3.1 LINEARIZING ALGEBRAIC FUNCTIONS

Often, the system contains elements described by nonlinear algebraic functions as shown in Figure 3-1. Typical examples are a spring which hardens in compression and softens in tension, or an orifice whose flow coefficient changes with pressure ratio.

The problem is to find a linear approximation to the output of the nonlinear device, and, as illustrated, the method consists of running a straight line

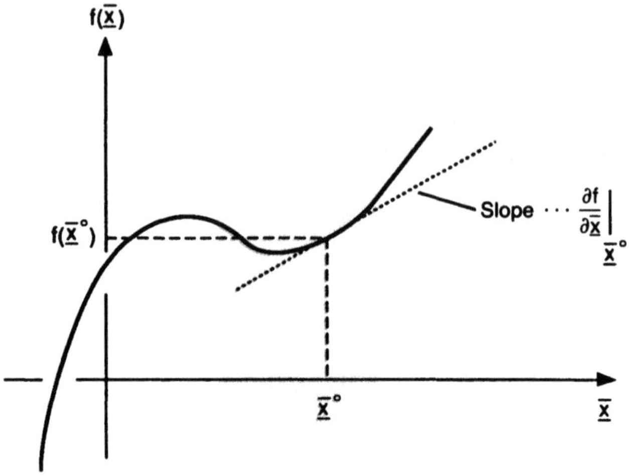

Figure 3-1. Nonlinear algebraic function in a single variable.

through the operating point whose slope best fits the region of interest. Ordinarily, the tangent is taken for analytic ease because of its presence in the Taylor expansion.

$$f(\underline{\bar{x}}) = f(\underline{\bar{x}}^0) + \frac{(\underline{\bar{x}} - \underline{\bar{x}}^0)}{1!} \cdot \left.\frac{\partial f}{\partial \underline{\bar{x}}}\right|_{\underline{\bar{x}}^0} + \frac{(\underline{\bar{x}} - \underline{\bar{x}}^0)^2}{2!} \cdot \left.\frac{\partial^2 f}{\partial \underline{\bar{x}}^2}\right|_{\underline{\bar{x}}^0} + \cdots$$

As long as the perturbation, $x = \underline{\bar{x}} - \underline{\bar{x}}^0$, is small we argue that powers x^2, x^3, etc., are infinitesimal and the Taylor series may be truncated to give

$$f(\underline{\bar{x}}) \approx f(\underline{\bar{x}}^0) + x \cdot \frac{\partial f}{\partial \underline{\bar{x}}}(\underline{\bar{x}}^0)$$

where $x \equiv \underline{\bar{x}} - \underline{\bar{x}}^0$. There are certain obvious restrictions on this method. First, there can be no ambiguity in $f(\underline{\bar{x}})$, which would occur if it were multivalued as in Figure 3-2(a). Such a case is multivalued in both the function and its argument, and is said to have memory because past history is needed to specify the output value. Secondly, if $f(\underline{\bar{x}})$ contains a discontinuity in the neighborhood of the operating point as shown in Figure 3-2(b), no suitable linearization is possible, because the function is not unique at the discontinuity and its derivatives are either enormous or nonexistent.

Methods of analysis of such functions include the phase plane and describing function, which are included in later chapters.

So far we have considered nonlinear functions of a single variable. Yet functions of two or more variables frequently occur, such as the torque of an ac power servo motor (Figure 3-3), which depends on both the voltage and the

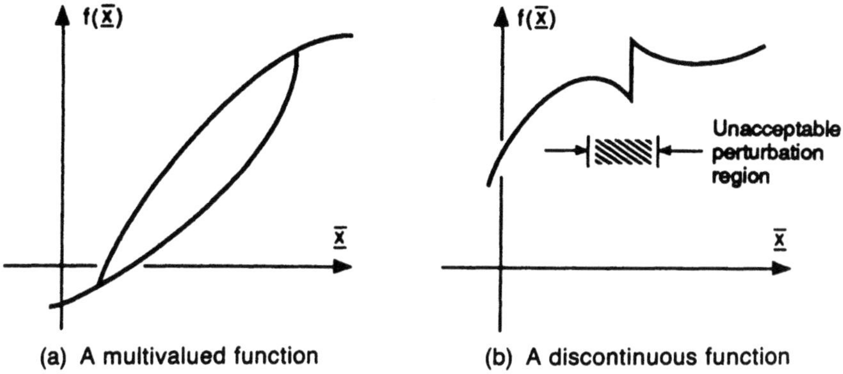

Figure 3-2. Algebraic functions that do not linearize.

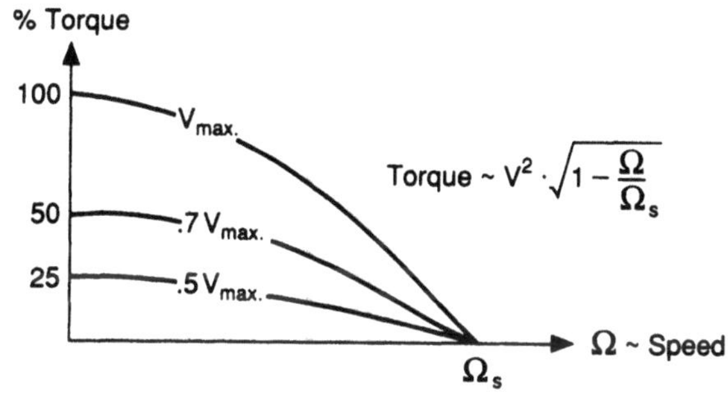

Figure 3-3. Torque speed of ac power servo motor; example of bivariant function.

motor speed. Thus, the torque, $\tau = f(V, \Omega)$, is expandable in a Taylor series about an operating point, as shown below.

$$\tau = f(V^0, \Omega^0) + (V - V^0) \cdot \frac{\partial f^0}{\partial V} + (\Omega - \Omega^0) \cdot \frac{\partial f^0}{\partial \Omega}$$

$$+ \cdots + \frac{(V - V^0)^2}{2!} \cdot \frac{\partial^2 f^0}{\partial V^2} + \frac{(\Omega - \Omega^0)^2}{2!} \cdot \frac{\partial^2 f^0}{\partial \Omega^2}$$

$$+ (V - V^0)(\Omega - \Omega^0) \frac{\partial^2 f^0}{\partial V \cdot \partial \Omega} + \cdots$$

Perturbations in V and Ω are again considered small so that their powers and products disappear,

$$\tau \approx f(V^0, \Omega^0) + v \cdot \frac{\partial f^0}{\partial V} + \omega \cdot \frac{\partial f^0}{\partial \Omega}$$

where $v \equiv V - V^0$ and $\omega \equiv \Omega - \Omega^0$, and the partials are evaluated at the operating point (V_0, Ω_0). When there are n variables in the argument, the approach yields,

$$y = f(\bar{\underline{x}}_1 + \bar{\underline{x}}_2, \cdots, \bar{\underline{x}}_n) = f(\bar{\underline{x}}])$$

$$y \approx f(\bar{\underline{x}}^0]) + x_1 \cdot \frac{\partial f^0}{\partial \bar{\underline{x}}_1} + x_2 \cdot \frac{\partial f^0}{\partial \bar{\underline{x}}_2} + \cdots + x_n \cdot \frac{\partial f^0}{\partial \bar{\underline{x}}_n}$$

and, if the output has more than one component, it can be described as a vector,

$$y] = \begin{matrix} f_1(\bar{\underline{x}}]) \\ f_2(\bar{\underline{x}}]) \\ \vdots \\ f_n(\bar{\underline{x}}]) \end{matrix} \Bigg] = \bar{f}(\bar{\underline{x}}])$$

and the linear approximation is,

$$y] = [c] \cdot x]$$

where $c_{ij} = \partial f_i^0 / \partial x_j$.

3.2 LINEARIZING A TRANSISTOR

At low frequencies the reactive characteristics of a bipolar transistor are negligible, and its resistive characteristics dominate. Thus, the characteristics contain neither capacitive nor inductive effects, and the device is not frequency-dependent. The common-emitter configuration is shown in Figure 3-4.

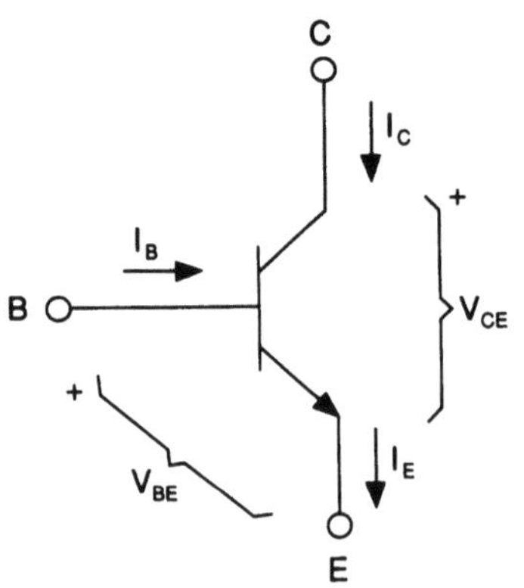

Figure 3-4. Bipolar transistor: common emitter.

TABLE 3.1. Steps in Linearizing a Transistor's Characteristics

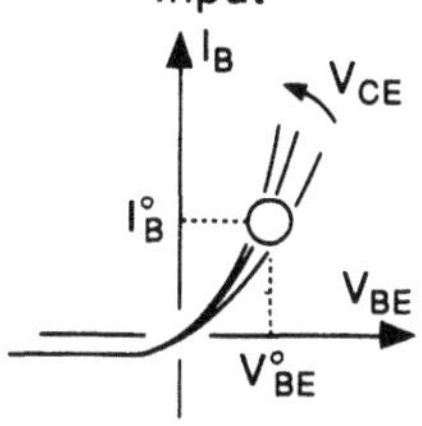

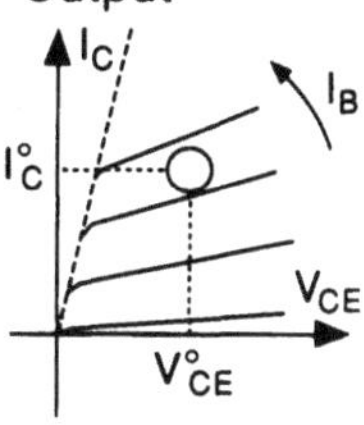

	INPUT CHARACTERISTICS	OUTPUT CHARACTERISTICS
GEN'L RELATION	$I_B = F(V_{BE}, V_{CE})$	$I_C = G(V_{CE}, I_B)$
DIFFERENTIATE	$dI_B = \frac{dF}{dV_{BE}} \cdot dV_{BE} + \frac{\partial F}{\partial V_{CE}} \cdot dV_{EE}$	$dI_C = \frac{\partial G}{\partial V_{CE}} \cdot dV_{CE} + \frac{\partial G}{\partial I_B} \cdot dI_B$
DEFINE OPERATING POINT	$\{V_{BE}^0, V_C^0, I_B^0, I_C^0\}$	
PERTURBATIONS	$v_{BE} \equiv V_{BE} - V_{BE}^0, \cdots$ etc. $\cdots$	
SMALL PERTURBATIONS	$v_{BE} \approx d_{BE}, \cdots$ etc. $\cdots$	
PERTURB'N RELATIONS	$i_B \approx \frac{\partial F}{\partial V_{BE}} \cdot v_{BE} + \frac{\partial F}{\partial V_{CE}} \cdot v_{CE}$	$i_C \approx \frac{\partial G}{\partial V_{CE}} \cdot v_{CE} + \frac{\partial G}{\partial I_B} \cdot i_B$
CONSTANT PARTIALS	$i_B \approx \left.\frac{\partial F}{\partial V_{BE}}\right\|_{OP} \cdot v_{BE} + \left.\frac{\partial F}{\partial V_{CE}}\right\|_{OP} v_{CE}$	$i_C \approx \left.\frac{\partial G}{\partial V_{CE}}\right\|_{OP} \cdot v_{CE} + \left.\frac{\partial G}{\partial I_B}\right\|_{OP} \cdot i_B$

Linear model

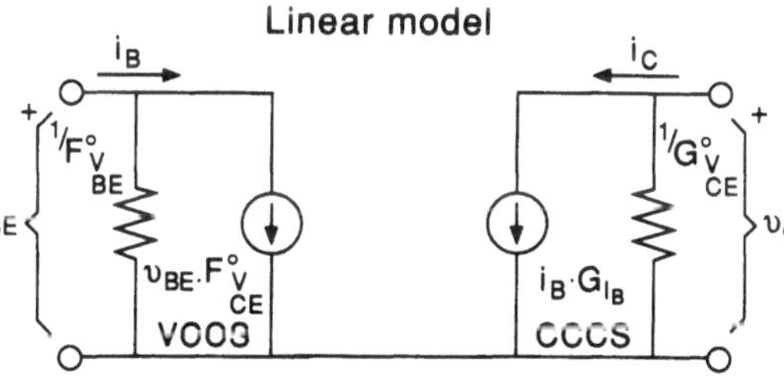

In this configuration the base-emitter terminals make up the input port, while the collector-emitter terminals make up the output port. By custom the input port is characterized by its I-V (current-voltage) relation with one of the output variables held constant, and a corresponding I-V relation characterizes the output. Table 3-1 depicts these relationships and tabulates the successive steps in the linearization, culminating in a small-signal model.

Manufacturer's data sheets provide values for the various partials as fol-

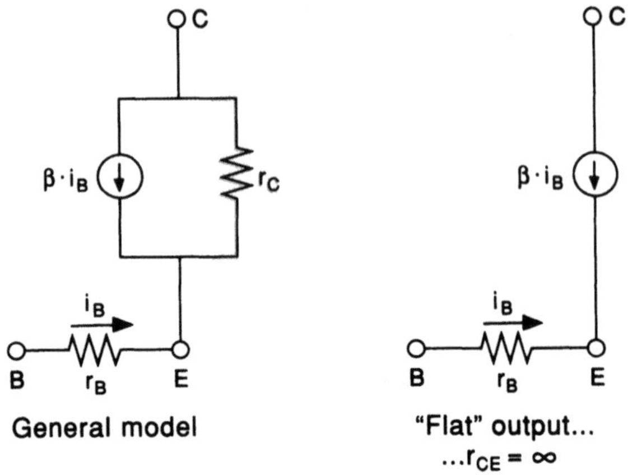

Figure 3-5. Linearized transistor models.

lows:

$$\text{base conductance: } \frac{\partial F^0}{\partial V_{BE}} = \frac{\partial I_B^0}{\partial V_{BE}} \equiv \frac{1}{r_B}$$

$$\text{negligible transconductance: } \frac{\partial F^0}{\partial V_{CE}} = \frac{\partial I_B^0}{\partial V_{CE}} \approx 0$$

$$\text{current gain: } \frac{\partial G^0}{\partial I_B} = \frac{\partial I_C^0}{\partial I_B} \equiv \beta$$

$$\text{collector conductance: } \frac{\partial G^0}{\partial V_{CE}} = \frac{\partial I_C^0}{\partial V_{CE}} \equiv \frac{1}{\nu_c}$$

The linearized models are shown in Figure 3-5.

3.3 LINEARIZATION OF DIFFERENTIAL FUNCTIONS

Linearization of differential functions, and thereby differential equations, is often encountered in engineering problems. Although succeeding chapters address other approximating methods, and the prior chapter addressed closed form solutions, it is important to have in one's bag of tricks a wide variety of methods of analysis, as one never knows, a priori, which is applicable.

In most approaches to studying the dynamic behavior of systems, it is assumed (because it is usually true) that some nominal behavior pattern will occur or that the system will come to rest if it is stable. On the assumption that such behavior is known (or can be found theoretically or in practice), we then ask the question, What happens if such normal behavior is disturbed? That is, what happens to the system if some external event perturbs the system so that the steady state is disrupted?

For example, consider the linear oscillator

$$x'' + x = 0$$

which has a steady-state oscillation expressed by

$$x = c_1 \sin t + c_2 \cos t$$

and if the initial position $x(0) = 0$ and velocity $x'(0) = 1$, the system dynamics become *for those initial conditions*

$$x = \cos t$$

Now supposing a perturbing vibration would change the system so that $x(t_1) = 1.2$ and $x'(t_1) = 0.6$, and this occurred when $t_1 = \pi$, then the system dynamics would oscillate according to

$$x = -0.6 \sin t - 1.2 \cos t$$

The above example was straightforward, because the system model was linear. In such cases, closed-form solutions are obtainable. What do we do when the model is nonlinear and/or a closed-form solution is not known? These issues are addressed below.

3.3.1 Small Perturbation Assumptions

Time domain linearization is illustrated by an ordinary nonlinear differential equation representing an ideal simple pendulum.

$$\ddot{\theta} + \sin \theta = 0$$

Because we can envision the pendulum's motion, it is easy to interpret our method of solution. In systems that are not as physically understandable, we require a methodology which allows for the insights to unfold. Therefore, we assume that a convenient and arbitrary operating point Θ^0 is selected. We next seek the effect of a small perturbation θ on the system. That is, we assume

$$\Theta \equiv \Theta^0 + \theta$$

The operating point Θ^0 is assumed fixed (i.e., stationary) and the resulting variable dynamics due to the perturbation variable θ. Then $\Theta' = \theta'$, $\Theta'' = \theta''$.

Substitution gives an equation in the perturbation variable and the operating point,

$$\ddot{\theta} + \sin \theta \cdot \cos \Theta^0 + \cos \theta \cdot \sin \Theta^0 = 0$$

and because our basic assumption is that the perturbation variable must be small,

$$\sin \theta \approx \theta, \qquad \cos \theta \approx 1$$

yielding the linear equation,

$$\ddot{\theta} + \theta \cdot \cos \Theta^0 = -\sin \Theta^0$$

which, using the convenient definition, $\omega_0^2 = \cos \Theta^0$, has the solution,

$$\theta(t) = \theta(0) \cdot \cos \omega_c t + \frac{\dot{\theta}(0)}{\omega_0} \cdot \sin \omega_0 t - \{1 - \cos \omega_c t\} \cdot \tan \Theta^0,$$

$$\text{for } \omega_0^2 > 0$$

$$\theta(t) = \theta(0) \cdot \cosh \alpha t + \frac{\theta(0)}{\alpha} \cdot \sinh \alpha t - \{1 - \cosh \alpha t\} \cdot \tan \Theta^0,$$

$$\text{for } \omega_0^2 = -\alpha^2 < 0$$

3.3.2 How Small Perturbations Can Fail

The above method works satisfactorily as long as the nonlinearity in the equation is only in the zeroth derivative such as the ideal pendulum, or $\ddot{\underline{X}} + \dot{\underline{X}} + \underline{X}^2 = 0$. However, difficulties will be encountered when the nonlinearity occurs in other than the zeroth derivative. Reexamining the procedure, note that we chose an operating point and defined the perturbation. However, we failed to state anything regarding the perturbation velocity

$$X = x_0 + x, \qquad \dot{X} = \dot{x}$$

We have forced too much on the procedure because it denies $\dot{x} \ll \dot{\underline{X}}$, and there is no guarantee of "smallness" in products and powers. This difficulty is illustrated by the Van der Pol equation,

$$\ddot{\underline{X}} - 2 \cdot \epsilon_0 \cdot \{1 - \beta \cdot \underline{X}^2\} \cdot \dot{\underline{X}} + \underline{X} = 0$$

choosing,

$$\underline{X} \equiv \underline{X}^0 + x \Rightarrow \dot{\underline{X}} = \dot{x} \qquad \text{and } \ddot{\underline{X}} = \ddot{x}$$

substituting,

$$\ddot{x} - 2 \cdot \epsilon_0 \cdot \{1 - \beta \cdot (\underline{X}^0)^2\} \cdot \ddot{x} + x$$
$$= -4\epsilon_0 \beta \underline{X}^0 \cdot x \cdot \dot{x} - 2\epsilon_0 \beta \cdot x^2 \cdot \dot{x} - \underline{X}^0$$

Specifying a small perturbation permits the discard of the right-hand nonlinear term. (Even this is questionable because the velocity may be extremely large.) However, we cannot justify that the left-hand nonlinear term be neglected, and so the method needs modification.

3.3.3 A Generalized Approach—State Variables

The above method works well for first-order systems but needs modification for n^{th}-order systems. Rather than dealing with a single n^{th}-order equation, the system is redescribed by n first-order equations. Each of these is the derivative of a conveniently arbitrarily chosen state, composed as a linear combination of all states. Each state is then considered to be made up of a constant portion established by the operating point or initial condition and a perturbation value.

To illustrate, the Van der Pol equation is chosen.

$$\ddot{\underline{\overline{X}}} - 2 \cdot \epsilon_0 \cdot (1 - \beta \cdot \underline{\overline{X}}^2) \cdot \dot{\underline{\overline{X}}} + \underline{\overline{X}} = 0$$

The choice of states is arbitrary; however, in engineering applications, successive derivatives (called phase variables) offer a clear display of the system's dynamic nature, so,

$$\underline{\overline{X}}_1 \equiv \underline{\overline{X}}, \qquad \text{and } \underline{\overline{X}}_2 \equiv \dot{\underline{\overline{X}}}$$

These two states are assumed to be perturbed from an operating point,

$$\underline{\overline{X}}_1 \equiv \underline{\overline{X}}_1^0 + x_1, \qquad \underline{\overline{X}}_2 \equiv \underline{\overline{X}}_2^0 + x_2$$

Upon substitution of the phase variables, Van der Pol's equation is transformed into two first-order equations,

$$\dot{\underline{\overline{X}}}_1 = \underline{\overline{X}}_2$$

$$\dot{\underline{\overline{X}}}_2 = -\underline{\overline{X}}_1 + 2 \cdot \epsilon_0(1 - \beta \cdot \underline{\overline{X}}_1^2) \cdot \underline{\overline{X}}_2$$

and further substitution, using $\dot{\underline{\overline{X}}}_1 = \dot{x}_1$ and $\dot{\underline{\overline{X}}}_2 = \dot{x}_2$, gives the perturbation equations,

$$\dot{x}_1 = x_2 + \underline{\overline{X}}_2^0$$

$$\dot{x}_2 = \underbrace{-x_1 + 2 \cdot \epsilon_0 \cdot x_2}_{\text{Linear Terms}} - \underbrace{\underline{\overline{X}}_1^0 + 2 \cdot \epsilon_0 \cdot \underline{\overline{X}}_2^0}_{\text{Constant Terms}}$$

$$\underbrace{- 2 \cdot \epsilon_0\beta(\underline{\overline{X}}_1^0 + x_1)^2 \cdot (\underline{\overline{X}}_2^0 + x_2)}_{\text{Nonlinear}}$$

Products and powers are discarded, which linearizes the system since these are precisely the nonlinear terms.

$$\dot{x}_1 = x_2 + \underline{\overline{X}}_2^0$$

$$\dot{x}_2 = -\{1 + 4\epsilon_0\beta \cdot \underline{\overline{X}}_1^0 \cdot \underline{\overline{X}}_2^0\} \cdot x_2 + 2 \cdot \epsilon_0 \cdot \{1 - \beta(\underline{\overline{X}}_1^0)^2\} \cdot x_2$$

$$+ \cdots + [-\underline{\overline{X}}_1^0 + 2\epsilon_0 \cdot \underline{\overline{X}}_2^0\{1 - \beta(\underline{\overline{X}}_1)^2\}]$$

which has the matrix form,

$$\dot{x}] = [A] \cdot x] + k]$$

where $[A]$ is the linearized system matrix and $k]$ is a constant vector composed of the operating point values. This equation can be solved by the usual linear methods to determine the time response "in the small" near the system operating point. Its eigenvalues can also be determined from the characteristic equation to show the nature of its motion. Thus,

$$|sI - A| = 0$$

is the characteristic equation, giving,

$$\begin{aligned} A_{1,2} = \epsilon_0 \cdot \{1 - \beta(\overline{\underline{X}}_1^0)^2 \overline{\underline{X}}_2^0\} \\ \pm \sqrt{\{\epsilon_0(1 - \beta(\overline{\underline{X}}_1^0)^2)\overline{\underline{X}}_2^0\}^2 + X_1^0(1 + 4\epsilon_0 \beta \overline{\underline{X}}_1^0 \overline{\underline{X}}_2^0)} \end{aligned}$$

which is oscillatory if the radicand is negative, aperiodic if it is positive, and unstable if either root has a positive real part. It is evident that the system is oscillatory if ϵ_0 and β are positive; in addition, it is unstable for small values of the operating point, while stable for large values. The conclusion that a limit cycle must exist is demonstrated in Chapter 4.

3.3.4 Discontinuities Defy Linearization

Since linearization limitations have been acknowledged, it is only fair to investigate some failures to demonstrate the cause. One is a mechanical clock with coulomb friction whose linearization is prevented by the bang-bang nature of the friction. Another is a relay-controlled integrating loop with the hysteresis in the relay causing a limit cycle. Finally, free-play in gears creates an effective hysteresis action when embedded in an integrating loop, as is the case of a motor-driven actuator.

3.3.4.1 Coulomb Friction. Coulomb friction opposes motion with a constant force, behaving as a relay with respect to velocity. The work done is converted to heat; therefore, coulomb friction is stabilizing as it extracts mechancial energy from a system. It is a candidate for smoothing oscillatory actuators. An example is the pendulum clock, a lightly damped second-order system with coulomb friction in its hinge. For small oscillations, its block diagram is shown in Figure 3-6.

In the figure, the procedure apparently breaks down because; while the $SGN(\overline{\underline{X}}_2)$ is a perfectly valid operation, it is one of those peculiar functions which doesn't have a derivative.

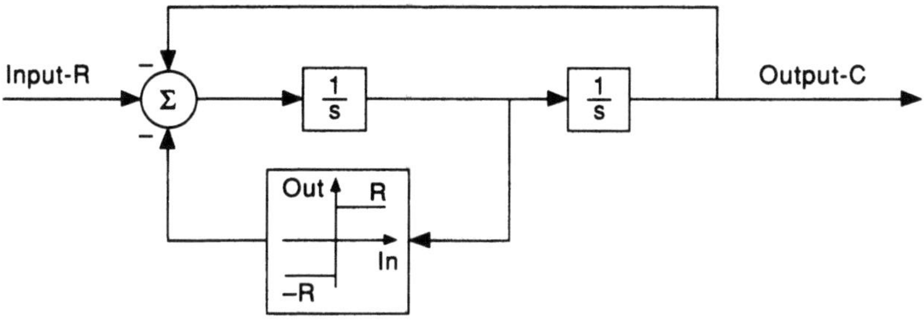

$$\ddot{c} + k \cdot \frac{\dot{c}}{|\dot{c}|} + c = R$$

$$\underline{\overline{X}}_1 \equiv c \;\&\; \underline{\overline{X}}_2 \equiv \dot{c}$$

$$\dot{\underline{X}}] = \begin{bmatrix} f_1(\underline{\overline{X}}_1, \underline{\overline{X}}_2) \\ f_2(\underline{\overline{X}}_1, \underline{\overline{X}}_2) \end{bmatrix} = \begin{bmatrix} \underline{\overline{X}}_2 \\ -\underline{\overline{X}} - k \cdot \mathrm{SGN}(\underline{\overline{X}}) \end{bmatrix}$$

$$d\dot{\underline{X}}] = \begin{bmatrix} \dfrac{\partial f_1}{\partial \underline{\overline{X}}_1} \cdot d\underline{\overline{X}}_1 + \dfrac{\partial f_1}{\partial \underline{\overline{X}}_2} \cdot d\underline{\overline{X}}_2 \\ \dfrac{\partial f_2}{\partial \underline{\overline{X}}_1} \cdot d\underline{\overline{X}}_1 + \dfrac{\partial f_2}{\partial \underline{\overline{X}}_2} \cdot d\underline{\overline{X}}_2 \end{bmatrix} \cdots$$

$$\cdots \text{ Which produces, } \frac{\partial f_2}{\partial \underline{\overline{X}}_2} = -k \cdot \frac{\partial}{\partial \underline{\overline{X}}} \{\mathrm{SGN}(X)\},$$

Figure 3-6. System with coulomb friction.

3.3.4.2 Relay with Hysteresis. Consider the bang-bang relay control system shown in Figure 3-7. The hysteresis nonlinearity depicted has memory (i.e., its output depends on what has gone on before). As an example, consider a room with thermostatically controlled temperature regulation. We presuppose a system operating point, say 70°F, at which we neither wish to heat nor cool. If the temperature rises to a preset upper limit, say 72°F, the cooling cycle activates, lowering the temperature. In Figure 3-7, the origin corresponds to 70°F, the actuating temperature 72°F corresponds to the 2°F incrementing tempera-

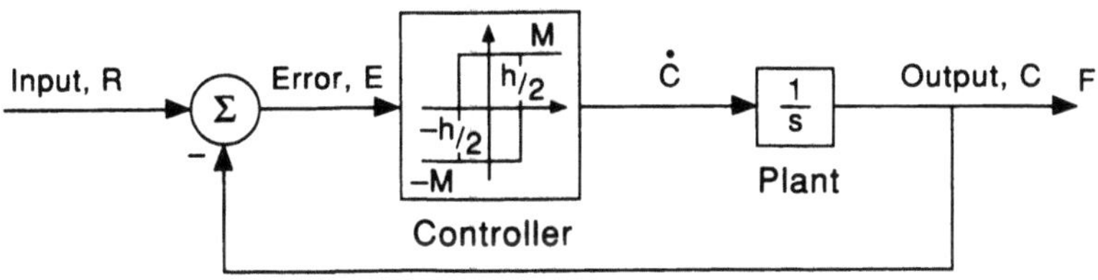

Figure 3-7. "Bang-bang" control with hysteresis.

ture $h/2$, and we presuppose a constant system output equal to M until the demand for cooling ceases. For ease of discussion, we will choose −2°F or 68°F as the trigger for heat.

Then, if the system controller were perfect, heating would begin and the lowest temperature reading would be 68°F; but we know that system inertia carries the temperature below 68°F. Moreover, although heat is called for, the system does not have the power to assure that we do not go below 68°F (and instead fall to, say, 66°F). The described behavior is shown in Figure 3-8.

At 66°F the demand for heat, which had been activated at 68°F, begins to dominate and the system temperature begins to rise. At 72°F the heating cycle shuts off and the cooling cycle is again initiated.

Thus, as shown in Figure 3-9, two distinct paths are traced for the nonlinearity's idealized output, each path dependent upon what happened in the

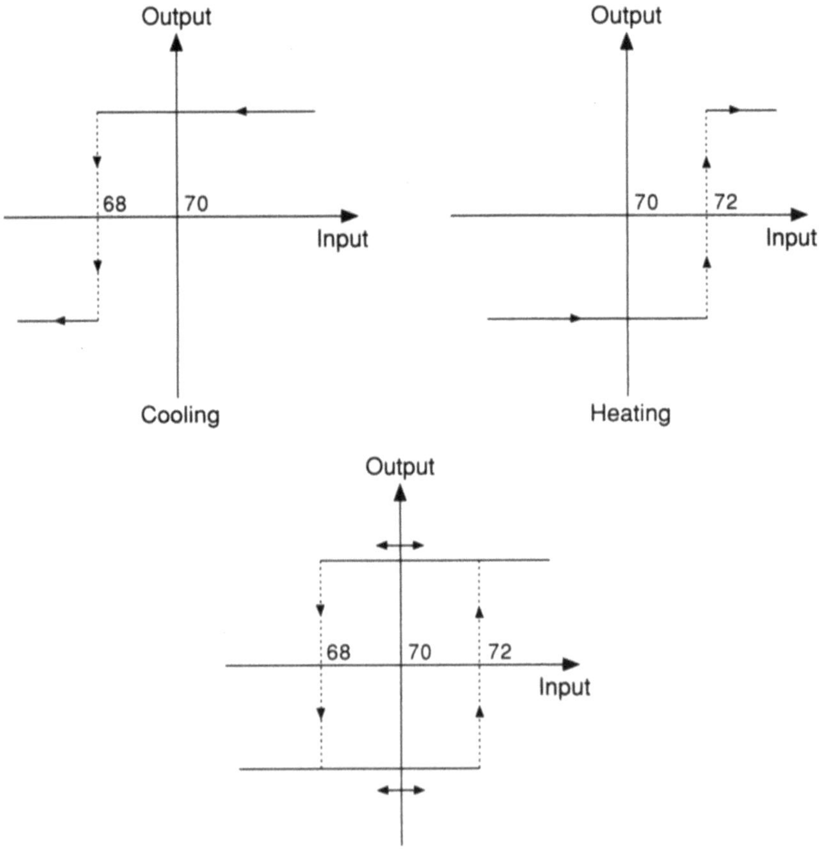

Figure 3-8. Hysteresis in thermostat controller.

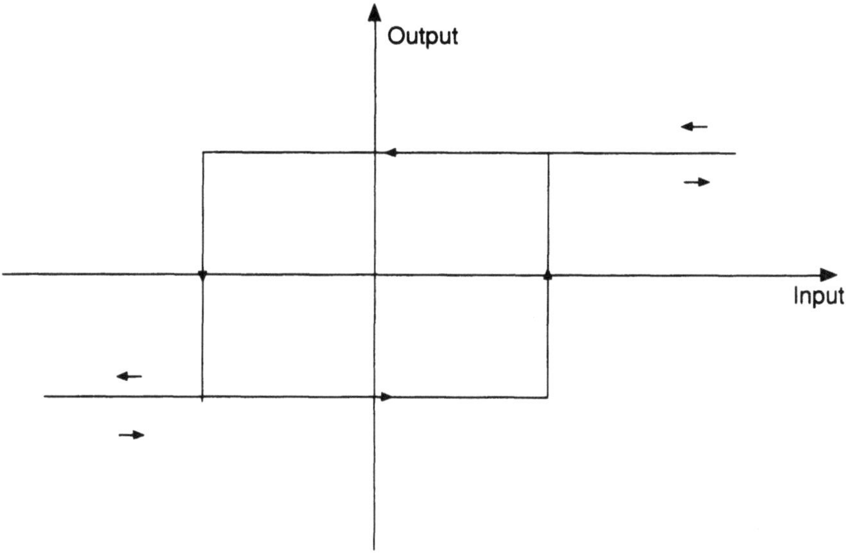

Figure 3-9. Hysteresis in controller.

immediate past. Thus, we say the nonlinearity has memory, as depicted in Figure 3-9.

Mathematically, this would be attempted to be described as

$$y = \begin{cases} +M, & e > h/2 \\ -M, & e < -h/2 \end{cases}, \qquad \text{which leaves out } (-h/2,\ h/2)$$

and at $-h/2 < e < h/2$,

$$y = \begin{cases} +M, & \text{if at last max } e > h/2 \text{ and at last min } e > -h/2 \\ -M, & \text{if at last max } e < -h/2 \text{ and at last min } e < h/2 \end{cases}$$

This description is cumbersome and not susceptible to common algebra. The system's solution can be found in pieces from an initial position, $C(0)$ despite the nonlinearity because of the simplicity of the dynamics. If the input is zero, it follows that

$$\nu(t) = 0, \qquad C(0) > 0$$

interval	$\dot{e}(t)$	$e(t)$
$0 < t < t_1$	$+M$	$M \cdot t - C(0) \Rightarrow e(t_1) = h/2$
$t_1 < t < t_2$	$-M$	$h/2 - M \cdot (t - t_1) \Rightarrow e(t_2) = -h/2$
$t_2 < t < t_3$	$+M$	$-h/2 + M \cdot (t - t_2) \Rightarrow e(t_3) = h/2$

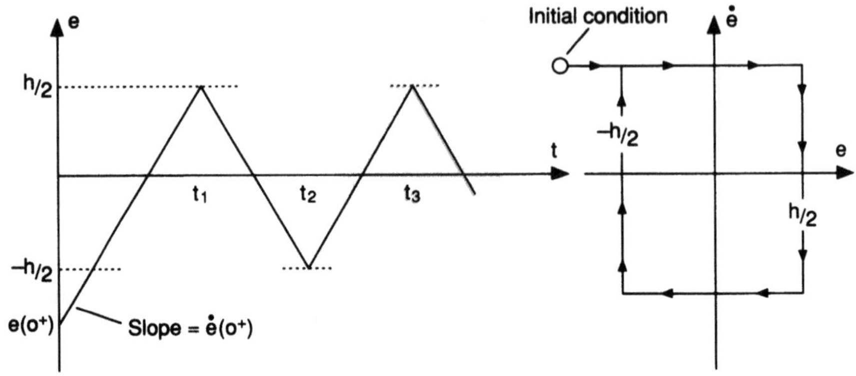

Figure 3-10. "Bang-bang" system response.

The solution is shown in Figure 3-10, where the motion develops into a limit cycle. With this knowledge, it is possible, by hindsight, to argue that the system is unstable for small errors (inside the hysteresis band) and stable for large errors (outside the hysteresis band). The loop has positive feedback for small values due to the sign inversion of the relay, and hence is unstable. For large values of the error, the relay causes no sign inversion and the loop is stable because of the negative feedback. But this type of argument is only true of first-order plants, and is inappropriate to higher-order plants which can always be unstable.

3.3.4.3 Free Play in Gears. Another failure of linearization is shown in Figure 3-11, a generalization of the frequently encountered free play between teeth in a gear, as shown in Figure 3-12. The only neighborhoods of valid linearized solutions lie entirely inside or entirely outside the deadband, and the salient features correspond to local gains of zero or one. Difficulty arises when the system error passes in and out of the deadband, for which the gain is some weighted average in between. This average would be the linearized approximation; however, finding it requires prior knowledge of the system's solution, which denies attempts at its linearization.

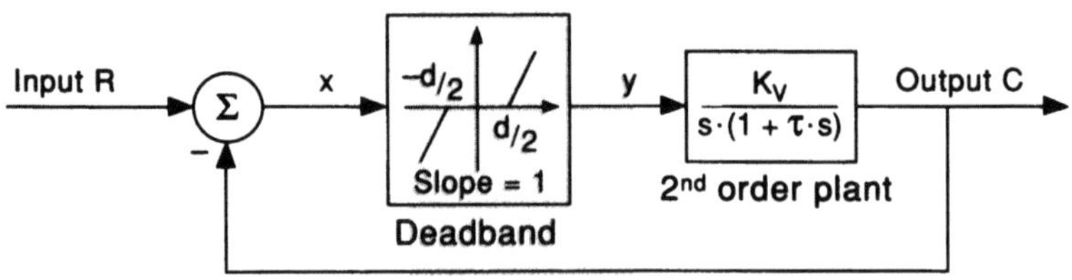

Figure 3-11. System with deadband.

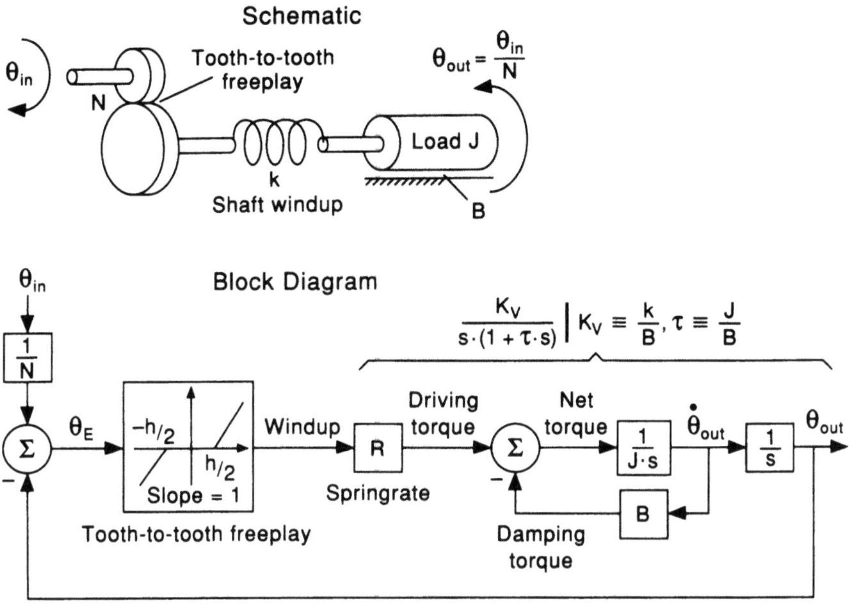

Figure 3-12. Gear pass: lumped parameters.

A useful by-product of this system is the nature of its overall input–output relation at low frequency. This could be determined experimentally from its Lissajous pattern by applying a sine wave as shown in Figure 3-13. In each cycle the error is driven back and forth through the single-valued deadband to produce, surprisingly, the multivalued hysteresis as the system's characteristic. The utility of this observation accrues to the system modeler seeking to simulate hysteresis in an actuator. Attempts to model multivalued functions explicitly are awkward, requiring memory to initiate their simulation—a feature unavailable to analog methods, and clumsy at best in digital methods.

The overall hysteresis characteristic shown by its Lissajous pattern in Figure 3-13 can be developed as an asymptotic result of the system of Figure 3-11 with $\tau = 0$ (ref. 1). The output is given piecewise, according to,

$$\dot{c} = \begin{cases} K_V \cdot (r - c + h/2), & x < -h/2 \\ 0, & -h/2 \leq x \leq h/2 \\ K_V \cdot (r - c - h/2), & x > h/2 \end{cases} \qquad \text{where } x \equiv r - c$$

By allowing the integrator gain, K_V to become large, the bracketed factors must attenuate to correspond. In the limit as $K_V \to \infty$, the bracketed factors must

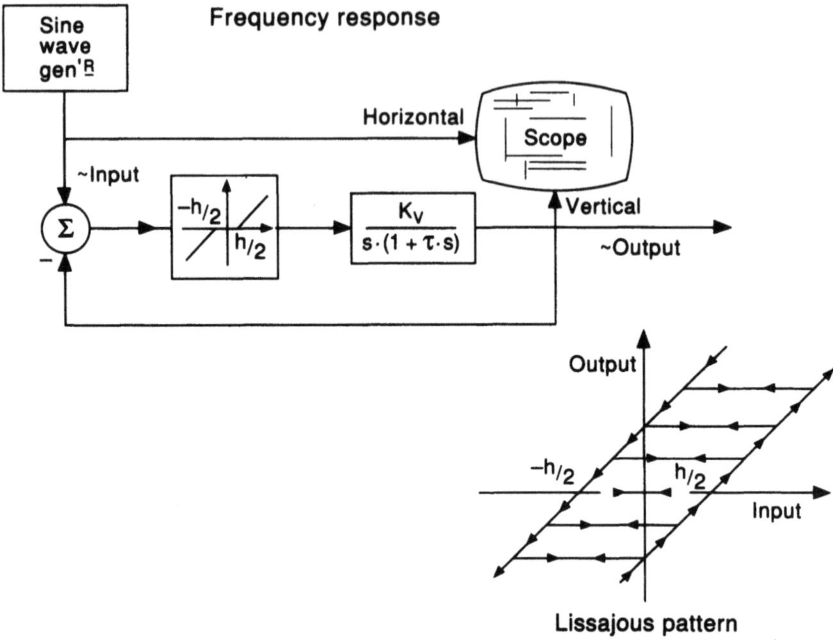

Figure 3-13. Hysteresis in gears.

vanish, giving,

$$c = \begin{cases} r + h/2, & x < -h/2 \\ \text{constant} = c_{\text{LAST}}, & -h/2 < x < h/2 \\ r - h/2, & x > h/2 \end{cases}$$

This example is evidence that a multivalued function can be created implicitly with feedback from a single-valued function. It is interesting to speculate on a generalized synthesis procedure yielding a single-valued function when embedded in an implicit feedback loop with arbitrarily specified multivaluedness. Such a procedure would be a reward to dynamic simulations, but the fact that the functions of most interest are nonlinear spells failure because the question of existence is unanswerable.

3.4 LINEARIZING SATELLITE MOTION

A spacecraft of mass m moves under the inverse square law in the presence of a central gravitational body (say, Earth) of mass M, as shown in Figure 3-14. With the assumption that $m \ll M$ it is reasonable to assume the central body

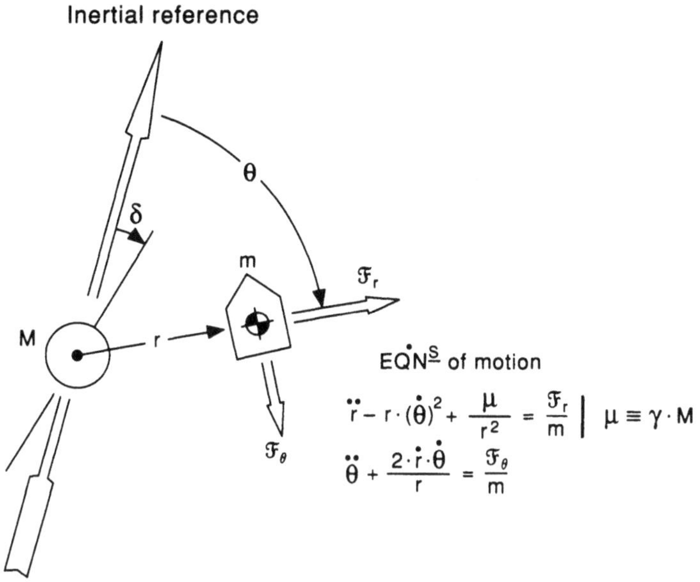

Figure 3-14. Single-body dynamics.

is fixed in space, and the equations of planar motion derive from either Newton's second law or Lagrange.

In most systems governed by nonlinear equations, the exact analytic solution is unknown, but this is a rare exception (ref. 2), where it is known to be a conic,

$$r = \frac{\rho}{1 - e \cdot \cos(\theta - \delta)}, \quad \text{giving} \begin{cases} e = 0\text{—circle} \\ 0 < e < 1\text{—ellipse} \\ e = 1\text{—parabola} \\ e > 1\text{—hyperbola} \end{cases}$$

$$\text{where } p \equiv \frac{(2 \cdot \dot{A})^2}{\mu} \text{ semi-latus rectum}$$

$$\text{and } e = \sqrt{1 + p^2\left(v_0^2 - \frac{2\mu}{r_0}\right)}, \qquad v_o \equiv r_o\dot{\theta}_o, \quad \dot{A} \equiv \frac{r_o^2\dot{\theta}_o}{2}$$

Because its solution is known, this example offers the benefit of exact comparison with the linearized approximation. The reality of engineering seldom allows any comparison whatsoever, so the engineer is never truly sure of his linear approximation.

The linear perturbation method selects an operating point and then perturbs the model. Since the exact solution is known to be a conic, and a circle is a conic, we select the operating point (r_0, θ_0) as a convenient point on the circular orbit. We then assume that either radial or tangential departures occur due to the perturbation to investigate the resultant motion of a spacecraft in circular orbit to understand how to attempt rendezvous or docking maneuvers. Four state variables are assigned, each a linear combination of the derivatives in the equations, the simplest such combination being,

$$x_1 \equiv r, \qquad x_2 \equiv \dot{r}, \qquad x_3 \equiv \theta, \qquad x_4 \equiv \dot{\theta}$$

corresponding inputs are,

$$u_1 \equiv 0, \qquad u_2 \equiv \mathfrak{F}_r, \qquad u_3 \equiv 0, \qquad u_4 \equiv \mathfrak{F}_\theta$$

and the equations for any craft or projectile of mass are,

$$\left.\begin{aligned} \dot{x}_1 &= x_2 + u_1{}^0 \\ \dot{x}_2 &= x_1 \cdot x_4^2 - \frac{\mu}{x_1^2} + \frac{u_2}{m \cdot x_1} \\ \dot{x}_3 &= x_4 + u_3{}^0 \\ \dot{x}_4 &= -\frac{2x_2x_4}{x_1} + \frac{u_4}{m \cdot x_1} \end{aligned}\right\} \dot{x}] = F(x], u])$$

To linearize, $F(x], u])$ could be expanded in a Taylor series around the operating point and higher-order terms discarded. The same result can be obtained by differentiation and the replacement of total differentials with finite increments.

$$d\{\dot{x}]\} = d\{F(x], u])\} \longrightarrow \Delta\dot{x}] \approx \Delta F(x], u])$$

Identifying total variables with a "bar" allows the operation to be symbolically expressed, leaving perturbation variables to be "unbarred," where

$$x_i = \bar{x}_i - x_i^0, \qquad i = 1, 2, 3, 4 \cdots$$

The resulting equations are not yet linear as can be seen below:

$$\dot{x}] = \begin{bmatrix} 0 & 1 & 0 & 0 \\ \bar{x}_4^2 - \dfrac{2\mu}{\bar{x}_1^3} & 0 & 0 & 2 \cdot \bar{x}_1 \cdot \bar{x}_4 \\ 0 & 0 & 0 & 1 \\ \dfrac{2 \cdot \bar{x}_2 \cdot \bar{x}_4 - u_4/m}{\bar{x}_1^2} & -\dfrac{2 \cdot \bar{x}_4}{\bar{x}_1} & 0 & -\dfrac{2 \cdot \bar{x}_2}{\bar{x}_1} \end{bmatrix} \cdot x] + \cdots$$

$$\cdots + \begin{bmatrix} 0 & 0 & 0 & 0 \\ 0 & 1/m & 0 & 0 \\ 0 & 0 & 0 & 0 \\ 0 & 0 & 0 & 1/m \end{bmatrix} \cdot u] + \begin{bmatrix} 0 \\ \bar{x}_1 \cdot \bar{x}_4^2 + \dfrac{\mu}{\bar{x}_1^2} + \dfrac{\bar{u}_2}{m} \\ 0 \\ \dfrac{\bar{u}_4}{m} \cdot \left(1 - \dfrac{1}{\bar{x}_1}\right) \end{bmatrix}$$

or $\dot{x}] = [\Phi] \cdot x] + [\Psi] \cdot u] + C]$

A corollary of small perturbation theory is that total variables change negligibly, rendering $[\Phi]$ and $[\Psi]$, both constant, thereby completing the linearization

$$\dot{x}] = [A] \cdot x] + [B] \cdot u]$$

where

$$[A] \equiv [\Phi^0] = \begin{bmatrix} 0 & 1 & 0 & 0 \\ 3 \cdot \omega_0^2 & 0 & 0 & 2r_0\omega_0 \\ 0 & 0 & 0 & 1 \\ 0 & -2\omega_0/r_0 & 0 & 0 \end{bmatrix}$$

and

$$[B] \equiv [\Psi^0] = \begin{bmatrix} 0 & 0 & 0 & 0 \\ 0 & 1/m & 0 & 0 \\ 0 & 0 & 0 & 0 \\ 0 & 0 & 0 & 1/mr_0 \end{bmatrix}$$

in which

$$\omega_0 = \dot{\theta}_0 = \sqrt{\mu/r_0^3}$$

A set of local coordinates is convenient when describing perturbation motion near the circular reference orbit (Figure 3-15). The coordinates we select are (V_L, h_L), representing localized vertical and horizontal variables, respectively. The final linearized equations of motion are,

$$\ddot{V}_L = 3\omega_0^2 V_L + 2\omega_0 \dot{h}_L + \frac{\mathfrak{F}_r}{m}$$

$$\ddot{h}_L = -2 \cdot \omega_0 \dot{V}_L + \frac{\mathfrak{F}_\theta}{m}$$

The analyst is cautioned to be concerned with the answer obtained when these linear equations are "solved." Yes, we have a model and, yes, we can

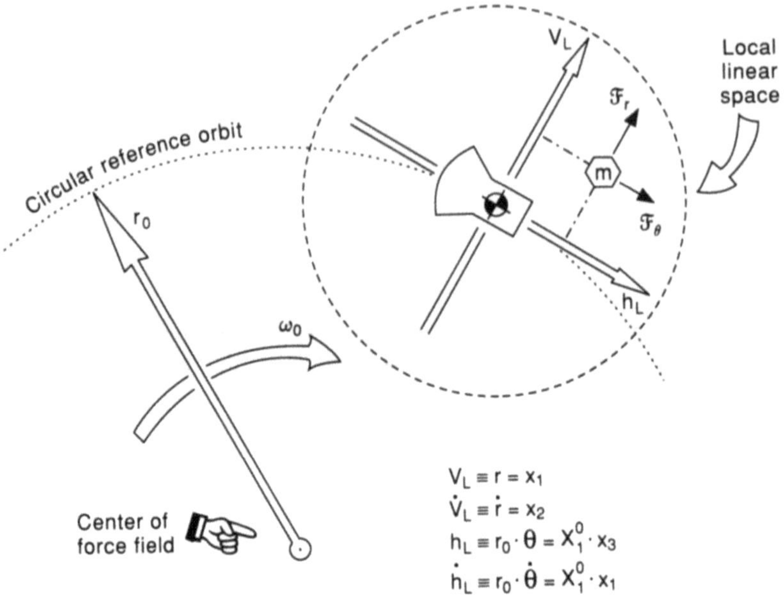

Figure 3-15. Linearized coordinates.

solve it. However, what relation *does* the solution of the model have to the dynamics of the physical system? We show in the next two sections why such caution is advisable.

3.4.1 Vertical Ejection

The linearized equations permit us to study the motion of a projectile ejected "vertically" at some velocity, $\dot{V}_L(0)$ relative to the spacecraft. Employing Laplace transforms,

$$V_L(s) = \frac{\dot{V}_L(0)}{\omega_0} \cdot \frac{\omega_0}{s^2 + \omega_0^2}, \qquad h_L(s) = -2 \cdot \dot{V}_L(0) \cdot \frac{\omega_0}{s(s^2 + \omega_c^2)}$$

which yields the linearized solutions as

$$V_L(t) = (\dot{V}_L(0)/\omega_0) \sin \omega_0 t$$

$$h_L(t) = -(2\dot{V}_L(0)/\omega_0)(1 - \cos \omega_0 t)$$

This solution can be reexpressed as an elliptical trajectory whose equation is

$$\left(\frac{V_L(t)}{\dot{V}_L(0)/\omega_0}\right)^2 + \left(\frac{h_L(t) + 2\dot{V}_L(0)/\omega_0}{2\dot{V}_L(0)/\omega_0}\right)^2 = 1$$

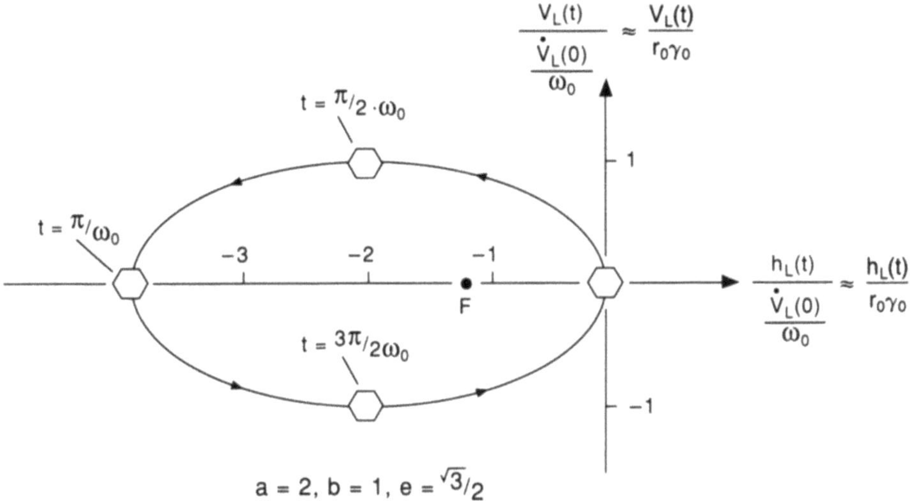

Figure 3-16. Vertical ejection, local linearized coordinates.

This is plotted in Figure 3-16, using normalized coordinates, having its major, minor axes 4, 2, respectively, and a corresponding eccentricity of $\sqrt{3}/2$.

A comparison of the projectile's linearized approximate orbit can now be made with its exact orbit by superimposing the local trajectory on the reference circular orbit. The exact orbit is established by the overall initial velocity. These results are presented in Figure 3-17 for a vertical ejection velocity at 25% of

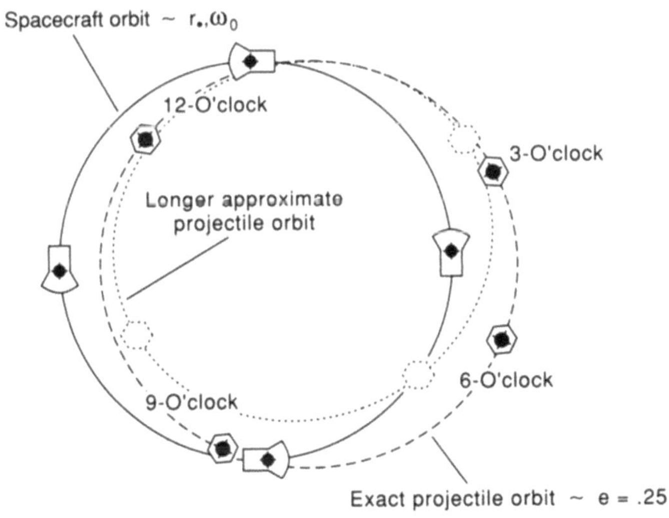

Figure 3-17. Vertical ejection, inertial coordinates, $\dot{V}(0) = v_0/4$.

circular orbit speed at "12 o'clock." Clearly, there is a substantial difference between the two orbits, although the linearized approximation appears to be accurate up to "3 o'clock."

3.4.2 Horizontal Ejection

An initial condition, $\dot{h}_L(0)$, represents a horizontal ejection of the projectile from the spacecraft. Employing Laplace transforms, the local coordinates are,

$$V_L(s) = 2 \cdot \dot{h}_L(0) \cdot \frac{\omega_0}{s(s^2 + \omega_0^2)}$$

$$h_L(s) = \frac{\dot{h}_L(0)}{\omega_0} \cdot \frac{\omega_0}{(s^2 + \omega_0^2)} - 3 \cdot \omega_0 \cdot \dot{h}_L(0) \cdot \frac{\omega_c}{s^2(s^2 + \omega_0^2)}$$

so

$$V_L(t) = 2 \cdot \frac{\dot{h}_L(0)}{\omega} \cdot (1 - \cos \omega_0 \cdot t)$$

and

$$h_L(t) = -3 \cdot \frac{\dot{h}_L(0)}{\omega_0} \cdot \omega_0 t + 4 \cdot \frac{\dot{h}_L(0)}{\omega_0} + \sin \omega_0 \cdot t$$

which are the parametric equations of a cycloid, plotted in Figure 3-18 using normalized coordinates.

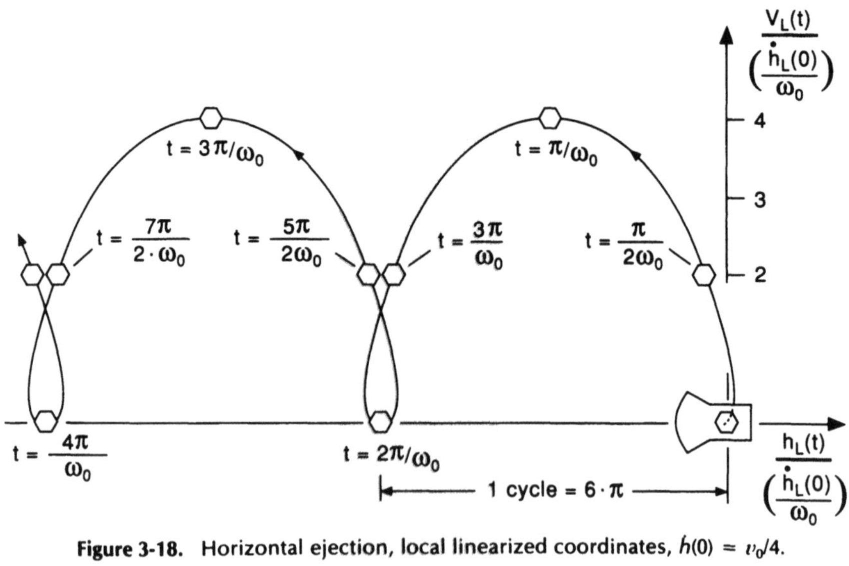

Figure 3-18. Horizontal ejection, local linearized coordinates, $\dot{h}(0) = v_0/4$.

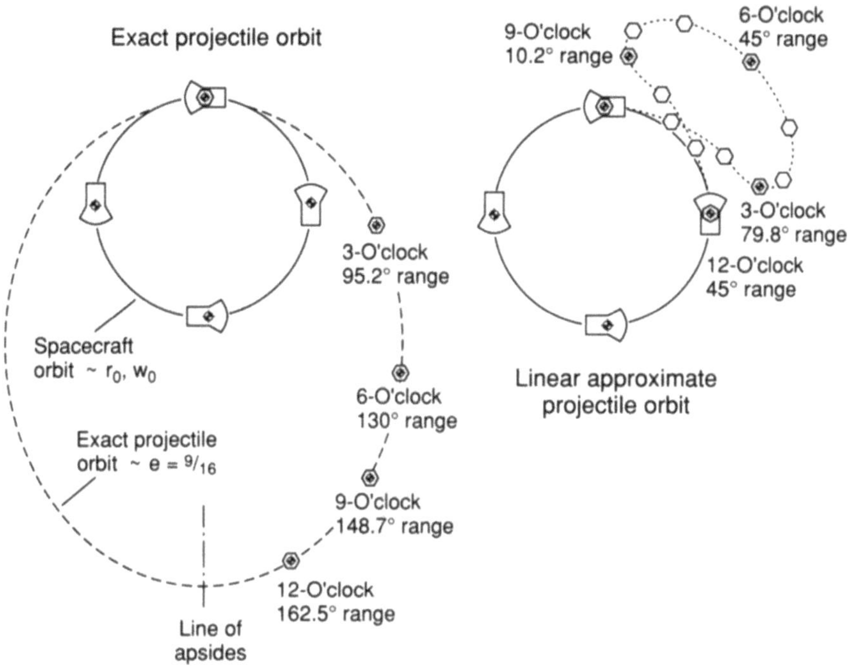

Figure 3-19. Horizontal ejection, inertial coordinates, $\dot{h}(0) = v_0/4$.

Superimposing this linearized local trajectory on the reference circular orbit provides the approximate orbit of the projectile as deduced from its linearized equations, while the exact orbit is established by the overall initial velocity. These results are presented in Figure 3-19 for a horizontal ejection velocity at 25% of circular orbit speed occurring at "12 o'clock."

3.5 CONCLUDING EXAMPLE

The cabin pressurization system for a commercial airliner is an example of controlling a nonlinear plant. As discussed in Chapter 10, its design proceeds by linearizing and then analyzing enough operating points to be sure that the design meets its performance requirement throughout the flight spectrum. Bleed air from the compressors of the main propulsion engines is conditioned to the right temperature by turbocompressors and heat exchangers. The air is then ducted to the cabin and, as it flows in, pressure builds up. When sufficient pressure is achieved, an outflow valve opens to allow as much air to drain to ambient (outside the aircraft) as is being supplied from the main propulsion engines. By modulating the area of the outflow valve (i.e., its discharge area), cabin pressure can be made to increase or decrease. The mechanics of the pressure change are found from the Ideal Gas law,

$$p = \frac{W \cdot R \cdot T}{V} = \rho \cdot g \cdot R \cdot T \qquad \text{IDEAL GAS law}$$

where p = gas pressure
W = weight of gas in vessel
V = volume of vessel
T = gas temperature
ρ = gas density
g = gravity
R = gas constant

The rate of change of pressure is found by differentiating the above expression with respect to time.

$$\dot{p} = \dot{\rho} \cdot g \cdot R \cdot T + \rho \cdot g \cdot R \cdot \dot{T}$$

If the cabin temperature were constant, the last term would disappear, but pressure fluctuations will cause some rapid temperature fluctuations which cannot be smoothed by the relatively slow temperature control. Thus, it is customary to assume an adiabatic cabin for which,

$$p = \rho^{\gamma} \cdot K$$

where γ = ratio of specific heats, and K = a constant. Equating the two expressions for p from the gas law and an adiabatic process gives a relation between density, ρ, and temperature, T, from which their rates are related by,

$$\dot{T} = \frac{(\gamma - 1) \cdot T \cdot \dot{\rho}}{\rho}$$

Substituting into the relation for pressure rate gives the relation for net inflow,

$$\dot{p} = \frac{\gamma \cdot R \cdot T}{V} \cdot \dot{W}_{\text{NET}}$$

The next step is to find the relation between cabin pressure and outflow valve area, which will be seen to be quite nonlinear. With reference to Figure 3-20,

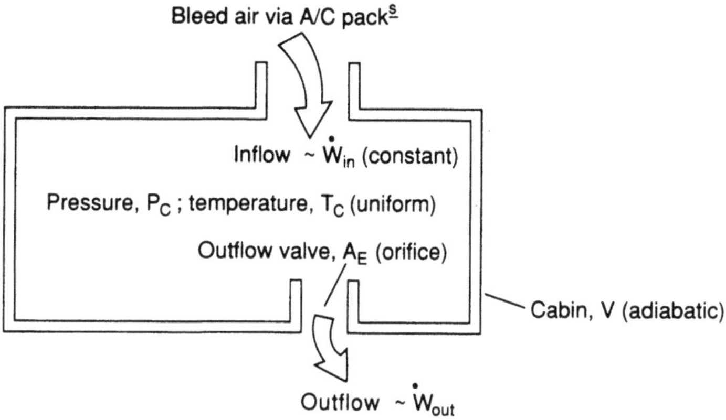

Figure 3-20. Aircraft cabin schematic.

the continuity equation is,

$$\dot{W}_{\text{OUT}} = \rho_2 \cdot A_{E_2} \cdot V_2 \cdot g$$

$$\begin{aligned} \text{where } A_{E_2} &= A_2 \cdot C_0 \text{ effective area} \\ A_2 &= \text{geometric area} \\ C_0 &= \text{discharge coefficient} \\ V_2 &= \text{velocity through valve} \end{aligned}$$

The pressure ratio across the valve is given by the compressible flow equation, a clearly nonlinear relation,

$$\frac{p_2}{p_1} = \left\{1 - \frac{\gamma - 1}{2} \cdot \left(\frac{V_2^2 - V_1^2}{a_1^2}\right)\right\}^{\gamma/\gamma - 1}$$

where $V_1 \approx 0$ since $A_1 >> A_2$, and

$$a_1 = \sqrt{\frac{\gamma \cdot p_1}{\rho_1}} = \sqrt{\gamma \cdot g \cdot R \cdot T}$$

This is solved for the gas speed V_2, through the valve and substituted into the continuity equation. Two cases arise in which V_2 is either subsonic or sonic (valve is choked). The outflow is given by the nonlinear equation,

$$\dot{W}_{\text{OUT}} = \frac{p_1 A_{E_2}}{\sqrt{T}} \cdot k \cdot \frac{p_2}{p_1}$$

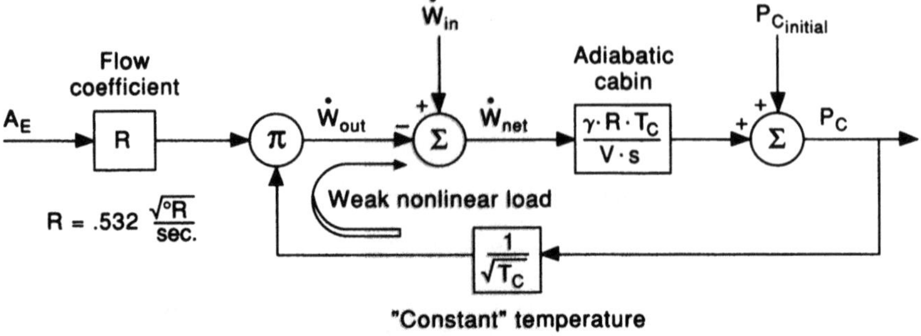

Figure 3-21. Cabin dynamics: cruise and choked valve.

Assuming the pressure vessel has neither leakage nor mechanical compliance (i.e., stretch of the enclosure walls is not allowed and the system is rigid) and that the temperature inside is everywhere the same and slowly varying, the plant can be described for both the choked and unchoked cases as shown in Figures 3-21 and 3-22 (from ref. 3), where the subscripts a and c refer to outside-ambient and cabin conditions, respectively. The transfer function relating changes in cabin pressure, Δp_c, from an operating point p_c^0, to changes in effective valve area, ΔA_E, from its operating point, A_E^0, is developed by considering the perturbations,

$$\Delta k = k - k^0$$

$$\Delta A_E = A_E - A_E^0$$

$$\Delta p_c = p_c - p_c^0 \Rightarrow \Delta \dot{p}_c = \dot{p}_c$$

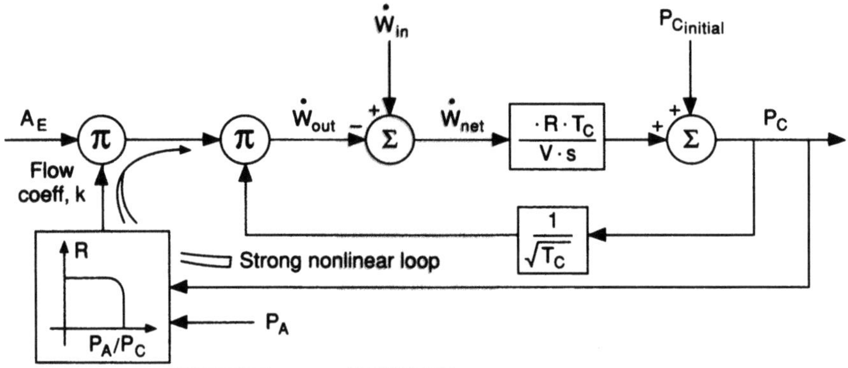

$$k = K \cdot \sigma^{1_\gamma} \sqrt{1 - \frac{\gamma - 1}{\phi^\gamma}} \Big| \; k \equiv \sqrt{\frac{2g}{R} \cdot \frac{\gamma}{\gamma - 1}}; \; \phi \equiv \frac{P_A}{P_C}$$

Figure 3-22. Cabin dynamics: unchoked valve.

Substitution gives,

$$\dot{p}_c = \frac{R \cdot T}{V} \cdot \dot{W}_{\mathrm{NET}} = \frac{R \cdot T}{V} \cdot (\dot{W}_{\mathrm{IN}} - \dot{W}_{\mathrm{OUT}})$$

$$= \frac{R \cdot T}{V} \cdot \dot{W}_{\mathrm{IN}} - \frac{R\sqrt{T}}{V} \cdot k \cdot A_E \cdot p_c$$

and, after discarding products and powers of the perturbation variables as negligibly small,

$$\Delta \dot{p}_c = \underbrace{\frac{R \cdot T}{V} \cdot \dot{W}_{\mathrm{IN}} - \underbrace{\frac{R \cdot \sqrt{T}}{V} \cdot k^0 \cdot A_E^0 \cdot p_c^0}_{\dot{W}_{\mathrm{OUT}}}}_{\text{these cancel for } \dot{p}_c = 0}$$

$$- \frac{R \cdot \sqrt{T}}{V} \cdot \{k^0 A_E^0 \Delta p_c + k^0 p_c^0 \Delta A_E + p_c^0 A_E^0 \Delta k\}$$

The flow coefficient, k, is only dependent on cabin pressure, p_c, for level flight since p_a is constant, thus,

$$\Delta k \approx \left. \frac{\partial k}{\partial p} \right|_{p_c^0} \cdot \Delta p_c$$

When this is substituted and the Laplace transform is taken, the transfer function is

$$\frac{\Delta p_c}{\Delta A_E}(s) = \frac{-\dfrac{R \cdot T}{V} \cdot k^0 \cdot p_c^0}{s + \left\{\dfrac{R \quad T \cdot A_E^0}{V} \cdot \left(k^0 + p_c^0 \cdot \left.\dfrac{\partial k}{\partial p_c}\right|_{p_c^0}\right)\right\}} = \frac{k}{s + a}$$

which is a dc gain and simple lag, both of which depend upon the operating condition. Use will be made of this relation in Chapter 10, where the design of a cabin pressure system is discussed. It is sufficient to note at this point that this linearized model yields usefully correlated results in practice. The solutions one obtains allow a dynamic range of values for the gain having a lower limit k_l and an upper limit k_u, such that $k_l < k < k_u$ and $k_u/k_l = 2{,}576$, while the lag a satisfies $a_u/a_l = 188$. These ratios hold throughout the flight conditions normally anticipated.

3.6 SUMMARY

Linearization is very helpful but, to the unwary, dangerous. We want to know what will happen to a dynamic system, and so we try to force the system to fit our methods of analysis. When one assumes too much, we distrust the system and thus can be lead wildly astray. Care must be taken to assume that the method applied is indeed applicable and that the conditions reached are meaningful. Thus, we suggest that when closed-form analytical solutions are not possible, more than one method of analysis should be applied to the problem and the results compared. Additional methods of analysis follow in subsequent chapters.

3.7 PROBLEMS

1. Eulerian axes are those which relate the angular orientation of a body to another set of space-fixed axes. As shown in Figure 3-23, the orientation is described by three rotations from space axes to body axes by the succession ψ about Z, then θ about the new Y-axis, and finally by ϕ about the new X-axis. The rate of change of these Eulerian angles is related to the components of the body's angular velocity p, q, and r about its x, y, z axes, respectively, by the nonlinear relations,

$$\dot{\phi} = p + \{q \cdot \sin\phi + r \cdot \cos\phi\} \cdot \tan\theta$$

$$\dot{\theta} = q \cdot \cos\phi - r \cdot \sin\phi$$

$$\dot{\psi} = \frac{r \cdot \cos\phi + q \cdot \sin\phi}{\cos\theta}$$

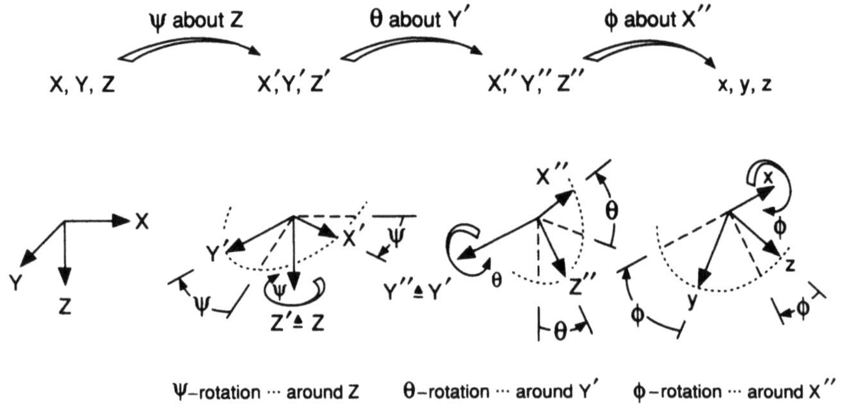

Figure 3-23. Eulerian angles.

For the case of level flight, the operating point is $\phi^0 = \theta^0 = \psi^0 = 0$. Find the linear relation between Eulerian angle rates and body rates. Repeat for the arbitrary operating point $\theta^0 = 90°$.

2. The angular velocity of a free body has components about its axes of principal inertia. They are related by the three nonlinear equations below, in which the J's are the polar moments of inertia.

$$\dot{\Omega}_1 = \frac{J_2 - J_3}{J_1} \cdot \Omega_2 \cdot \Omega_3$$

$$\dot{\Omega}_2 = \frac{J_3 - J_1}{J_2} \cdot \Omega_3 \cdot \Omega_1$$

$$\dot{\Omega}_3 = \frac{J_1 - J_2}{J_3} \cdot \Omega_1 \cdot \Omega_2$$

(a) Linearize these equations about an arbitrary operating point.
(b) Show that if the body is spinning about just one axis, its motion is unstable if this is an axis of intermediate inertia (i.e., greater than one and less than the other).

3. With reference of Figure 3-22, ambient pressure fluctuations, such as those which occur during rotation at lift-off, can propagate into the cabin because the valve is unchoked. Linearize the system and find the transfer function relating changes in cabin pressure to changes in ambient pressure, Δp_a. What can you deduce about the rapidity that ambient pressure fluctuations propagate into the cabin at low cabin-to-ambient pressure differentials, say, on the order of an inch of water?

4. With respect to Figure 3-24, the motion of a dissipationless unrolling rug

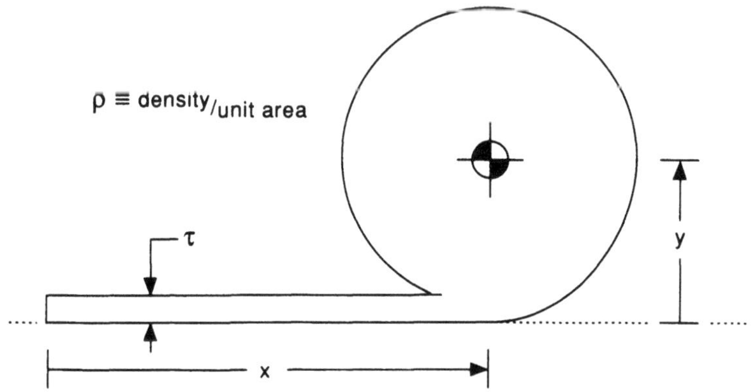

Figure 3-24. Unrolling rug.

is given by,

$$\left[3 \cdot (M - \rho\tau x) + \frac{\rho\tau^2}{2\pi}\right] \cdot \ddot{x} - \frac{3\rho\tau}{2} \cdot (\dot{x})^2 - 3g\tau\sqrt{\frac{\rho}{\pi} \cdot (M - \rho\tau x)} = 0$$

Select state variables and linearize the problem. Examine it for the two cases.

(a) x very small (initial motion).

(b) x very large (terminal motion)

5. The general schematic for a dc motor is shown in Figure 3-25, with system equations described.

   Show that:

   (a) When run with a constant flux ϕ, and excited by the armature voltage, the motor is a linear system.

   (b) When run with constant armature voltage, and excited by the field voltage, the motor is nonlinear. In this case linearize the system equations and compare the resulting transfer function with that of the constant field configuration of part (a).

   (c) When both the armature and field are excited from the same source, the motor is said to be run in shunt. Show that this is a nonlinear system. Linearize its equations, and determine the transfer function.

   (d) The armature circuit can be connected in series with the field circuit, in which configuration the motor is said to be run in series. Show that this is a nonlinear system. Linearize its equations and determine the transfer function.

6. The body axes commonly employed to analyze the motion of an aircraft

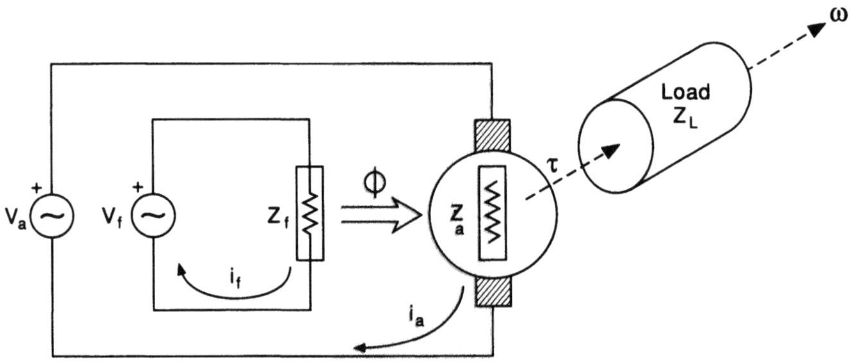

$V_a = \phi \cdot \omega \cdot K_b + Z_a \cdot i_a$ ... armature voltage

$V_f = Z_f \cdot i_f$ ... field voltage

$\tau = \phi \cdot i_a \cdot K_t$ ... electromagnetic torque

$\phi = K_\phi \cdot i_f$ ... flux

$\omega = \tau / Z_L$ ... shaft speed

Figure 3-25. Motor schematic (dc).

are shown in Figure 3-26. For the case of constant roll rate and forward velocity, the appropriate equations of motion are,

$$\dot{\alpha} = q - p_0 \cdot \beta$$

$$\dot{q} = \omega_\theta^2 \cdot \alpha - 2\zeta_\theta\omega_\theta \cdot q - (p_0^2 - r^2) \cdot \frac{I_{zx}}{I_{yy}} - p_0 \cdot r \cdot \frac{I_{xx} - I_{zz}}{I_{yy}}$$

$$\dot{\beta} = -r + p_0 \cdot \alpha$$

$$\dot{r} = \omega_\psi^2 \cdot \beta - 2\zeta_\psi\omega_\psi \cdot r - q \cdot r \cdot \frac{I_{zx}}{I_{zz}} - p_0 \cdot q \cdot \frac{I_{yy} - I_{xx}}{I_{zz}}$$

where ζ_θ and ζ_ψ are the damping factors in pitch and yaw, and ω_θ and ω_ψ are the nonrolling undamped frequencies in pitch and yaw.

As long as the airplane does not roll, $p_0 = 0$, there is no coupling between the first two equations and the last two provided the product of inertia, I_{zx}, is zero. (This implies no coupling between lateral and longitudinal modes of motion.)

In this problem make the following assumptions: linearize the equations, and show that in a steady horizontal roll, $p = p_o$ at constant speed, there can be an aperiodic instability that will occur for low values of ζ_θ and ζ_ψ at a rate intermediate to ω_θ and ω_ψ, provided the aircraft is long and slender, $I_{xx} << I_{yy}$, I_{zz} and $I_{yy} \simeq I_{zz}$.

7. Show in the statement of problem that if the aircraft has isoinertia (i.e., $I_{xx} = I_{yy} = I_{zz}$) that there will be no instability.
8. With reference to Figure 3-26, consider the aircraft to be performing a steady level turn in which its roll rate is zero but its pitch and yaw rates

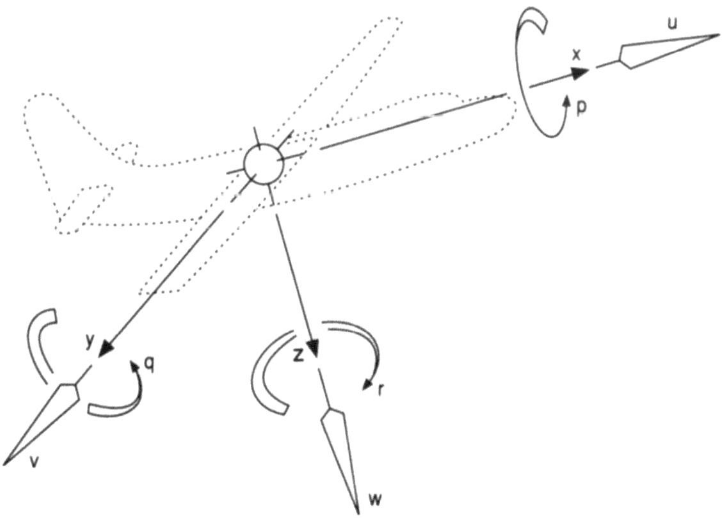

Figure 3-26. Eulerian axes (x, y, z).

are $q_0 + q$ and $r_0 + r$, respectively. Consider the case where the product of inertia is negligible, $I_{zx} = 0$, and show that there can be no coupling between lateral and longitudinal modes of motion.

9. Repeat problem 8 without neglecting the product of inertia (i.e., $I_{zx} \neq 0$), but assuming a "flat" turn of low bank angle in which $r_0 >> q_0$. Does the product of inertia tend to couple or decouple the lateral and longitudinal modes of motion.
10. A circular ring carries a bead without friction, as shown in Figure 3-27. The ring is rotated about a vertical axis at constant speed. The bead is unimpeded in passing the gimbal pins. Its equation of motion is,

$$\ddot{\theta} = \frac{\Omega_0^2}{2} \sin 2\theta - \frac{g}{r} \sin \theta$$

Linearize the system about an operating point and characterize its motion for the cases: (a) $\theta^0 = 0°$; (b) $\theta^0 = 90°$; (c) $\theta^0 = 180°$.

11. An nth order differential equation is to be linearized about an operating point using phase variables.

$$\overset{(n)}{y} = \frac{d^n y}{dt^n} = f(\overset{(n-1)}{y}, \cdots, \dot{y}, y, u)$$

nonlinear function — system input

$$\underline{\bar{x}}_1 = y, \quad \underline{\bar{x}}_2 = \dot{y}, \cdots, \underline{\bar{x}}_n \equiv \overset{(n-1)}{y} \qquad \text{phase variables}$$

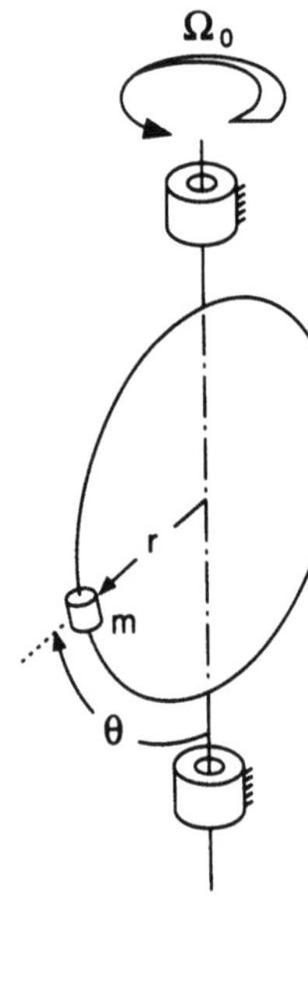

Figure 3-27. Rotating circular ring carries sliding bead at no loss.

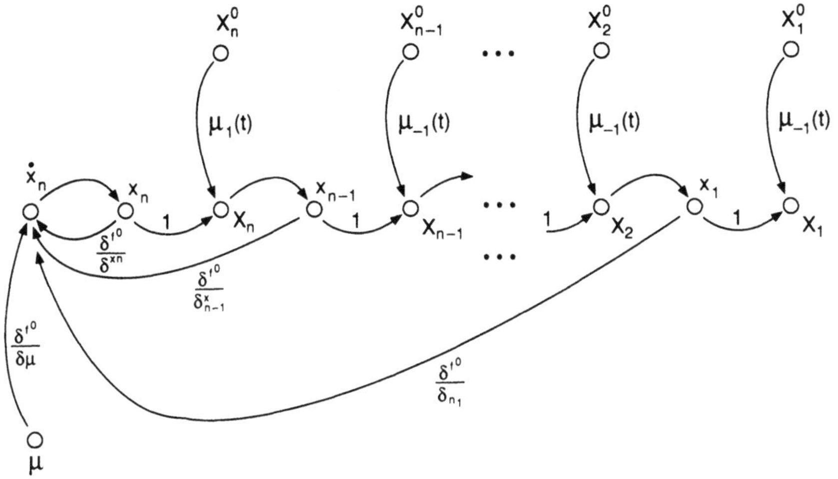

Figure 3-28. Signal flow graph of linearized nth order ordinary differential equation.

Show that the operating point values become the initial conditions on each integrator as shown in Figure 3-28. How does the initial condition, $\underline{\bar{x}}_1(0) = \underline{\bar{x}}_1^0$ enter the problem?

3.8 REFERENCES AND RELATED LITERATURE

3.8.1 References

1. Gibson, John E., *Nonlinear Automatic Control*, McGraw-Hill, New York (1963).
2. Kolk, W. R., Modern Flight Dynamics, Prentice-Hall, Englewood Cliffs, New Jersey (1961).
3. Kolk, W. R., Cabin pressure control—An exercise in nonlinear design, Transactions for First Purdue Control Engineering Conference (April 1974).

3.8.2 Related Literature

1. Brogan, William L., *Modern Control Theory*, Quantum Publishers, New York (1974).
2. Eveleigh, Virgil W., *Introduction to Control Systems Design*, McGraw-Hill, New York (1972).
3. Gille, J. C., Pelegrin, M. J., and Decaulne, P., *Feedback Control Systems*, McGraw-Hill, New York (1967).
4. Schultz, John G., and Melsa, James L., *State Functions and Linear Control System*, McGraw-Hill, New York (1967).
5. Truxal, John G., Editor, *Control Engineers' Handbook*, McGraw-Hill, New York (1958).

Chapter 4
The Describing Function

4.1 DESCRIBING FUNCTION

While working on constant speed propellers for the Curtiss Electric Propeller Co. during World War II, Ralph Kochenburger became interested in their cyclic behavior—a limit cycle, in which the propeller would periodically change speed due to the action of its "bang-bang" controller. The controller commanded an electric motor to rull full-tilt either forward or reverse to change the pitch of the propeller, enabling it to take a larger or smaller "bite" of the air and, correspondingly, decrease its speed.

The core of the problem was the relay action of the controller, which despite its simplicity, was about the only electrical gear capable of withstanding the harsh environment of an aircraft engine. So, instead of controlling the propeller's speed to a constant value, it would cycle around that value—which was undesirable, but the best that could be done. In those days there was no method of analysis to adequately model the phenomenon and the press of war left no time to study it. The end of the war, however, led Kochenburger back to MIT for a Ph.D. where his dissertation was a brilliant linearization known as the describing function which addressed the existence of limit cycles.

A closely related problem is the design of an oscillator, the difference being that its cyclic behavior is the goal rather than the difficulty. Barkhausen (ref. 2) established a criterion for sustained oscillations in a feedback system which formally holds true only in linear systems, although it is a matter of observation that nonlinear oscillators faithfully comply as well. An appreciation of why is demonstrated with Kochenburger's Describing Function, which rests on the proposition that a cyclic signal can be adequately "described" by its first harmonic. (In fact, concurrent analyses in Europe, but unknown to him, called it the "First Harmonic Approximation," see Refs. 3, 4, and 5.) Barkhausen argues that to sustain a cyclic output, the loop gain of a feedback system must simultaneously have an n-cycle delay ($2\pi n$ phase loss), and a magnitude of unity.

Put in these terms, the system must be linear as the loop gain is simply the impulse response of a system governed by linear ordinary differential equations with constant real coefficients (LODE/CRC). What Kochenburger acknowledges is the presence of nonlinearity, for which he prescribed an "equivalent gain," and in so doing, the augmented loop gain becomes a function of not only frequency but amplitude (of the first harmonic) as well.

The system of Figure 4-1 is assumed to be in a fully developed, steady-state limit cycle, which is self-sustaining (having zero input). Although it is not necessary, the nonlinearity is usually taken to be algebraic and symmetric (i.e., an odd function), the latter guaranteeing zero dc-mean in all cyclic signals.

We now take the provocative step of assuming that the nonlinearity's output, $\tilde{y}$, can be approximated by its first harmonic, which is determined by truncating all but the first terms in its Fourier series (note that the dc-term is already zero because the nonlinearity is odd). The merit of this step is defended by noting that a truncated Fourier series is the "best" approximation in the sense of least squares (ref. 6).

This step is critical because, by discarding all the harmonics in the cyclic signal, the linear dynamics are subjected to a pure sine wave. Thus, their only influence is to produce a sine wave of like frequency but different amplitude and phase, according to their frequency response, $L(j\omega)$. This sine wave is fed back, inverted, and applied to the nonlinearity as in Figure 4-1, which suggests the following interpretation:

- The nonlinearity is excited by, and produces, sine waves at the same frequency (i.e., the first harmonic approximation), the two waves differing only in phase and amplitude. Mathematically, we then state, for a sinusoidal input

$$\tilde{x} = x_m \sin \omega t$$

- The output is defined as the first harmonic approximation

$$\tilde{y} \equiv \tilde{y}_1 = A_1 \cos \omega t + B_1 \sin \omega t$$

Stated this way, the nonlinearity performs exactly (albeit approximately) as do the linear dynamics, according to their transfer function. So, it is natural

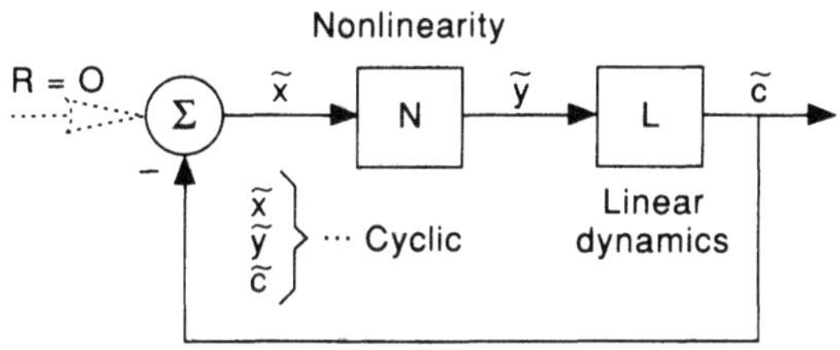

Figure 4-1. System configuration for describing function analysis.

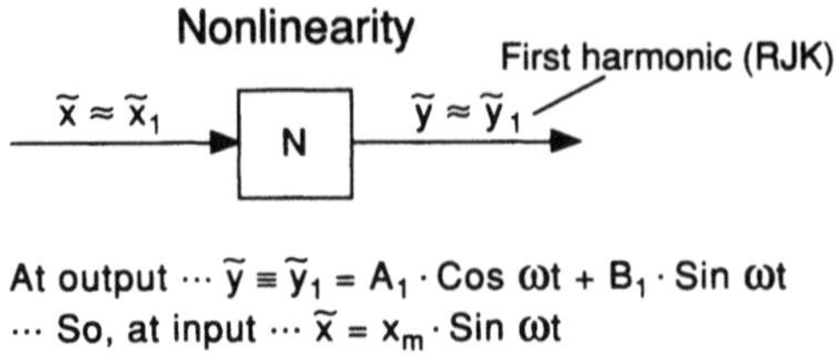

Figure 4-2. Concept of the describing function.

to imagine a "transfer function" which describes this action—Kochenburger pragmatically called it the Describing Function or $N_1(x_m, j\omega)$.

Note that the nonlinearity, with sine waves of like frequency at both input and output, can be described by the quotient of their Laplace transforms, exactly as a system of linear ordinary differential equations with constant real coefficients (LODE/CRCs) is described by its transfer function as shown in Figure 4-2.

The describing function is formally defined as

$$N_1(x_m, j\omega) \equiv \left. \frac{\mathcal{L}\{\tilde{y}_1(t)\}}{\mathcal{L}\{\tilde{x}_1(t)\}} \right|_{\substack{IC^s=0 \\ s=j\omega}} \equiv \left. \frac{\mathcal{L}\{A_1 \cos \omega t + B_1 \sin \omega t\}}{\mathcal{L}\{x_m \sin \omega t\}} \right|_{\substack{IC^s=0 \\ s=j\omega}}$$

with initial conditions set at zero and evaluated at $s = j\omega$. Substituting and evaluating we obtain N_1, the frequency response function based on the first harmonic of the nonlinearity's output.

$$N_1(x_m, j\omega) = \left. \frac{A_1 \dfrac{s}{s^2 + \omega^2} + B_1 \dfrac{\omega}{s^2 + \omega^2}}{x_m \dfrac{\omega}{s^2 + \omega^2}} \right|_{s=j\omega} = \frac{A_1 j + B_1}{x_m}$$

As N_1 is a complex term and, in general, its argument is both frequency- and amplitude-dependent, according to the following rules:

Nonlinear function	Describing function
Algebraic	$N_1 = N_1(x_m)$
Differential equation	$N_1 = N_1(x_m, j\omega)$
Odd and single-valued	N_1 = Real (i.e., $A_1 = 0$)
Odd and multivalued	N_1 = Complex
Odd + dc	N_1 = Asymmetric describing function (pbm 11)
Even	NO GAME!

Kochenburger's analysis, focusing on the odd and single-valued relay, led to a real describing function, with only real terms, which naturally became

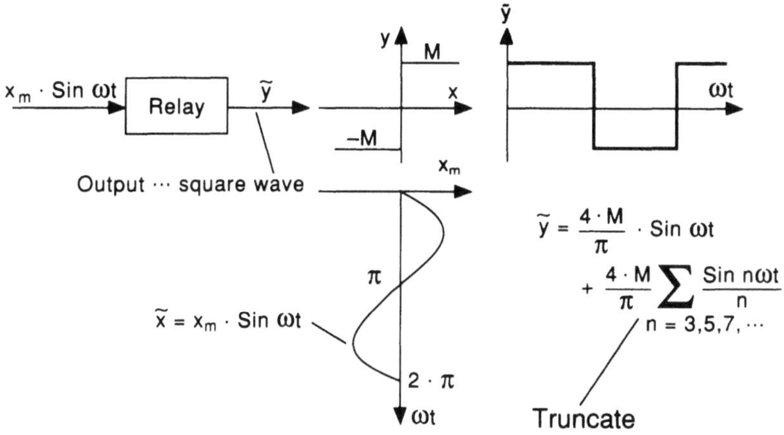

Figure 4-3. Describing function for ideal relay.

known as an "equivalent gain" and is calculated by assuming a sine wave input. The output is a square wave (Figure 4-3), whose Fourier series consists of only the odd harmonics.

Since the Fourier series is fundamental to the computation of the describing function, we summarize it here. The function, $f(t)$, must be absolutely convergent and satisfy Dirichlet's conditions over the interval, $[t_0, t_0 + P]$. Its Fourier coefficients are

$$A_n = \frac{2}{P}\int_{t_0}^{t_0+P} f(t)\cos n\omega t\, dt \qquad n = 1, 2, 3, \ldots$$

$$B_n = \frac{2}{P}\int_{t_0}^{t_0+P} f(t)\sin n\omega t\, dt \qquad n = 0, 1, 2, 3, \ldots$$

$$\omega \equiv \frac{2\pi}{P}$$

A cyclic rendition, $\tilde{f}(t)$, of the original function, $f(t)$, over the interval, $[t_0, t_0 + P]$, is recovered from the Fourier coefficients by

$$f(t) = \frac{A_0}{2} + \sum_{n=1}^{\infty} (A_n \cos n\omega t + B_n \sin n\omega t)$$

The two operations can be considered a transform pair if the coefficients are defined for the cyclic function, and, should the period be allowed to grow to infinity, this pair become the Fourier transform and its inverse.

In our example the Fourier coefficients of the square wave for the relay are found to be

$$A_0 = 0, \quad A_1 = \frac{4M}{\pi}, \quad A_n = \frac{4M}{\pi n} \ (n \text{ odd}), \quad A_n = 0 \ (n \text{ even})$$

$$B_n = 0$$

and the relay's square-wave output is

$$\tilde{y} = \frac{4M}{\pi} \sum_{n \text{ odd}}^{\infty} \frac{\sin n\omega t}{n}$$

and truncating the series for all harmonics higher than the first we define

$$\tilde{y} \equiv y_1 = A_1 \sin \omega t = \frac{4M}{\pi} \sin \omega t$$

and the describing function becomes

$$\frac{y_1}{x_1} = \frac{\frac{4M}{\pi} \sin \omega t}{x_m \sin \omega t} = \frac{4M}{\pi x_m} \equiv N_1(x_m)$$

The notion of an equivalent gain clearly emerges from the relay's describing function by considering the time average of gain during one cycle of operation. Each point on the relay's characteristic is an instantaneous gain, defined by the tangent of its coordinates. At zero input it is infinite, but then decreases as the input increases, ultimately reaching zero. Averaging this value over one cycle of an input sine wave gives an average, or effective, gain (see Problem 1).

Another common linearity often contributing to limit cycles is saturation (Figure 4-4), which, being odd, has a real describing function, depending only on amplitude.

A type of deadband (Figure 4-5) is still another familiar nonlinearity. Its describing function can be found algorithmically from the Fourier coefficients (see Problem 2) or synthesized by subtracting a linear gain from the already known saturation. Since the describing function is a linearizing process, it distributes over addition; hence, the describing function of a sum is the sum of the describing functions.

$$N_1 = N_1^a - N_1^b$$

where $N_1^a = k$, and $N_1^b = (k/\pi)\,(2\theta_1 - \sin 2\theta_1)$,

$$N_1 = \begin{cases} k[1 - (2\theta_1 - \sin 2\theta_1)/\pi], & x_m > x_{db} \\ 0, \text{ otherwise} \end{cases}$$

The importance of the describing function lies in its algebraic make-up, allowing it to be used as a multiplicative gain, exactly as if it were a transfer

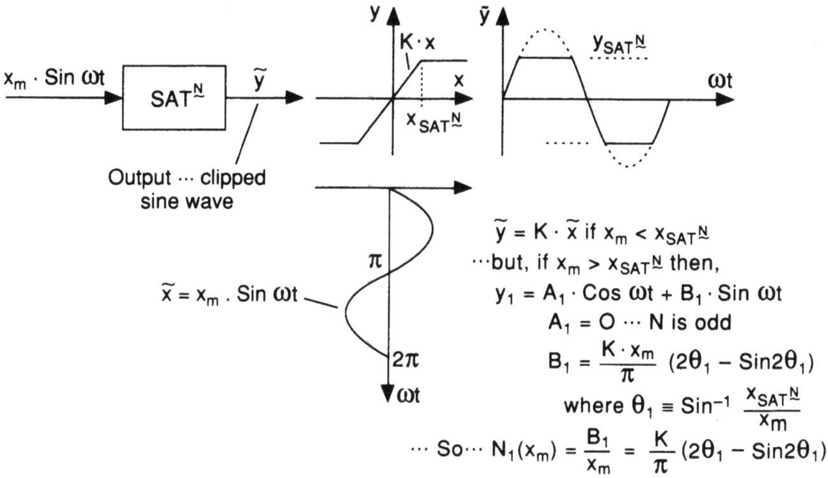

Figure 4-4. Describing function for saturation.

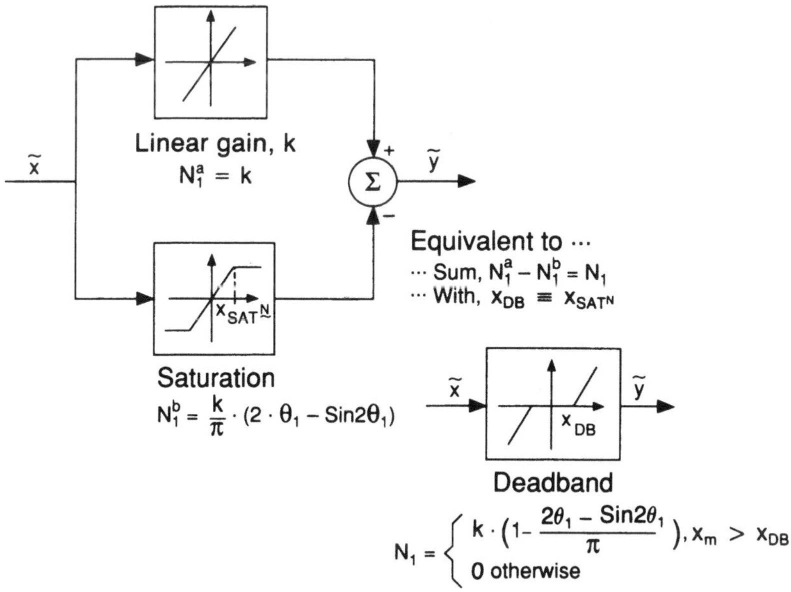

Figure 4-5. Describing function for deadband.

function. Thus, the routine use of common algebra for solving linear systems with Laplace transforms is left unaltered, with the provision that it only applies to an unforced, steady cyclic system.

Justification for Kochenburger's approximation is more qualitative than quantitative. The linear dynamics are ultimately "low-pass" in nature, having more poles than zeros—so, $L(j\omega)$ filters out the higher harmonics. This is especially true when only the odd harmonics are present in $\tilde{y}$, typically when the nonlinearity produces a wave with half-wave symmetry (always true of odd functions). Further justification for the approximation is found in the "smoothness" of $\tilde{y}$, for the smoother it is, the more rapidly its Fourier series converges [i.e., the term-by-term attenuation in successive coefficients (ref. 7, p. 226)]. A measure of this alternative is the number of derivatives in $\tilde{y}$ that are continuous (meaning single-valued). If $\tilde{y}$ is discontinuous at its Nth derivative, its Fourier coefficients, A_n and B_n, attenuate as $n^{(-1-N)}$. Some examples include:

$\tilde{y}$	N	Attenuation at
Square wave	0	$1/n$
Triangular wave	1	$1/(n^2)$
Sin (t)	∞	Instantaneous! [i.e., sin (t) is so smooth it is its own Fourier series!]

Despite these virtuous qualitative arguments, the proof of the pudding rests with a convincing quantitative argument. Attempts have been made over the years to put numbers together in a clear, convincing story (ref. 8). Whether it is convincing, we leave the reader to judge. Some examples give rather good agreement (see Problem 5), yet the lack of an exact solution leaves the question moot.

4.1.1 Existence of Limit Cycles—Using the Describing Function to Find Them

Since the describing function replaces the nonlinearity with an algebraic multiplier, the output of an unforced feedback system depends implicitly on itself, as shown in Figure 4-6. The transfer function analysis yields

$$C = \tilde{y}_1 L$$

$$\tilde{y}_1 = \tilde{x} N_1$$

$$\tilde{x} = R - C$$

and from transfer function algebra, $C/R = 1/(1 + N_1 L)$. The characteristic equation is thus

$$1 + N_1 L = 0$$

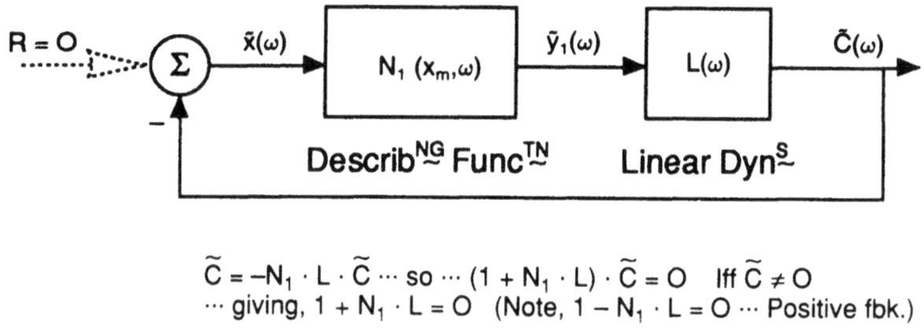

Figure 4-6. Condition for existence of a limit cycle.

Any means to satisfy the characteristic equation $1 + N_1 L = 0$ is fair game, and while there are several, the most common is to equate

$$-1/N_1(x_m, \omega) = L(j\omega)$$

at $\omega = \omega_{lc}$ and $x_m = x_{lc}$, where the subscript lc refers to limit cycle. The solution is found graphically by locating the intersection of the two loci in the complex plane. An example is the system of Figure 4-7, where a relay controls a three-pole plant.

Using the previously found describing function for the relay, its negative reciprocal is a locus at "9 o'clock" (−180° corresponding to a real negative number), moving out the negative real axis as in Figure 4-8. The frequency response function, $L(j\omega)$, of the linear dynamics crosses this only once, and thus the existence of a limit cycle is predicted where its phase is $-\pi$ (i.e., −180°).

To determine the intersection of the locus of the linear dynamics $L(j\omega)$ with that of the nonlinear dynamics $N_1(x_m, \omega)$, we recognize that $L(j\omega)$ is

$$L(j\omega) = \frac{1}{(j\omega + 1)^3} = \frac{(1 - 3\omega^2) - j\omega(3 - \omega^2)}{(1 - 3\omega^2)^2 + (3 - \omega^2)\omega^2}$$

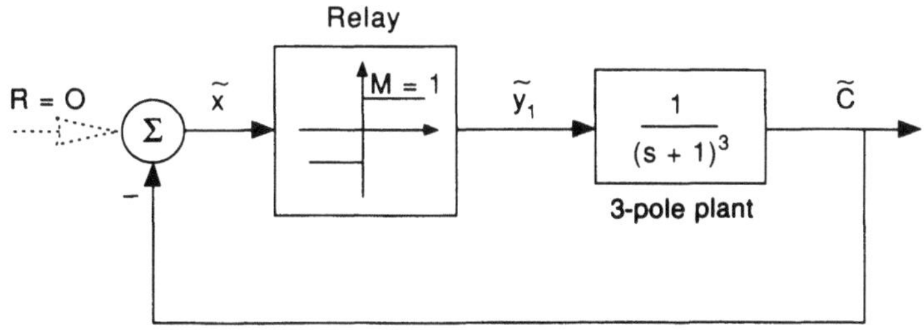

Figure 4-7. A system which limit cycles.

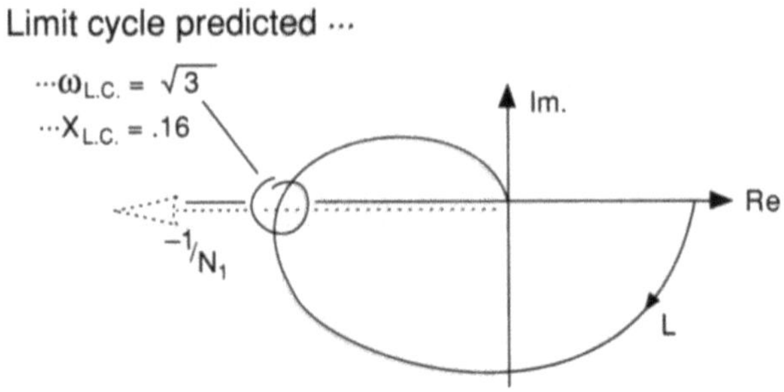

Figure 4-8. Prediction of a limit cycle.

which can be expressed as

$$L(j\omega) = A + jB$$

and the modulus of the radius vector is given by $(A^2 + B^2)^{1/2}$, and the instantaneous angle of the radius vector by

$$\underline{/L(j\omega)} = \tan^{-1} B/A$$

Thus, we set $L(j\omega) = \pi$, obtaining $\omega_{lc} = 3^{1/2}$. At this value of ω_{lc}, the modulus is found to be 1/8.

Since $N = 4M/\pi x_m = |L(j\omega_{lc})| = 1/8$, for $M = 1$, a limit cycle is predicted at an output amplitude of 0.16 and a period of 3.6 seconds (period $= 2\pi/\omega$). The three-pole attenuation qualitatively seems sufficient to meet the legacy of the first harmonic approximation.

While this analysis is reasonable, the fact remains that, lacking an exact solution, it is just speculation. Digital simulation provides some comfort by offering an approximate solution, the validity of which can be made credible, if not convincing, by changing both the parameters and the system model. Using the method described in Chapter 8, in which "samplers" and "holds" are inserted (Figure 4-9), the system is approximated with a set of pulse filters, each having its own recursion equation. A familiarity with the use of Z transforms

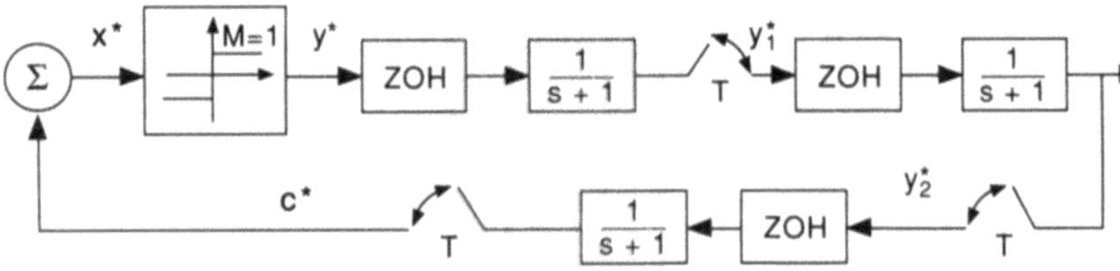

Figure 4-9. Simulation of a limit-cycling system.

can be gained in Chapter 8, and we assume the reader is familiar with those methods for the following discussion.

The recursion formula for a relay is

$$y^{(n)} = \begin{cases} 1 & x^{(n)} > 0 \\ -1 & x^{(n)} < 0 \end{cases}$$

and for each lag by use of Z transforms, we obtain

$$y_2(z)/y_1(z) = \mathcal{Z}\{(1 - e^{-sT})/s(s + 1)\} = (1 - e^{-T})/(z - e^{-T})$$

Therefore, the resulting recursion formula is

$$y_2(n + 1) = e^{-T}y_2(n) + (1 - e^{-T})y_1(n)$$

Choosing the sample time, T, is central to the accuracy of the simulation and, as a guide, the sample frequency should be about 100 times smaller than that of the limit cycle. The describing function predicts a limit cycle with a period of 3.6 seconds, so the sample time should be around .036 seconds, as shown in Figure 4-10.

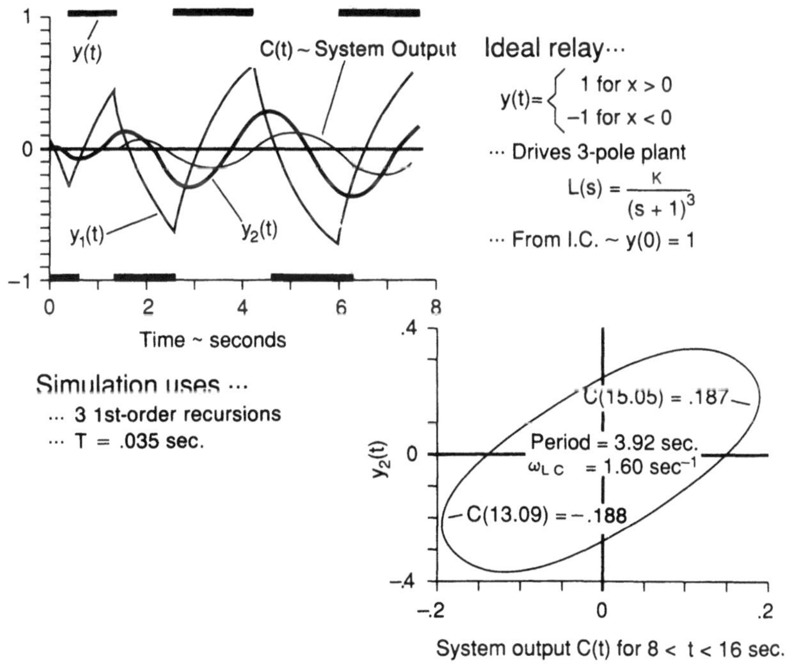

Figure 4-10. Simulation results for limit-cycling system.

4.1.2 Stability of Limit Cycles

Thus far in the analysis, the question of limit-cycle stability has not been asked, for indeed, we sought only the necessary conditions under which a cyclic solution could exist, and not the additional conditions that it persist. What brings this into focus is the possibility that the $(-1/N_1)$-locus intersects the $L(j\omega)$-locus more than once, as in Figure 4-11, where the analysis honestly predicts two limit cycles. Can the system cycle at both intersections? Or will it choose one over the other—and why?

Before addressing these questions, it is interesting to consider the stability of the limit cycle, a property which we will define as causing a system to be attracted to, or repelled from, a cyclic solution. This is a far more intriguing question than simply its existence. The following definitions are offered . . .

STABLE LIMIT CYCLE System is *attracted* toward a cyclic solution (i.e., the system will always return to the cyclic solution no matter how it is disturbed).

UNSTABLE LIMIT CYCLE System is *repelled* from a cyclic solution (i.e., the system with either "grow" in an otherwise unstable manner or "subside" in an otherwise stable manner, but will not return to the cyclic solution from which it was perturbed).

NEUTRAL LIMIT CYCLE System *passes through* a cyclic solution (i.e., the system will either "grow" through a cyclic solution or "subside" through a cyclic solution). Note that this is the case of tangency between L and $-1/N_1$ at the confluence of an even number of intersections.

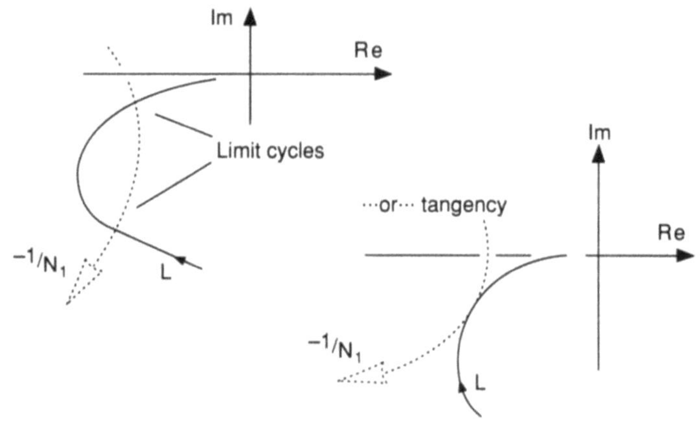

Figure 4-11. Intersection of $-1/N_1$ and L loci.

To test for stability, the system is perturbed from its assumed limit cycle, and it either returns (to a stable limit cycle) or departs (from an unstable limit cycle).

4.1.3 A Naive Criterion That Works!

Most accounts of limit-cycle stability properly suppose the limit-cycle amplitude to be perturbed by an arbitrary Δx. This, however, is taken along the $(-1/N_1)$-locus. Then the condition that it disappears (or grows) follows the argument of Nyquist's criterion by somehow asserting that all points along the $(-1/N_1)$-locus magically take on the same significance as the $(-1, j0)$-point in his classic analysis. With reference to Figure 4-12, based on an orthodox, stable open-loop, we recall that if the Nyquist plot circles the -1 point, we have an unstable linear system. Thus, while the frequency increases and the locus approaches the origin, the question of limit-cycle stability is seemingly resolved in a clean, algorithmic way.

Beginning at the steady-state condition, which is the limit cycle with parameters (x_{lc}, ω_{lc}), perturb x_{lc} by Δx along $-1/N_1$ to point A (or B) or C (or D). If the perturbation returns to the intersection of L and $-1/N_1$, the limit cycle is termed stable. If the perturbation increases in amplitude and does not return to the intersection, then the limit cycle is termed unstable.

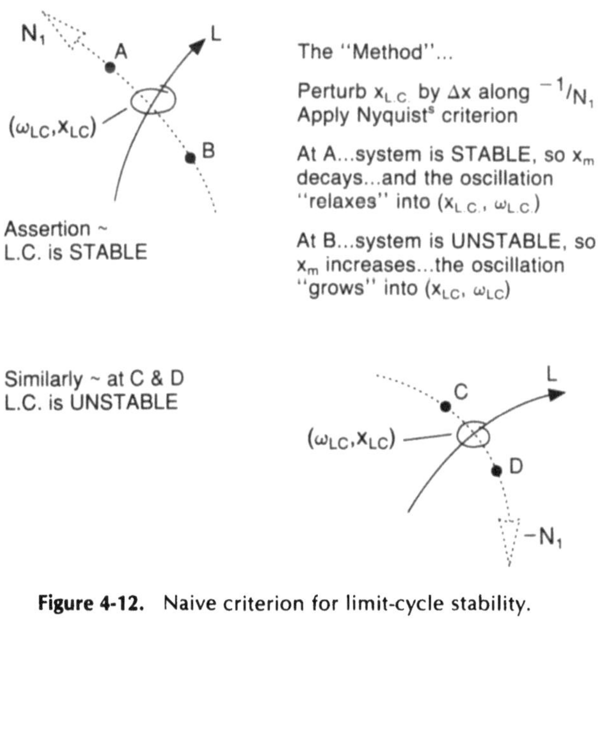

Figure 4-12. Naive criterion for limit-cycle stability.

The argument, however, is flawed by a contradiction; the system being assumed *both* cyclic and noncyclic. Surely, the perturbation is unsuitable for the cyclic analysis of a system which regularly waxes and wanes. But the intriguing particular is that the conclusions drawn are *exactly right*, as is shown next using a more perfect approach.

4.1.4 A Proper Criterion for Limit-Cycle Stability

Provided the nonlinearity is independent of frequency, the analysis is straightforward, and, as most systems fall into this category, we favor this assumption because of the clarity it lends. The existence of a limit cycle can be expressed as

$$L(s) + 1/N_1(x_m) = 0$$

where the limit cycle exists at $s = \sigma_{lc} + j\omega_{lc}$, $x_m = x_{lc}$. The limit cycle, subject to the first harmonic assumption, is a sine wave

$$\tilde{x}(t) = \tilde{x}_1(t) = x_{lc} \sin \omega_{lc} t$$

so applying Laplace transforms

$$X_1(s, x_{lc}) = L(\tilde{x}_1) = x_{lc}\omega_{lc}/(s^2 + \omega_{lc}^2)$$

We now perturb the limit cycle, $X_1(s, x_{lc})$, by perturbing its entire argument

$$ds = d\sigma + jd\omega, \qquad \text{and } dx$$

But what is dx? We have assumed we are dealing with a system whose operating point is a function of time, and we are willing to examine its behavior if we move an arbitrary "distance" made up of dx, $d\sigma$ and $d\omega$ from its operating point. Just as $d\sigma$ and $d\omega$ are perturbations of the limit-cycle decay-rate and frequency, so dx is the perturbation on the limit-cycle amplitude, x_{lc}, and all three are algebraic.

Notice that the perturbation causes a *transient* solution *near* the limit cycle given by

$$\tilde{x}_1 + dx = (x_{lc} + dx)e^{d\sigma t} \sin [(\omega_{lc} + d\omega)t]$$

The transient perturbation will either grow, causing an unstable limit cycle ($d\sigma > 0$), or disappear, causing a stable limit cycle ($d\sigma < 0$).

A Constraint: Since the system lies in the neighborhood of the original cyclic solution, it must obey

$$\phi = L(s) + 1/N_1(x_m) = 0,$$

and, consequently,

$$d\phi = 0$$

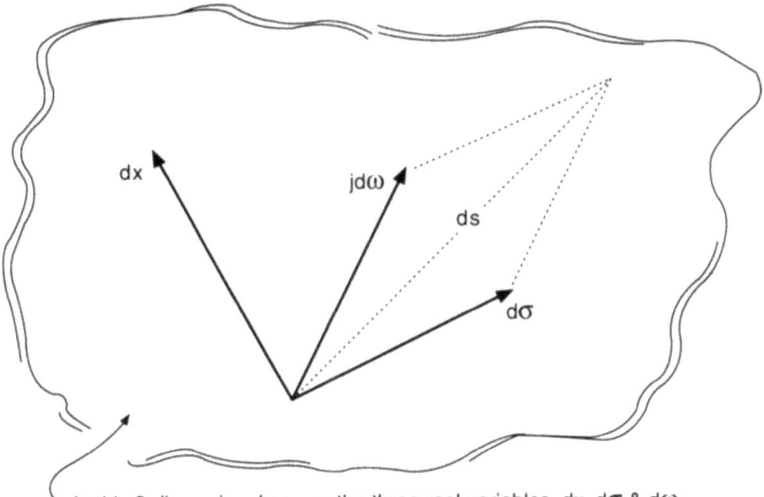

In this 3-dimensional space, the three real variables, dx, dσ & dω, are functionally related by the complex function, ϕ ···

$$\phi = L + 1/N_1 = 0 \cdots \text{so} \cdots d\phi = 0$$

which gives two conditions ··· dReϕ = 0, dImϕ = 0, both of which contain the three variables.

Figure 4-13. Limit cycle perturbations.

Now ϕ is complex, having both real and imaginary parts as does its differential, both of which must vanish, yielding two separate equations. Each equation contains the differentials of the argument of ϕ (i.e., $d\sigma$, $d\omega$, and dx), which you may replace with "delta quantities" if you wish. Being arbitrary, these are the unknowns, and we have the two equations relating three unknowns, of which any one can be removed by substitution, leaving just the other two related by a single equation, see Figure 4-13. We are interested in learning if dx decays (or inflates) with time, which is mostly dictated by $d\sigma$; in fact, a reasonable statement of stability is $dx/d\sigma < 0$. This is shown below.

Since

$$\phi = L(s) + 1/N_1(x) = 0$$

$$d\phi = d\{L(s) + 1/N_1(x)\} = 0$$

However, the bracketed term is complex, so that we can write

$$d\phi = d\{Re + jI_m\}$$

$$= \frac{\partial Re}{\partial x}dx + \frac{\partial Re}{\partial s}ds + j\frac{\partial Im}{\partial x}dx + j\frac{\partial Im}{\partial s}ds = 0$$

Now $\Sigma Re = 0$ and $\Sigma Im = 0$, which yields two equations in three unknowns. And since $s = \sigma + j\omega$, $ds = d\sigma + jd\omega$. After substitution, we get

$$\frac{\partial Re}{\partial x} dx + \frac{\partial Re}{\partial \sigma} d\sigma - \frac{\partial Im}{\partial s} d\omega = 0$$

$$\frac{\partial Im}{\partial x} dx + \frac{\partial Im}{\partial \sigma} d\sigma + \frac{\partial Re}{\partial s} d\omega = 0$$

We eliminate $d\omega$ (Why? See Problem 13)

$$\left\{\frac{\partial Re/\partial s}{\partial Im/\partial s} + \frac{\partial Im/\partial s}{\partial Re/\partial s}\right\} d\sigma = -\left\{\frac{\partial Im/\partial x}{\partial Re/\partial s} + \frac{\partial Re/\partial x}{\partial Im/\partial s}\right\} dx$$

Defining

$$\mathfrak{D} \equiv (\partial Im/\partial x)\,(\partial Im/\partial s) + (\partial Re/\partial s)\,(\partial Re/\partial x)$$

and thus we obtain

$$\{(\partial Re/\partial s)^2 + (\partial Im/\partial s)^2\}\, d\sigma = -\mathfrak{D}\, dx$$

Since the bracketed term is positive, for $d\sigma$ to be of opposite sign as dx, $\mathfrak{D} > 0$ is required.

4.1.5 Geometric Intepretation—and the Irony!

A geometric interpretation (refs. 10 and 11), developed below and shown in Figure 4-14, shows that crossing the L-locus (actually its tangent) into the $(-1/N_1)$-locus (also its tangent), if positive, represents a stable limit cycle and, if negative, represents an unstable limit cycle.

Since $L(j\omega)$ and $-1/N_1(x_m)$ are complex, we can rewrite (assuming N_1 is algebraic)

$$L(j\omega) = A(j\omega) + jB(j\omega), \qquad -1/N_1(x_m) = C(x_m) + jD(x_m)$$

Thus,

$$Re(L + 1/N_1) = A(j\omega) - C(x_m)$$

$$Im(L + 1/N_1) = B(j\omega) - D(x_m)$$

Let us now *define* the discriminant $\mathfrak{D}$ as the *determinant*

$$\mathfrak{D} = \begin{vmatrix} -B' & C' \\ A' & D' \end{vmatrix} = -B'D' - A'C'$$

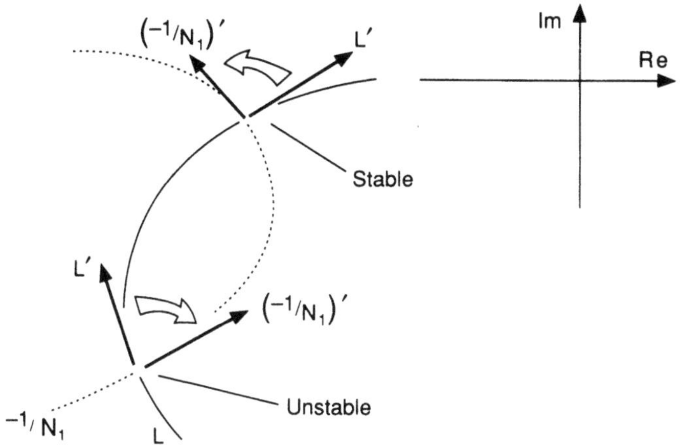

Figure 4-14. Proper criterion for limit-cycle stability.

where ( )′ denotes differentiation with respect to the variable's augment; that is, for $s = j\omega$

$$A' = \partial Re/\partial s, \qquad B' = \partial Im/\partial s, \qquad C' = -\partial Re/\partial x_m, \qquad D' = -\partial Im/\partial x_m$$

Now if we identify vectors tangent to L and $1/N_1$ as

$$dL/d\omega = j\partial L/\partial\omega = -B' + jA' \equiv L'$$

$$d(-1/N_1)/dx_m = C' + jD' \equiv (-1/N_1)'$$

We observe that the determinant is the value of the vector-product of the tangents and thus provides a basis for a geometric interpretation, shown in Figure 4-14.

The irony of this elegant analysis is that, while it is quite properly derived, it gives the same answer that the flawed use of Nyquist's criterion gave with a lot less work.

4.2 FREQUENCY-DEPENDENT DESCRIBING FUNCTIONS

Just as algebraic nonlinearities produce amplitude-dependent describing functions, nonlinear differential equations give rise to frequency-dependent describing functions, which is slightly ambiguous, because their argument contains not only frequency but amplitude as well. Their calculation is straightforward, although possibly cumbersome, yet there is a common case where the calculation simplifies. This is in equations whose nonlinearity is confined exclusively to the input variable, the circumstance most commonly encountered. Admissible operations are $y = x'^3$ or $y = \sin^2 x$, but not $y = x - y'^2$ or $y = x - \tan y$. The simplification consists of assigning the phase angle of the nonlinearity to the input variable, which, being linear, reduces the algebra. The calculations are:

$$\text{For } N_1(x, \omega) = |N_1|e^{j\angle N_1}$$

$$\text{Let } \tilde{x}_1 = x \sin(\omega t - \angle N)$$

$$= x_1 \sin \omega t + x_2 \cos \omega t$$

$$\text{And } \tilde{y} = \tilde{y}_1 = y \sin \omega t$$

Therefore,

$$\cos \angle N_1 = x_1/x, \qquad \sin \angle N_1 = -x_2/x$$

where

$$\angle N_1 = \tan^{-1}(-x_2/x_1)$$

and

$$|N| = |y/x| = |y|/(x_1^2 + x_2^2)^{1/2}$$

Substituting yields

$$-1/N_1 = -x_1/y - jx_2/y$$

A counterexample follows, in which we allow the output to be nonlinear with a linear input. Consider a nonlinear differential equation that yields a frequency-dependent describing function

$$(\dot{y})^3 + y = x$$

a form of a saturating integrator. Noting that this is linear in its input, x, the phase angle is associated with $x(t)$, and the describing function shown in Figure 4-15 is found as follows:

$$\tilde{y}_1 = Y \sin \omega t, \qquad \text{and } \dot{\tilde{y}}_1 = \omega Y \cos \omega t$$

$$\tilde{x}_1 = x_1 \sin \omega t + x_2 \cos \omega t$$

Direct substitution, using $\cos^3 \omega t = (3/4) \cos \omega t + (1/4) \cos 3\omega t$ yields

$$(\omega^3 y^3/4)(3 \cos \omega t + \cos 3\omega t) + \omega y \sin \omega t = x_1 \sin \omega t + x_2 \cos \omega t$$

Discarding the third harmonic term in line with the describing function's first harmonic approximation, and equating like coefficients

$$y = x_1$$

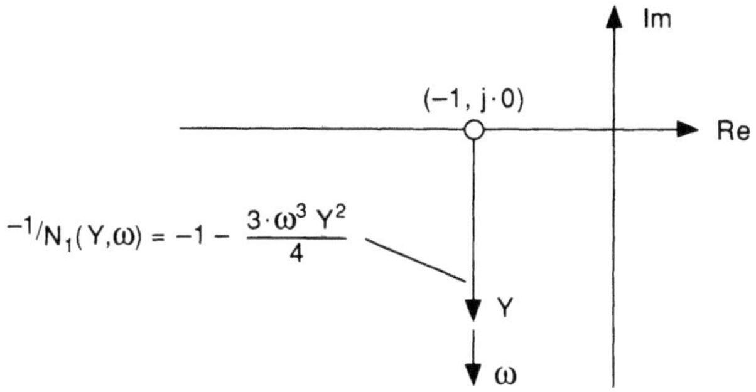

Figure 4-15. A frequency dependent describing function for $\dot{y}^3 + y = x$.

or

$$Re(-1/N_1) = -1$$

$$3\omega^3 y^3/4 = x_2$$

or

$$Im(-1/N_1) = -3\omega^3 y^2/4$$

yielding

$$-1/N = -1 - 3\omega^3 y^2/4$$

4.2.1 Example of Frequency-Dependent Describing Function

The existence of a limit cycle at the intersection of the $-1/N_1$ and $L(j\omega)$ loci is illustrated in the following example, which provides some evidence to support the first harmonic assumption. The system consists of a feedback loop in which the nonlinearity drives the linear plant, consisting of either a single-pole or a double-pole, as in Figure 4-16.

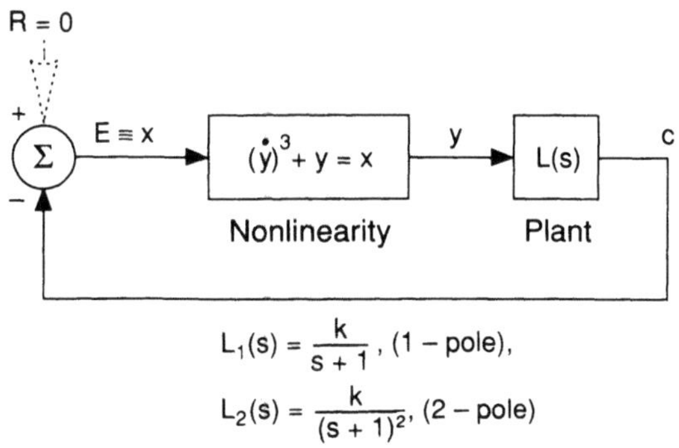

Figure 4-16. System with nonlinear differential controller.

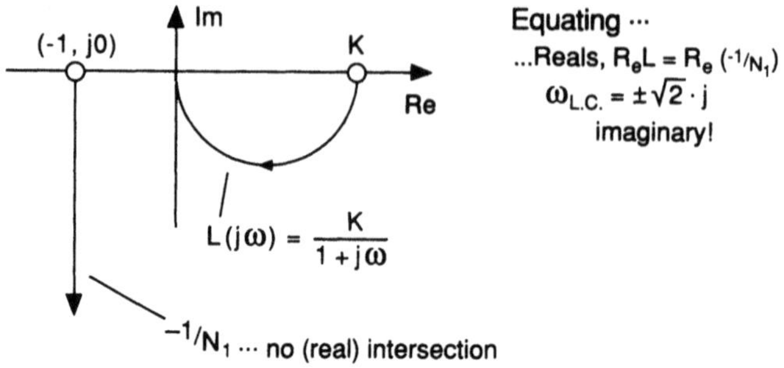

Figure 4-17. No limit cycle properties for $L_1 = \frac{K}{s+1}$.

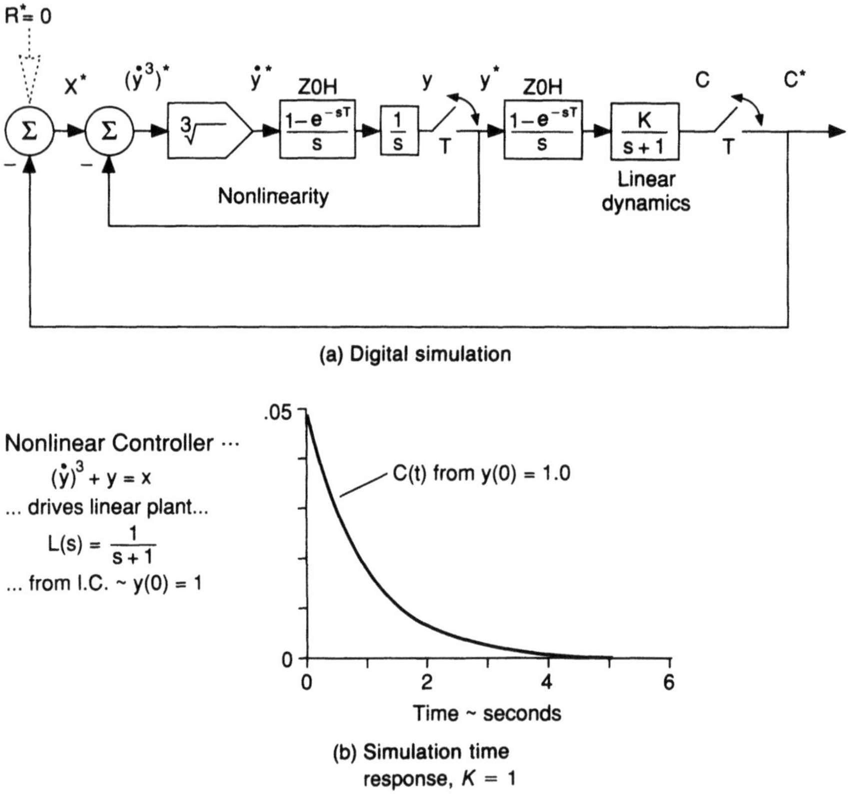

Figure 4-18. Simulation of system with nonlinear differential controller driving $L_1 = \frac{K}{s+1}$.

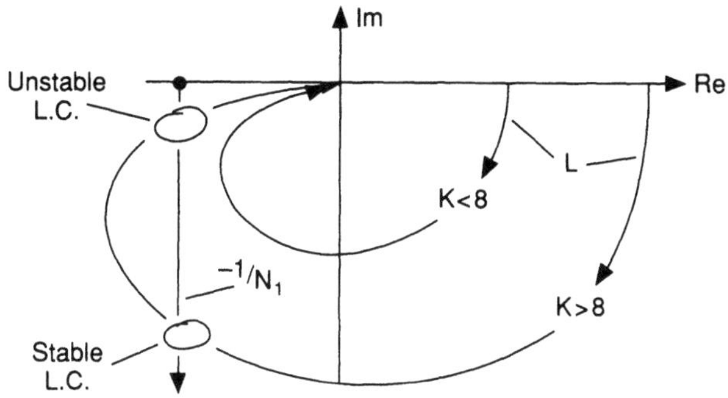

Figure 4-19. Limit cycles predicted for $L_2 = \frac{K}{(s + 1)^2}$.

In the case of $L_1(s)$, consisting of a single pole, there is no intersection of the describing locus and the linear locus because the latter never escapes from the fourth quadrant, as shown in Figure 4-17. By analytically forcing the equality, the "intersection" is shown to occur at an imaginary amplitude and frequency, which we know is a physical impossibility and must be discarded.

Furthermore, a digital simulation of the system, based on the method discussed in Chapter 8, is shown in Figure 4-18 along with its response from an initial condition, which is seen to be aperiodically damped, giving credence to the assertion that no limit cycle exits as obtained above.

The second case, with a two-pole plant (see Figure 4-16) gives the describing function further evidence of validity. Here the frequency response function occupies both the third and fourth quadrants (Figure 4-19), and depending on its dc gain, K, there will be either no intersection (i.e., no limit cycle) or two intersections (i.e., two limit cycles—one stable and the other unstable). At a gain in excess of that which gives a real part more negative than -1 (i.e., $K > 8$), there are two intersections. We leave it to the reader to verify that at the point of tangency $\partial L/\partial\omega = \infty$, and the real part of $L = -1$ yields $K = 8$.

4.3 DIGITAL SIMULATION VERIFIES DESCRIBING FUNCTION ANALYSIS

A digital simulation based on the method discussed in Chapter 8 is shown in Figure 4-20. At $K < 8$, the system should be free of limit cycles as is the case shown in Figure 4-21, where $K = 5$, the system responding to an initial condition on the nonlinearity output in an oscillatory but damped manner.

With the dc gain set at $K > 8$, there are two intersections, indicating a stable and an unstable limit cycle. The presence of the former is easy to verify

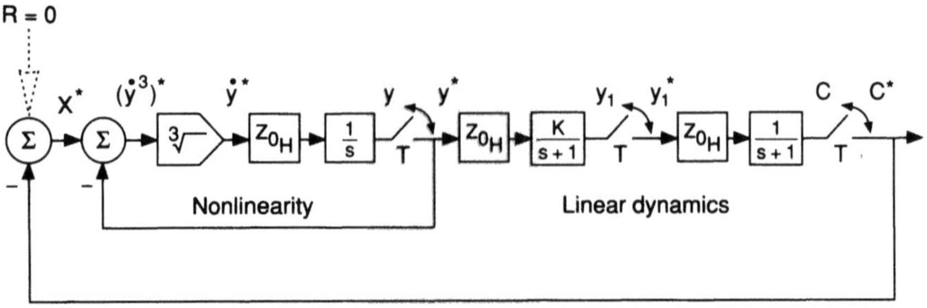

Figure 4-20. A digital simulation of $L = \frac{K}{(s + 1)^2}$.

in a simulation because the system naturally relaxes into it. This is shown in Figure 4-22.

The time response [Figure 4-22(a)] indicates that stimulation by an initial condition causes the system to relax into a stable limit cycle, frequency and amplitude of which can be measured peak-to-peak in the figure. Another way to view the limit cycle is to plot the two outputs—that of the nonlinearity and that of the system—with time being a running parameter [Figure 4-22(b)]. In the steady state, both variables are cyclic, so, by plotting one against the other, a closed path is found (also shown in the figure). This represents the stable limit cycle (predicted in Figure 4-19); the comparison of its amplitude and frequency with that predicted by the describing function is left as an exercise for the student. The validity of this comparison should be recognized to rest with the fidelity of the simulation, which in turn is heavily influenced by the choice of sample time.

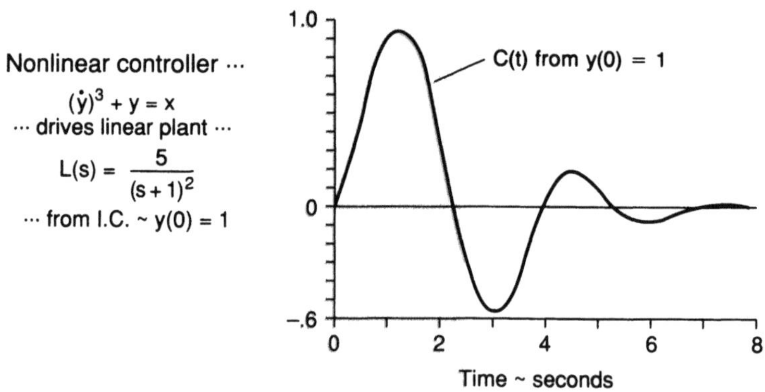

Figure 4-21. Simulated response for $K < 8$ verifies no limit cycle.

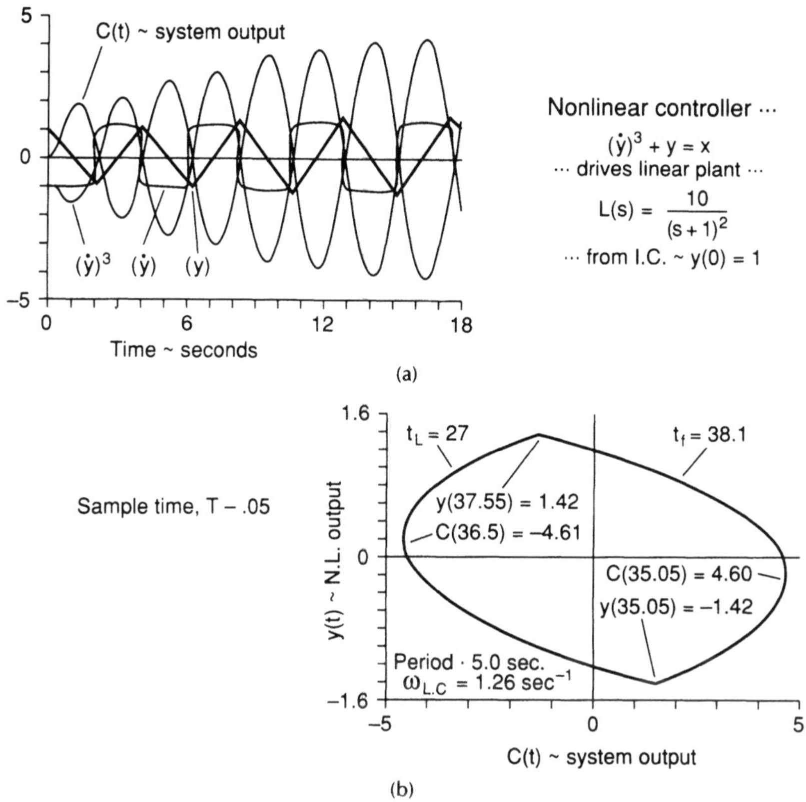

Figure 4-22. Simulation for $K > 8$ verifies limit cycle.

It is also possible to use the simulation to investigate the unstable limit cycle by guessing a set of initial conditions and tracing out one cycle to see if it gets smaller or larger. Eventually, a set of initial conditions can be found that very nearly gives a closed path. This not only determines the frequency and amplitude of the unstable limit cycle, but demonstrates clearly why it is unstable, there being a closed path from which all trajectories are repelled when the slightest departure from it is made.

4.4 ASYMMETRIC DESCRIBING FUNCTIONS

While nonlinearities are often symmetric (having odd-quadrant symmetry), an occasional asymmetric nonlinearity is encountered—the most common being an on–off switch. The sinusoidal response of such a device is offset from zero by a dc term, adding some complication to the calculation of its describing func-

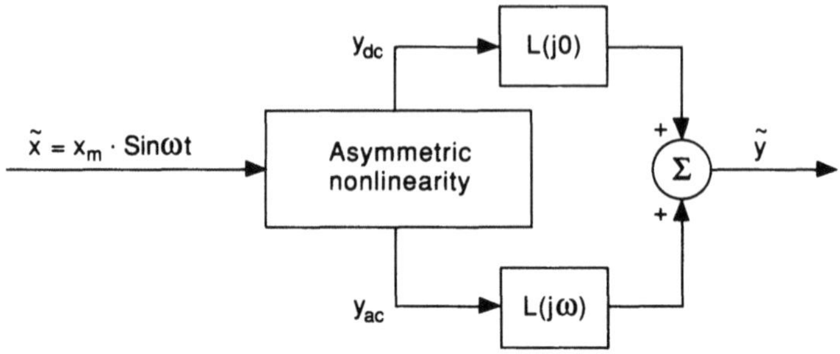

Figure 4-23. ac and dc components of an asymmetric nonlinearity.

tion. How this dc term is handled in the analysis of limit cycles is discussed here and is based on ref. 12. Being a dc value, it can excite (in steady state) a linear filter, $L(s)$, at zero frequency, by an amount proportional to $L(j0)$. Likewise, a sine-wave at frequency, ω, excites the linear filter by an amount proportional to $L(j\omega)$. Therefore, we are tempted to consider two paths from the nonlinearity: one, the usual ac path driving the frequency response function; and the other, a dc path driving the dc gain of the filter (as in Figure 4-23).

The implication of the dual paths in Figure 4-23 is that the asymmetric nonlinearity can discriminate between the ac and dc paths of a signal, which suggests that its describing function takes on a duality (i.e., N_{ac} and N_{dc}), acting on their corresponding signals. It is this capability of discriminating between ac and dc signals which is fundamental to the use of the asymmetric describing function (as shown in Figure 4-24).

The existence of a limit cycle, as proposed by Kochenburger, requires that the "characteristic equation" be satisfied; and, since there are both the ac

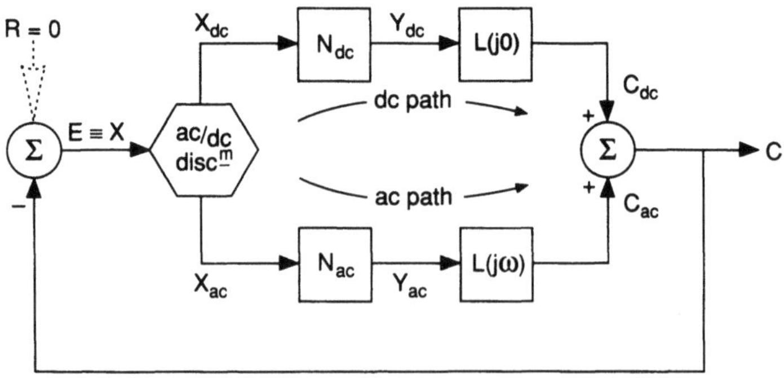

Figure 4-24. Embedding the asymmetric describing function.

and dc conditions to satisfy, two relations result:

$$1 + L(j\omega)\, N_1(x_m, \omega) = 0; \qquad \omega = \omega_{lc},\ x_m = x_{lc}$$

$$1 + L(j0)\, N_1(x_0, 0) = 0$$

The first is a phasor (complex) relation whose real and imaginary parts each provide conditions; while the second is a scalar (real) relation providing a third. Altogether, the three conditions allow the solution of the three characteristics of the limit cycle (i.e., its amplitude x_{lc}, its frequency ω_{lc}, and its dc component).

$$L(j\omega) = -1/N_{dc}$$

$$L(j\Omega) = -1/N_{ac} \Rightarrow |L(j\omega)| = |-1/N_{ac}|,\ \underline{/L(j\omega)} = \underline{/(-1/N_{ac})}$$

The asymmetric relay is used as an example of the asymmetric describing function (Figure 4-25).

Its rectangular-wave output is approximated by its truncated Fourier series

$$y(t) = A_0/2 + A_1 \cos \omega t + B_1 \sin \omega t$$

$$= y_0 + y_{ac}$$

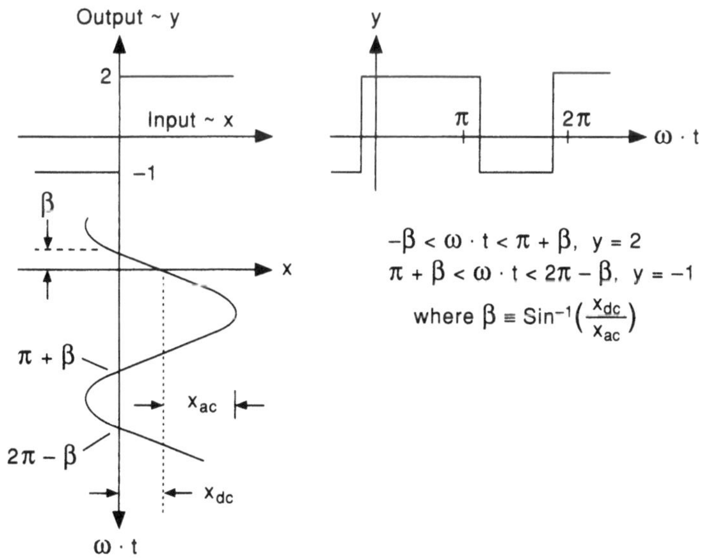

Figure 4-25. Characteristic of asymmetric relay.

where its Fourier coefficients are

$$A_0 = (M_1 - M_2) + 2\beta(M_1 + M_2)/\pi$$

$$A_n = \begin{cases} 0, & n \text{ even but } \neq 0 \\ -[2(M_1 + M_2)/n\pi] \sin\beta, & n \text{ odd} \end{cases} \qquad \beta \equiv \sin^{-1}(-x_{dc}/x_{ac})$$

$$B_n = \begin{cases} 0, & n \text{ even} \\ +[2(M_1 + M_2)/n\pi] \cos\beta, & n \text{ odd} \end{cases}$$

And the asymmetric describing function is

$$N_{dc} = y_0/x_{dc} = \{(M_1 - M_2)/2 + [(M_1 + M_2)/\pi] \sin(-x_{dc}/x_{ac})\}/x_{dc}$$

$$N_{ac} = |y_0|/|x_{ac}| = (B_1 + jA_1)/|x_{ac}|$$

4.4.1 Example of Asymmetric Describing Function

An example of the asymmetric relay is shown in Figure 4-26, driving a third-order integrating plant.

An apparent difficulty arises when the plant contains an integration, making the dc value of its transfer function infinite. The way around this is to force the dc part of the describing function to vanish, which produces a relation between x_{dc} and x_{ac}, as is illustrated in the example of Figure 4-26.

$$A_0 = 2, \qquad \beta = -(\pi/2)(M_1 - M_2)/(M_1 + M_2)$$

The ac component is

$$N_{ac} = (B_1 + jA_1)/|x_{ac}|, \qquad \text{evaluated at } a_1 = 0$$

$$= 2[(M_1 + M_2)/\pi|x_{ac}|] \cos\beta$$

The ac describing locus $-1/N_{ac}$ is plotted, along with the linear frequency response function, $L(j\omega)$, in Figure 4-27, which shows an intersection and, therefore, predicts a limit cycle.

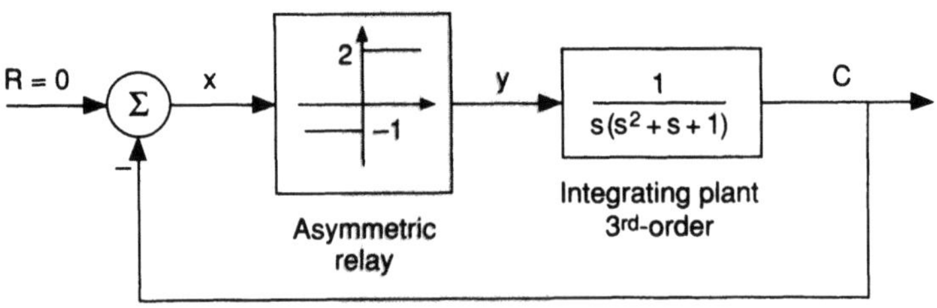

Figure 4-26. System controlled by asymmetric relay.

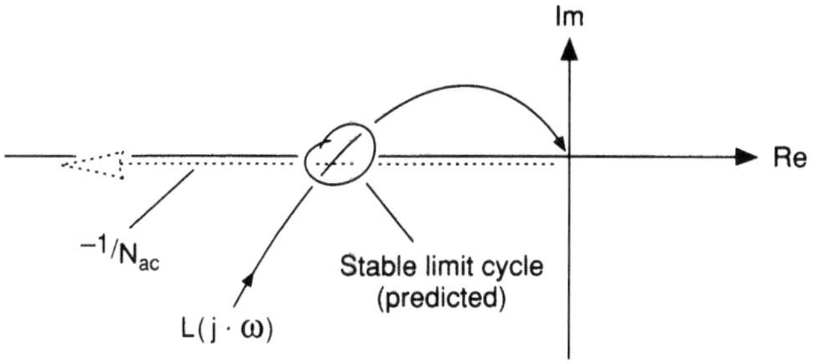

Figure 4-27. Limit cycle predicted for asymmetric relay.

4.4.2 Digital Simulations Verify Asymmetric Describing Function

These describing function results are verified by means of digital simulations: one using the Continuous System Modeling Program (CSMP) (ref. 13), shown in Figure 4-28, and the other using the method of Chapter 8, shown in Figure 4-29. The results obtained from the CSMP simulation are tabulated in Table

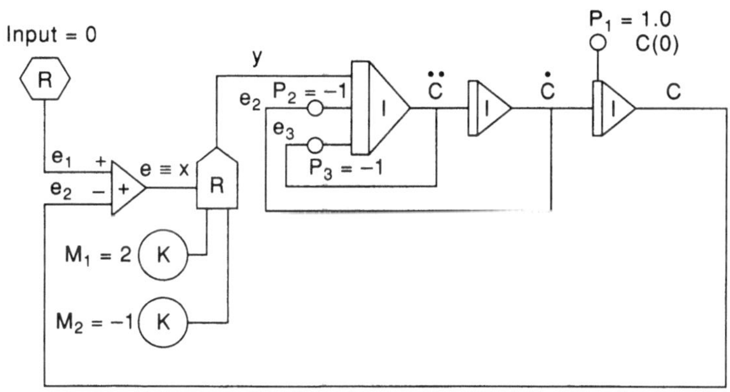

Figure 4-28. CSMP simulation.

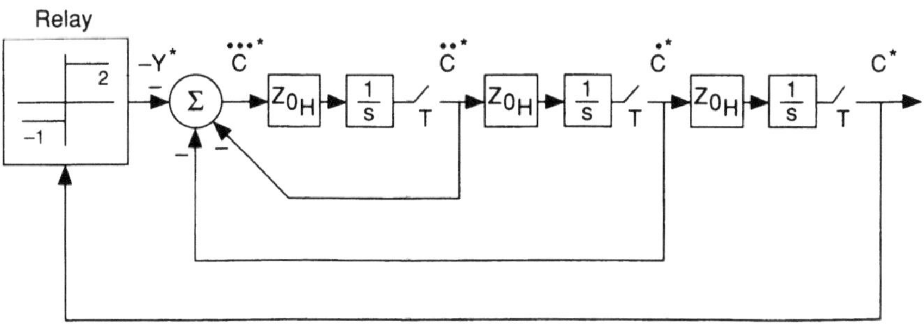

Figure 4-29. Use of samplers and holds for digitally simulating the system with asymmetric relay.

TABLE 4-1. CSMP Simulation Results

Time	Y Relay Output 2	$\ddot{C}$ Accel^N Output 3	$\dot{C}$ Vel^{TY} Output 4	C Position Output 5
100.000	2.0000	1.3962	1.5521	−0.2484
100.100	2.0000	1.2992	1.6869	−0.0862
100.200	−1.0000	1.1990	1.8119	0.0890
100.400	−1.0000	0.4541	1.9743	0.4714
100.600	−1.0000	−0.1717	1.9996	0.8720
100.800	−1.0000	−0.6775	1.9117	1.2658
101.000	−1.0000	−1.0669	1.7344	1.6324
101.200	−1.0000	−1.3468	1.4904	1.9563
101.400	−1.0000	−1.5271	1.2006	2.2264
101.600	−1.0000	−1.6193	0.8839	2.4354
101.800	−1.0000	−1.6362	0.5566	2.5795
102.000	−1.0000	−1.5908	0.2324	2.6582
102.200	−1.0000	−1.4960	−0.0774	2.6733
102.400	−1.0000	−1.3644	−0.3642	2.6285
102.600	−1.0000	−1.2073	−0.6219	2.5291
102.800	−1.0000	−1.0351	−0.8464	2.3814
103.000	−1.0000	−0.8567	−1.0357	2.1923
103.200	−1.0000	−0.6798	−1.1892	1.9690
103.400	−1.0000	−0.5104	−1.3080	1.7184
103.600	−1.0000	−0.3534	−1.3941	1.4474
103.800	−1.0000	−0.2121	−1.4502	1.1622
104.000	−1.0000	−0.0889	−1.4798	0.8686
104.200	−1.0000	0.0152	−1.4867	0.5714
104.400	−1.0000	0.0998	−1.4747	0.2748
104.600	2.0000	0.1654	−1.4477	−0.0177
104.800	2.0000	0.7532	−1.3524	−0.3008
105.000	2.0000	1.2068	−1.1532	−0.5537

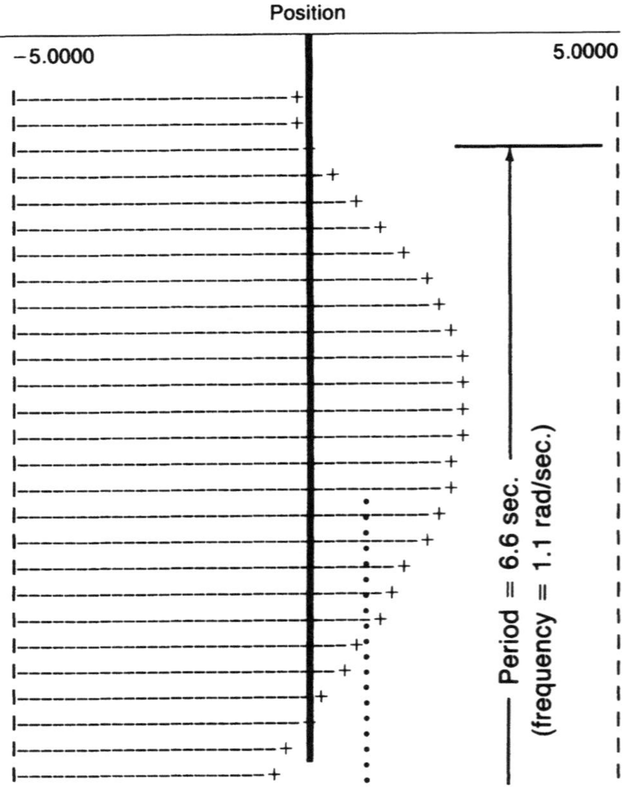

105.200	2.0000	1.5343	−0.8760	−0.7583
105.400	2.0000	1.7466	−0.5451	−0.9016
105.600	2.0000	1.8572	−0.1824	−0.9749
105.800	2.0000	1.8805	0.1935	−0.9740
106.000	2.0000	1.8314	0.5663	−0.8978
106.200	2.0000	1.7250	0.9233	−0.7483
106.400	2.0000	1.5755	1.2543	−0.5299
106.600	2.0000	1.3962	1.5521	−0.2484
106.800	−1.0000	1.1990	1.8119	0.0890
107.000	−1.0000	0.4541	1.9743	0.4714
107.200	−1.0000	−0.1717	1.9996	0.8720
107.400	−1.0000	−0.6775	1.9117	1.2658
107.600	−1.0000	−1.0669	1.7344	1.6324
107.800	−1.0000	−1.3468	1.4904	1.9563
108.000	−1.0000	−1.5271	1.2006	2.2264
108.200	−1.0000	−1.6193	0.8839	2.4354
108.400	−1.0000	−1.6362	0.5566	2.5795
108.600	−1.0000	−1.5908	0.2324	2.6582
108.800	−1.0000	−1.4960	−0.0774	2.6733
109.000	−1.0000	−1.3644	−0.3642	2.6285
109.200	−1.0000	−1.2073	−0.6219	2.5291
109.400	−1.0000	−1.0351	−0.8464	2.3814
109.600	−1.0000	−0.8567	−1.0357	2.1923
109.800	−1.0000	−0.6798	−1.1892	1.9689
110.000	−1.0000	−0.5104	−1.3080	1.7184

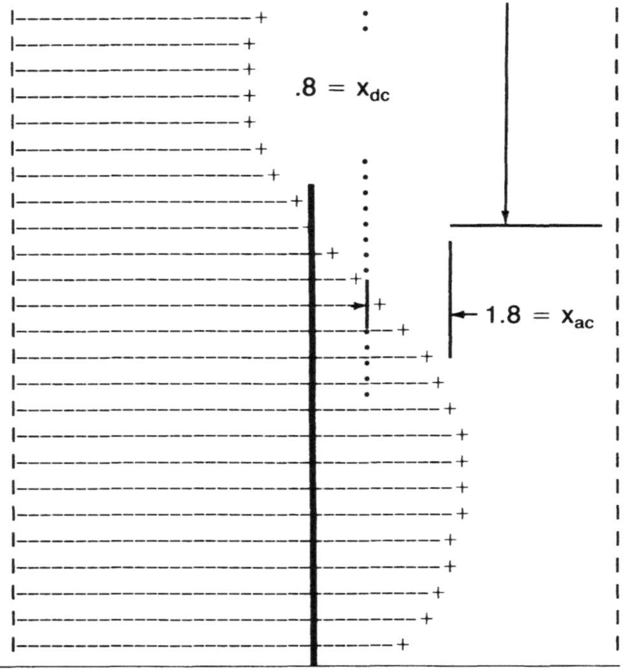

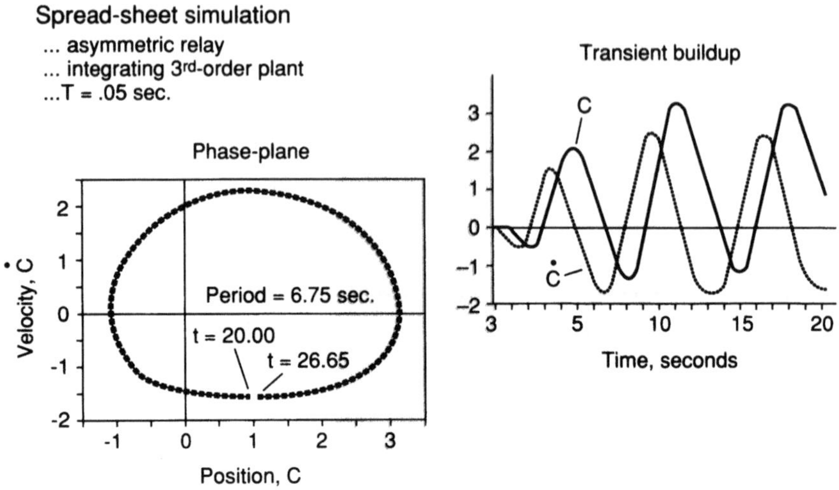

Figure 4-30. Spread sheet simulation results.

TABLE 4-2. Comparison of Describing Function/Simulation for Asymmetric Relay Driving Third-Order Integrating Plant

		x_{dc}	$\lvert x_{ac} \rvert$	$\omega_{L,C}$
Describing function		−.8	1.7	1.0
CSMP at $T = 1$		−.8	1.8	1.1
LOTUS at	$T = .1$	−1.2	2.4	1.1
	$T = .05$	−1.0	2.1	1.1
	$T = .02$	−.9	1.9	1.0

4-1 which indicate a dc offset of .8, an ac amplitude of 1.8, and a limit cycle period of 6.6 seconds.

The system can also be simulated using the method of Chapter 8 by inserting samplers and holds as indicated in Figure 4-29 (see also Problem 14 for another possible simulation). Its solution is shown in Figure 4-30 for one sample time, T, as simulated on a spread sheet. A comparison of the CSMP and spread-sheet results is given in Table 4-2.

4.5 PROBLEMS

1. Find the average gain of a saturation as described in Figure 4-4 by assuming a sine wave applied to its input and then averaging its output over one period. The ratio of average output to average input is the average gain, which will depend on the input amplitude. Plot this relation and show that the average gain falls off as the input grows.

2. Find the describing function for deadband (Figure 4-5) by formally determining the Fourier coefficients of its first harmonic. Determine also the Fourier coefficients of the third harmonic and show that they attenuate as indicated in the text.
3. Find the describing function for the following nonlinearities:
   (a) $y - x^{(\mathrm{odd})}$
   (b) $y = |x| x^{(\mathrm{odd})}$, i.e., an ''even''-powered odd function
   (c) $y = \sin x$

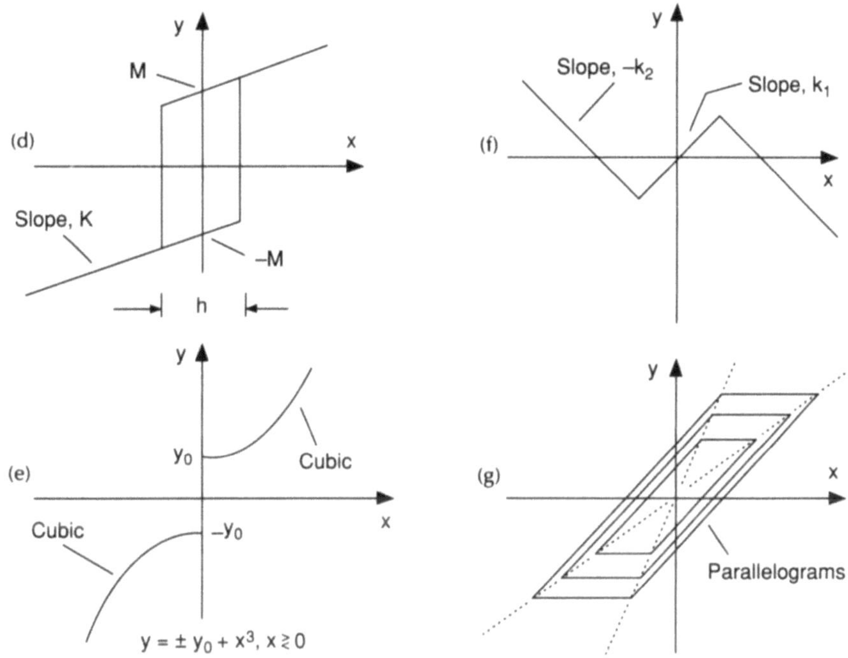

Figure 4-31. Nonlinearities for which describing functions are sought.

4. Find the Fourier series expansions for the two functions shown in Figure 4-32 and determine by the rate of attenuation in their successive terms which is smoother.

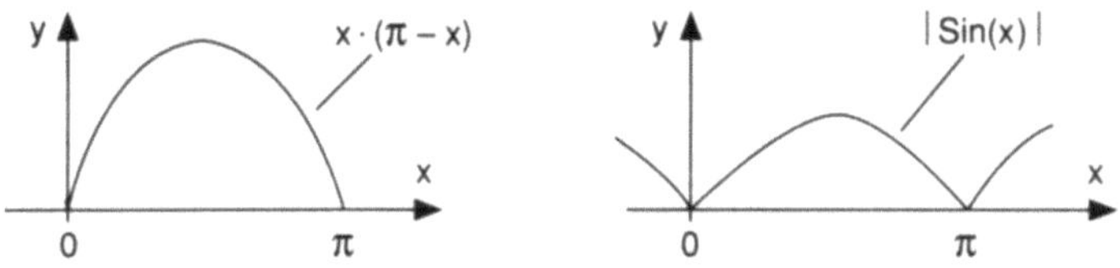

Figure 4-32. Wave shapes to be expanded in Fourier series.

5. In the system of Figure 4-33, an ideal relay drives a double integration. Using the describing function for the relay, (Figure 4-3), determine the amplitude and frequency of the limit cycle. This problem can be solved as a piecewise-continuous linear system consisting of steps applied to the integrators. Using this procedure, compute the exact solution from an initial condition, determine the exact frequency and amplitude of its limit cycle, and compare the result with that of the describing function. Would you expect the results to compare favorably?

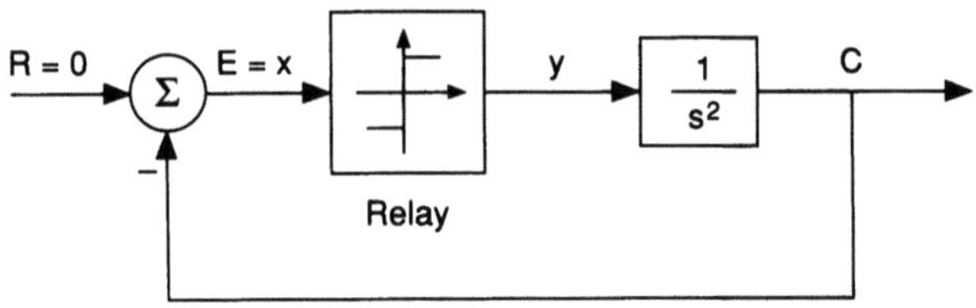

Figure 4-33. Relay with double integration.

6. Simulate a relay driving three poles using one ZOH and two samplers, Figure 4-34 so that the entire linear dynamics is simulated by just one recursion formula. Compare the results for several values of sample time, T, to determine an acceptably small value. Compare the results with those in the text to give some evidence of fidelity to digital simulation (at least in this particular example).

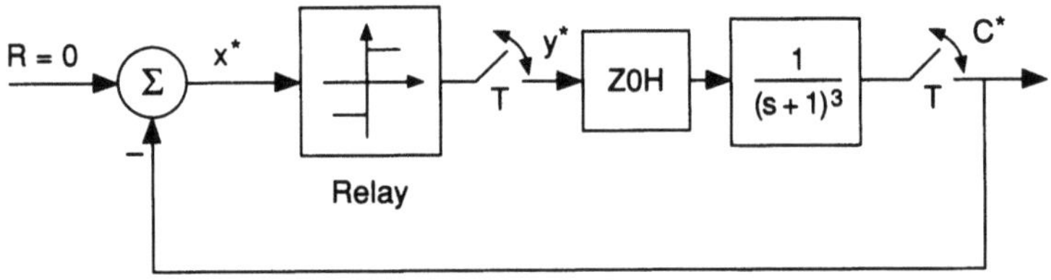

Figure 4-34. Relay simulation with 3 pole plant.

7. You can verify the idea that the describing function, $N_1(x_m)$, is an "equivalent gain" in the following system, which also lends credence to the first harmonic approximation.

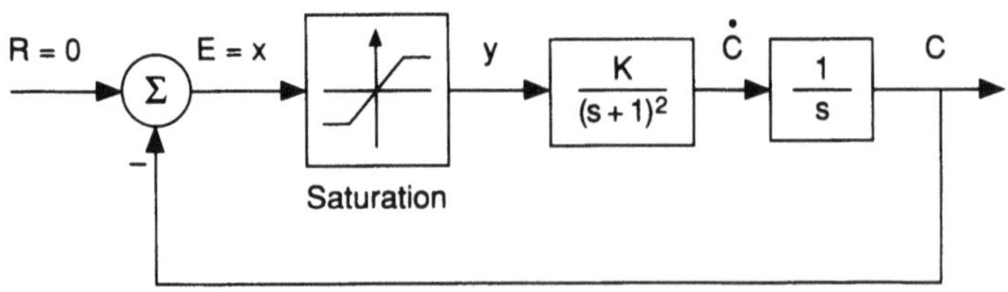

Figure 4-35. Saturating controller.

For $x_{sat} = \infty$ (a linear gain unity), the Routh test shows insipient instability will occur at $K = 2$.

Now suppose $x_{sat} = 1$. Find the value of N_1 and correlate the product KN_1 against the value of 2.

8. Verify the limit cycles predicted in Problem 7 with a digital simulation using a spreadsheet.
9. Using any available simulation package (i.e., AXSL, CSMP, CONTROL C MATRIX X, or whatever), compare its results with your spreadsheet results of Problem 8.
10. Why is the limit cycle predicted in the system of Figure 4-27 stable?
11. In Figure 4-36, an asymmetric relay drives a *non*integrating plant.

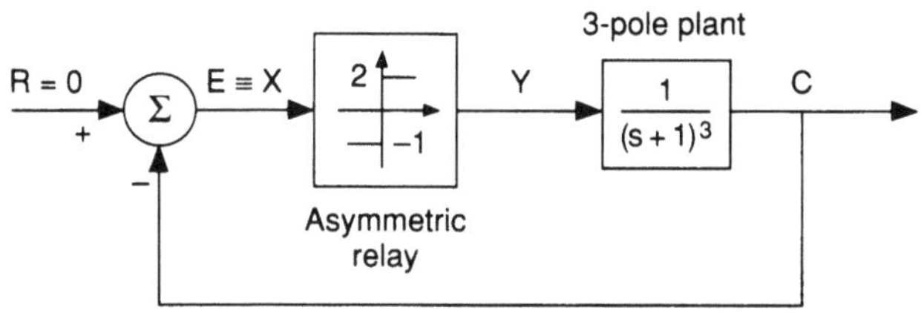

Figure 4-36. Asymmetric relay driving 3-pole plant.

Show that the three basic relations which must hold are:

(1) $L(j0) = -1/N_{dc}$ yielding x_{dc}
$= |x_{ac}| \sin \{-[\pi/(M_1 + M_2)]$
$[x_{dc}/L(j0) + (M_1 - M_2)/2]\}$

(2) $|L(j\omega)| = |-1/N_{ac}|$yielding $|x_{ac}| = 2(M_1 + M_2) \cos \beta/\pi(1 + \omega^2)^{3/2}$

(3) $\underline{/L(j\omega)} = \pi - \underline{/N_{ac}}$ yielding $\tan^{-1} [-\omega(3 - \omega^2)] = \pi$, since $\underline{/N_{ac}} = 0$.

What is the predicted amplitude and frequency of the limit cycle?

12. Develop a digital simulation using the method of Chapter 8, and verify the existence, frequency, and amplitude of the limit cycle.
13. With reference to the section where the stability of a limit cycle was developed, two relations between the three variables, dx, $d\omega$, and $d\sigma$ were found. It was decided to eliminate $d\omega$, thus leaving one relation between $d\sigma$ and dx. Why? Suppose you were to repeat the procedure, but instead of eliminating $d\omega$, you were to eliminate dx (or $d\sigma$). What would you be able to conclude about the limit cycle stability in these cases?
14. Another way to simulate the asymmetric relay problem discussed earlier is to lump the quadratic term (Figure 4-37), for which a single second-order recursion equation results. Program the simulation and compare the limit cycle to those of Table 4-1.

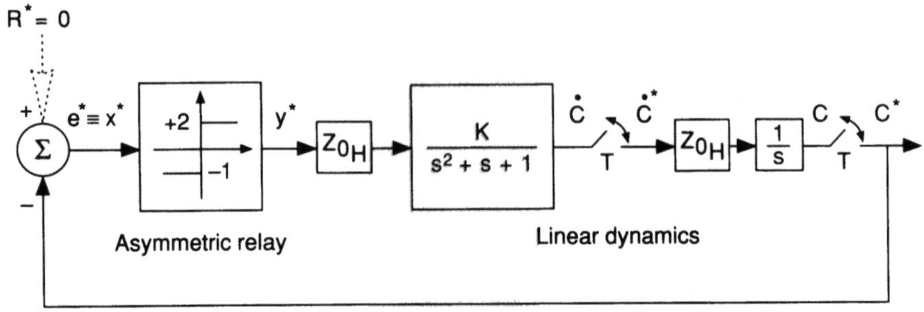

Figure 4-37. Simulation with asymmetric relay.

4.6 REFERENCES

1. Kochenburger, Ralph J., A frequency response method of analyzing and synthesizing contactor servomechanisms, *Transactions of the AIEE*, **69**, 270–284 (1950).
2. Members of E. E. Dept. of MIT, *Applied Electronics*, John Wiley & Sons, New York (1943).
3. Goldfarb, L. C., Concerning some nonlinear phenomena in control systems, *Automatica and Telemechanica* **8**, 344–383 (1947).
4. Tustin, A., The effects of backlash and of speed dependent friction on the stability of closed cycle control systems, *Journal of the Institute of Electrical Engineering (British)* **94**, pp. 143–151 (May 1974).
5. Oppelt, W., Locus curve method for regulators with friction, *Journal of the Institute of Electrical Engineers* (London), **94**, Part IIA, 1 and 2 (May 1947).
6. Aleksandrov, A. D., et al., *Mathematics: Its Content, Methods, and Meaning*, MIT Press, Cambridge, Massachusetts (1965), pp. 300–301.
7. Wylie, C. Ray, *Advanced Engineering Mathematics*, Fourth edition, McGraw-Hill, New York (1961).
8. Truxal, John G., *Automatic Feedback Control System Synthesis*, McGraw-Hill, New York (1955), pp. 601–611.
9. Stern, Thomas E., *Theory of Nonlinear Networks and Systems*, Addison-Wesley, Reading, Massachusetts (1965).
10. Atherton, Derek P., *Stability of Nonlinear Systems*, Wiley, New York (1981).
11. Gille, J. C., Pelegrin, M. J., and Decaulne, P., *Feedback Control Systems*, McGraw-Hill, New York (1959), pp. 419–420.
12. Lerman, Robert A., and Rosen, Fred K., *The Describing Function and Asymmetrical Nonlinearities*, Hamilton Standard Report Publication (May 19, 1966).
13. Anon., *1130 Continuous System Modelling Program*, IBM Applications Program, H20-0282-1 (circa 1964).

Chapter 5
Some Properties of Nonlinear Systems

5.0 INTRODUCTION

Nonlinear systems are modeled by use of nonlinear differential equations. More often than not, these equations do not admit to closed-form analytic solutions. Furthermore, many that do are not in terms of the elementary functions.

This leads to significant problems in discerning the behavior of the system. Consequently, a body of interpretative approaches has developed. It is the purpose of this chapter to provide some insights into periodic behavior, singular points, and the behavior of the system without actually solving the differential equations.

5.1 LINEAR SYSTEM CHARACTERISTICS

It is assumed that the reader is familiar with this subject; thus, we will only sketch the concepts here.

N^{th}-order linear differential equations with constant coefficients can be transformed into a polynomial equation by means of Laplace transforms or operators. This equation, termed the characteristic equation, can be factored (theoretically) to yield n roots, termed the characteristic roots or eigenvalues. Investigating the nature of the roots leads to an understanding of the characteristics of the system. One such characterization is the Routh–Hurwitz criterion.

5.1.1 Routh–Hurwitz

A linear differential equation with constant coefficients

$$b_0 + b_1x + b_2dx/dt + \cdots + b_ndx^n/dt^n = 0$$

is associated with its characteristic equation

$$0 = b_0\lambda^0 + b_1\lambda^1 + \cdots + b_n\lambda^n$$

and by examining the nature of the roots of the characteristic equation, we can discern certain properties about the behavior of the solution of the differential equation.

For example, if

(a) the characteristic equation contains terms with all powers of λ up to (but not including) n;
(b) the coefficients are all of the same sign;
(c) all successive determinants are positive, $D_0, \cdots, D_n > 0$, where

$$D_0 = b_1, \qquad D_1 = \begin{vmatrix} b_1 & b_0 \\ b_3 & b_2 \end{vmatrix}, \qquad D_2 = \begin{vmatrix} b_1 & b_0 & 0 \\ b_3 & b_2 & b_1 \\ b_5 & b_4 & b_3 \end{vmatrix}, \text{ etc.}$$

then the solution to the differential equation is stable. In vector-matrix form this can be expressed as

$$dx_i/dt = \dot{x}_i = f_i(x_1, \cdots, x_n), \qquad i = 1, \cdots, n$$

$$d\mathbf{x}/dt = \dot{\mathbf{x}} = F(\mathbf{x}) = B\mathbf{x}$$

Where the solution to the vector equation components is

$$x_i = \sum_{j=1}^{n} c_{ij} e^{\lambda_j t}$$

and the eigenvalues, λ_j, are defined by the equation

$$|B - \lambda I| = b_0\lambda_0 + b_1\lambda_1 + \cdots + b_n = 0, \qquad I = \text{unit matrix}$$

Form the matrix B

$$B = \begin{bmatrix} b_1 & b_0 & 0 & 0 & 0 & \cdot & \cdot & \cdot & 0 & 0 \\ b_3 & b_2 & b_1 & b_0 & 0 & \cdot & \cdot & \cdot & 0 & 0 \\ b_5 & b_4 & b_3 & b_2 & b_1 & b_0 & \cdot & \cdot & 0 & 0 \\ & & & & & & & & & \\ 0 & 0 & \cdot & 0 & 0 & \cdot & \cdot & \cdot & b_{n-1} & b_{n-2} \\ 0 & \cdot & \cdot & \cdot & \cdot & \cdot & \cdot & \cdot & 0 & b_n \end{bmatrix}$$

which yields the same characteristics as in the nonmatrix format.

$$b_1 > 0, \quad \begin{vmatrix} b_1 & b_0 \\ b_3 & b_2 \end{vmatrix} > 0, \quad \begin{vmatrix} b_1 & b_0 & 0 \\ b_3 & b_2 & b_1 \\ b_5 & b_4 & b_3 \end{vmatrix} > 0, \text{ etc.}$$

Note that the Routh–Hurwitz criterion is a technique for evaluating the characteristic equation without factoring it. Later in this chapter, we shall assume we know the nature of the roots of the characteristic equation and thereby establish conditions under which we can estimate further solution properties.

5.1.1.1 Periodic Properties. Some solutions to linear equations are periodic. An elementary example is the simple oscillator $x'' + \psi^2 x = 0$, which has the solution $x = C_1 \sin(\psi t + \phi)$.

Of interest is the prediction of repetitive events, such as the swing of a frictionless pendulum or the orbit of the moon about the Earth. Mathematically, we wish to examine the motions of the pendulum or of the moon so that we obtain solutions of the form

$$y(x) = y(x + \omega)$$

and, therefore, we can restrict (i.e., narrow) our investigation to the characteristics of the system's equation during one cycle without losing any information.

By definition and convention, such functions are termed periodic. The circular functions, sine and cosine, are the most familiar. Fourier series, being composed of such terms, are often periodic, but need not be. To be periodic, the series must *converge* to the function it represents on the interval. That is, the trigonometric series of $f(x)$ on the interval $(-L, L)$

$$f(x) = \sum_{n=0}^{\infty} a_n \cos nx + b_n \sin nx \tag{1}$$

where

$$a_n = 1/L \int_{-L}^{L} f(x) \cos nx \, dx$$

$$b_n = 1/L \int_{-L}^{L} f(x) \sin nx \, dx$$

is the Fourier series of $f(x)$.

Sufficiency conditions for convergence of a Fourier series include such conditions as follows. If

(a) $f(x)$ satisfies $f(x + L) = f(x)$, and
(b) $f(x)$ is piecewise continuous in the interval $(-L, L)$,
(c) then the Fourier series converges to $\frac{1}{2}[f(x + 0) + f(x - 0)]$ at every point where $f(x)$ has a left- and right-hand derivative.

The concept of periodic solutions leads to very interesting results, and we turn to some of the issues developed by Floquet. We recall that a linear second-order differential equation with constant coefficients

$$ay'' + by' + cy = 0, \qquad a, b, c = \text{constant}$$

has a general solution in terms of exponential functions of the form

$$y = c_1 e^{Ax} + c_2 e^{Bx}$$

where the values of $r = A, B$ are the system's eigenvalues (real or complex) and are solutions of the characteristic equation $ar^2 + br + c = 0$.

Now if the system has periodic solutions, then, by the definition of periodicity, the solutions must satisfy $y(x) \equiv y(x + \omega)$, where ω is a constant. We note that if $y(x)$ is periodic, and if $y(x)$ is also analytic, all of its derivatives exist and they are periodic. Thus, we can write

$$\sum_{n=0}^{\infty} \omega^n y^{(n)}/n! = 0$$

where $y^{(n)}$ is the nth order derivative of $y(x)$.

Therefore, every periodic analytic function $y(x)$ is a solution of this linear differential equation. Furthermore, the original differential equation, which we are studying and which has $y(x)$ as a solution, need not be linear. Hence, we observe that if a nonlinear differential equation has a periodic, analytic solution, $y(x)$, then that solution is also a solution of this infinite linear differential equation.

Since the solution of the linear equation is $y = e^{aix}$, where $a = 2\pi n/\omega$, $n = 0, \pm 1, \pm 2, \cdots$, then any sum of such terms is also a solution. For example,

$$y(x) = A_0 + \sum_{m=1}^{n} A_m \cos m\pi x/\omega + B_m \sin m\pi x/\omega,$$

$$A_m, B_m = \text{arbitrary constants}$$

5.1.2 Floquet Theory

These considerations lead us to the work of Floquet. Floquet was concerned with the existence of a periodic solution of a *linear* differential equation, where the coefficients of the equation's terms include periodic functions, all with the same period.

$$a(x)y'' + b(x)y' + c(x)y = 0, \qquad a(x), b(x), c(x) = \text{periodic}$$

Hill's equation

$$u'' + p(x)u' + q(x)u = 0$$

is such an example, which upon substituting $y = \exp[A(x)/2]u$ and $p = dA/dx$ becomes

$$y'' + Q(x)y = 0$$

where $Q = -0.5dp/dx - p^2/4 + q$.

It is therefore reasonable to continue our discussions with respect only to Hill's equation

$$y'' + Q(x)y = 0, \qquad (\ )' = d(\ )/dx$$

where we impose the conditions that x is defined for all values of x, that $Q(x)$ is piecewise continuous in every finite interval, and that $Q(x)$ is periodic with a minimum period π.

Using these characteristics, Floquet was able to show that if the differential equation has two continuously differentiable solutions $y_1(x)$ and $y_2(x)$, which are uniquely determined by the *normalized* conditions

$$y_1(0) = 1, \qquad y_1'(0) = 0, \qquad y_2(0) = 0, \qquad y_2'(0) = 1$$

and in this set of circumstances the eigenvalues are determined from the *characteristic* equation

$$r^2 - [y_1(\pi) + y_2'(\pi)]r + 1 = 0$$

having roots r_1, r_2

$$r_1 = e^{i\alpha x}, \qquad r_2 = e^{-i\alpha x}$$

Note that $r_1 r_2 = 1$ and $2\cos\alpha\pi = y_1(\pi) + y_2'(\pi)$.
This leads to the following.

5.1.2.1 Floquet's Theorem. *If the roots $r_l \neq r_2$ of the characteristic equation, then Hill's equation*

$$y'' + Q(x)\, y = 0$$

has two linearly independent solutions

$$f_l(x) = e^{i\alpha x}p_l(x), \qquad f_2(x) = e^{-i\alpha x}p_2(x)$$

where $p_l(x)$ and $p_2(x)$ are periodic with period π.

If $r_l = r_2$, Hill's equation has a nontrivial solution with period π or 2π. Let $p(x)$ be such a periodic solution and $y(x)$ another solution linearly independent of $p(x)$. Then

$$y(x + \pi) = r_l y(x) + kp(x), \qquad k = \text{constant}$$

and k = 0 is equivalent to

$$y_1(\pi) + y_2'(\pi) = \pm 2, \qquad y_2(\pi) = 0, \qquad y_1'(\pi) = 0$$

The significance of Floquet's theorem lies in the notion that for every solution $y(x)$ of Hill's equation (and, therefore, for virtually all second-order linear differential equations), there is an upper bound M for the absolute value of $|y(x)|$, provided α is real and r_1 and r_2 are unequal; furthermore, the solutions depend only on the initial conditions; and again, the solution is periodic. In the case where α is not real, then there exists a nontrivial unbounded solution of the equation. If the roots are equal, $r_1 = r_2$, then a necessary and sufficient condition that the solution be bounded is

$$y_1(\pi) + y_2'(\pi) = \pm 2, \qquad y_2(\pi) = 0, \qquad y_1'(\pi) = 0$$

These concepts naturally lead to the concept of stability in terms of bounded solutions. Because every solution of Hill's equation $y'' + Q(x)y = 0$ is bounded under these conditions, it is reasonable to assert the condition of boundedness to imply stability.

In sum, and for our purposes, however, we are content to know that for linear equations, conditions for periodicity and boundedness exist, that a great body of literature exists with respect to equations with either constant or periodic coefficients, and that solutions are possible with techniques of analysis well-defined. We also know that if a function can be expanded in a Fourier series, an almost periodic function (see Section 5.4.4) also can be expressed as a Fourier series to within an arbitrarily small constant.

A key issue here is that Floquet theory applies to *linear* differential equations and thus once a nonlinear system has been linearized, the theory can be applied; but if we are working in the nonlinear realm, we must resort to other means.

5.1.3 Solution Behavior Characteristics

It is convenient to consider a linear second-order system, investigate the system behavior by means of interpreting the equation on the y–y' plane, then extending these characterizations to higher-order systems and then to nonlinear systems.

To illustrate, let us consider the linear second-order system with constant coefficients. The characteristic equation has eigenvalues λ_1, λ_2 which are either:

(a) both real
(b) equal
(c) both complex
(d) purely imaginary

The solution to the linear differential equation has the solution for cases (a), (c), and (d)

$$y = c_1 e^{\lambda_1 x}, \qquad y' = c_2 e^{\lambda_2 x}$$

and for case (b)

$$y = c_1 e^{\lambda_1 x} + c_2 x e^{\lambda_1 x}, \qquad y' = [c_2 + \lambda_1(c_1 + c_2)]e^{\lambda_1 x}$$

These various cases lead to the tabulated results, which we will prove.

Eigenvalues	Solution characterizations
λ_1, λ_2 Real	Node
Both positive	Unstable node
Both negative	Stable node
Opposite signs	Saddle point
λ_1, λ_2 Complex conjugates	Focus
Real part negative	Stable focus
Real part positive	Unstable focus
λ_1, λ_2 Imaginary	Center

To derive these conditions, consider the system

$$x_1' = \lambda_1 x_1$$

$$x_2' = \lambda_2 x_2$$

Then, forming

$$dx_2/dx_1 = x_2'/x_1' = (\lambda_2/\lambda_1)(x_2/x_1) = kx_2/x_1$$

We can examine the nature of the resultant solution trajectories on the x_1-x_2 plane. See Problem 5 for confirmation that this system represents a general linearized second-order system with constant coefficients.

Let λ_1, λ_2 be real, unequal, and have the same sign, and consider three cases: $0 < k < 1$, $k = 1$, $k > 1$. In each case, integration of the equation $dx_2/dx_1 = kx_2/x_1$ yields $x_2 = x_1^k$.

Thus, if $k = 1$, a straight line results passing through the origin and if $k > 1$, a series of curves tangent to the x_1 axis at the origin. If $0 < k < 1$, curves are tangent to x_2 axis (Figure 5-1).

Note that if $\lambda_1, \lambda_2 > 0$, unequal and real, $e^{\lambda_1 x}$, $e^{\lambda_2 x}$ increase without bound and the system exhibits unstable solutions. Thus, in each of these cases, the trajectories move away from the origin and eventually exceed any bound that may be imposed.

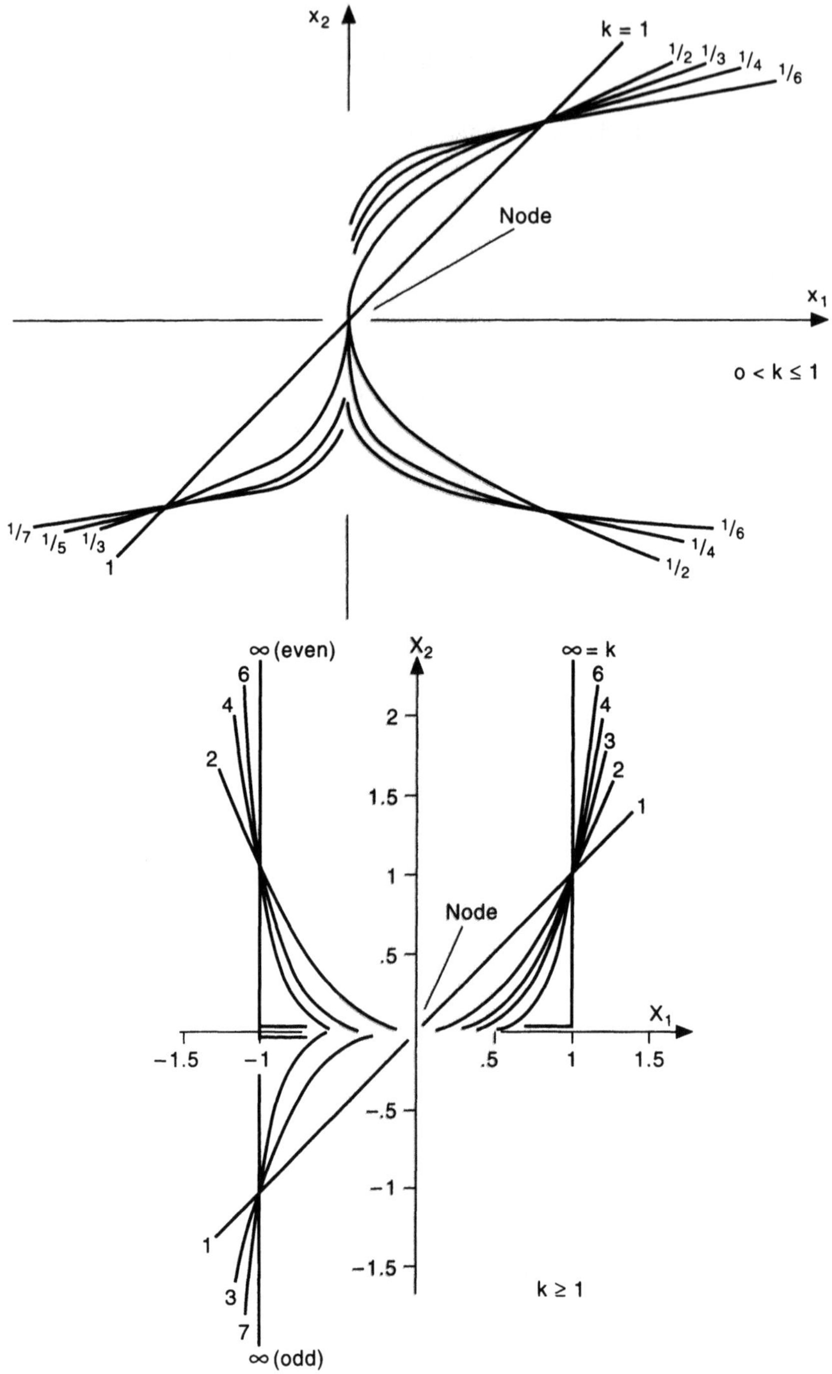

Figure 5-1. Node: $\lambda_1 \neq \lambda_2$, real.

If $\lambda_1, \lambda_2 < 0$, unequal and real, the trajectories move toward the origin and are stable.

Now suppose, λ_1, λ_2 are real, and of opposite sign. Therefore,

$$dx_2/dx_1 = (\lambda_1/\lambda_2)(x_2/x_1) = -kx_2/x_1$$

and upon integration

$$x_2 x_1^k = C$$

which represents a series of hyperbolas asymptotic to the axes, except the lines $x_1 = 0$ and $x_2 = 0$ pass through the origin and are unstable. This configuration is termed a saddle point (Figure 5-2).

In the case when the roots are complex conjugates, $\lambda_1 = m + jn$, $\lambda_2 = m - jn$, we can show that the system $x_1' = \lambda_1 x_1$, $x_2' = \lambda_2 x_2$ can be transformed using

$$x_1 = u + jv$$

$$x_2 = u - jv$$

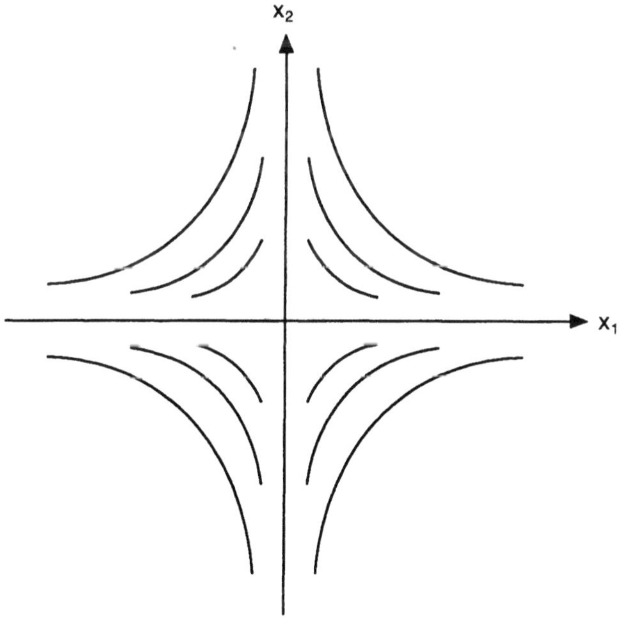

Figure 5-2. Saddle point.

and following a similar analysis results in

$$dv/du = \frac{mv + nu}{mu - nv} = \frac{u + rv}{ru - v}, \qquad r = m/n$$

It is convenient to introduce polar coordinates with $u = \rho \cos \theta$, $v = \rho \sin \theta$. Then,

$$du = -\rho \sin \theta \, d\theta + \cos \theta \, d\rho$$

$$dv = \rho \cos \theta \, d\theta + \sin \theta \, d\rho$$

$$\frac{dv}{du} = \frac{\rho \cos \theta \, d\theta + \sin \theta \, d\rho}{-\rho \sin \theta \, d\theta + \cos \theta \, d\rho} = \frac{\rho \cos \theta + r\rho \sin \theta}{r\rho \cos \theta - \rho \sin \theta}$$

Cross-multiplying and simplifying yields

$$r\rho d\theta = d\rho$$

and upon integration

$$\rho = Ce^{r\theta}$$

which is the equation of a logarithmic spiral ($r = m/n$, $u = \rho \cos \theta$, $v = \rho \sin \theta$), see Figure 5-3.

Note that if $r = 0$, then $dv/du = -u/v$ and if a proportionality constant is introduced, so that

$$dv/du = k^2 u/v$$

an ellipse results, and if $k = 1$, a circle is the solution (Figure 5-4).

It is important to recall that we have been working with linear systems.

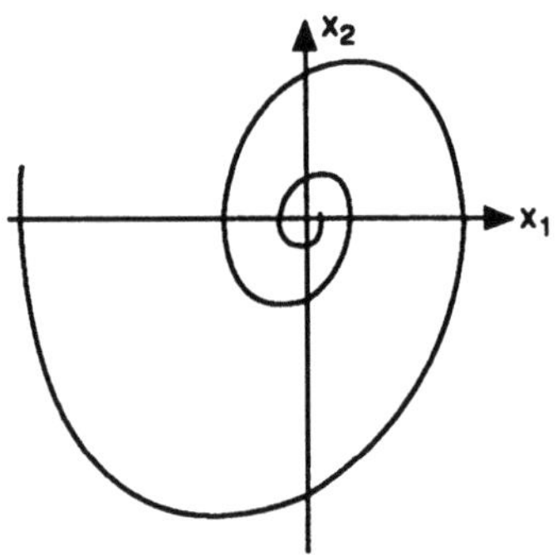

Figure 5-3. Spiral (focus).

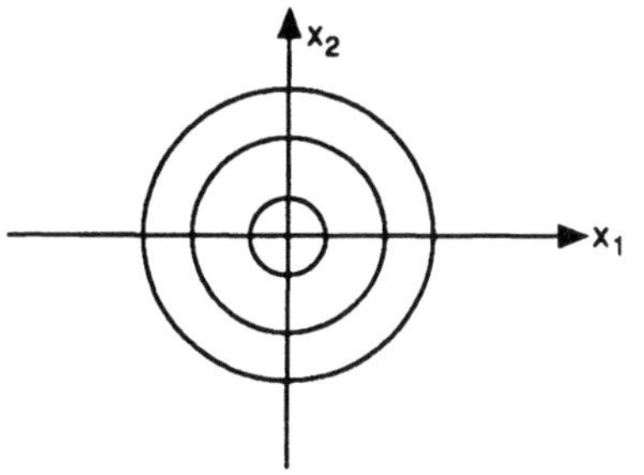

Figure 5-4. Center.

And the direction of the resultant trajectories, whether they are bounded or not, allows us to consider concepts of stability of the system. The Liapunov Stability Theorem (to be discussed in Chapter 6) allows us to examine the notion that the same characteristics of linear systems are valid in a neighborhood of the singularity of a nonlinear equation.

If one generalizes these concepts, it is easy to see that the x_1–x_2 plane can be extended into n-dimensional space, and the use of vectors creates the analogies we need. Thus, nodes, centers, foci, or saddle points all are useful concepts in higher-order systems, and we will rely on these thoughts as we consider new properties.

Let us now extend these concepts to nonlinear equations. Consider the general second-order system

$$dy/dt = F(x, y)$$

$$dx/dt = G(x, y)$$

Let us translate the origin of this system by means of the substitutions $x = w + p$ and $y = u + q$, where p, q are constants, and let F, G be analytic (i.e., continuous and differentiable). Thus, with the analyticity property, we can expand our system in a Taylor series to obtain

$$F = F(p, q) + wF_x(p, q) + uF_y(p, q) + \left(\frac{1}{2!}\right)[p^2F_{xx}(p, q) + 2pqF_{xy}(p, q) + q^2F_{yy}(p, q)] + \cdots$$

$$G = G(p, q) + wG_x(p, q) + uG_y(p, q) + \left(\frac{1}{2!}\right)[p^2G_{xx}(p, q) + 2pqG_{xy}(p, q) + q^2G_{yy}(p, q)] + \cdots$$

where the subscripts x, y refer to partial derivatives of F or G.

However, singular points of the system exist where F and G intersect at $F(p, q) = 0$, $G(p, q) = 0$, and the singular point (p_0, q_0) exists at

$$wF_x(p_0, q_0) + uF_y(p_0, q_0) + \text{higher-order terms} = 0$$

$$wG_x(p_0, q_0) + uG_y(p_0, q_0) + \text{higher-order terms} = 0$$

By linearizing this expansion, it is possible to solve for u and w. This linear solution will yield eigenvalues, and their interpretation will follow the same lines of reasoning as before.

Consequently, we conclude that within a properly defined circle of convergence (as defined by the Taylor series), nonlinear system behavior around a singularity can be interpreted by means of centers, foci, spirals, nodes, and saddlepoints.

5.2 NONLINEAR EQUATIONS WITH PERIODIC SOLUTIONS

We now turn to the concept of periodic solutions of nonlinear equations. By observation, we know that the simple frictionless pendulum will exhibit periodic motion. Thus, it will satisfy the relationship $y(x) = y(x + \omega)$. What about others? The following theorems establish conditions under which the equations

$$y'' + f(y)y' + g(y) = 0$$

$$y'' + f(y, y')y' + g(y) = 0$$

admit periodic solutions.

5.2.1 Lienard's Theorem

If $f(y)$ and $g(y)$ are continuous, integrable, and satisfy

(a) *$f(y)$ is an even function, and $g(y)$ is an odd function, where $yg(y) > 0$*

(b) *$F(y) = \int_0^y f(s)\, ds$, $G(y) = \int_0^y g(s)\, ds$ tend toward ∞ as y approaches ∞*

(c) *$F(y)$ has a single positive zero, $y = y_0$. In the interval $(0, y_0)$, $F(y) < 0$, but for $y > y_0$, $F(y) > 0$ and increases monotonically*

then

$$y'' + f(y)y' + g(y) = 0 \quad (\ )' = d(\ )/dx$$

has a periodic solution which is unique to within an arbitrary constant T of x.

Example

The examination of the limit-cycle behavior of Van der Pol's equation involves Lienard's theorem. Van der Pol's equation can be written as

$$y'' - a(1 - y^2)y' + by = 0, \qquad a, b > 0$$

We note that

$$f(y) = -a(1 - y^2)$$

$$g(y) = b$$

Clearly, condition (a) is satisfied since f and g are continuous, integrable, and f is even [i.e., $f(y) = f(-y)$], g is odd [i.e., $g(y) = -g(-y)$], and satisfies $yg(y) > 0$; condition (b) is satisfied since $F(y) = -a(y - y^3/3)$ and $G(y) = by$; and since $F(y)$ has only one positive root, $y = 3^{1/2}/3$, condition (c) is satisfied.

Thus, by application of Lienard's theorem, Van der Pol's equation has a unique periodic solution.

5.2.2 McHarg's Theorem

If $f(y)$ and $g(y)$ are odd functions [$f(y) = -f(-y)$, $g(y) = -g(-y)$], integrable, continuous, and positive for all values of y positive, [$f(y) > 0$, $g(y) > 0$, $y > 0$], and if k and y_l are positive constants such that

$$f(y) < kg(y), \qquad 0 < y < y_l$$

then every solution of the differential equation

$$y'' + f(y)y' + g(y) = 0$$

satisfying the initial conditions, $y = 0$, $y' = v_0$, is periodic, where

$$v_0 < \min\{1/k, [2G(y_l)]^{1/2}\}$$

and $G(y) = \int_0^y g(s)\, ds.$

Example

The equation

$$y'' + ayy' + by = 0, \qquad a, b > 0$$

satisfies the conditions of this theorem for $f(y) = ay$, $g(y) = by$.

5.2.3 Levinson–Smith's Theorem

Let $yg(y) > 0$ for $|y| > 0$ and $f(y, y')$, $g(y)$ be continuous and integrable, and let

$$G(y) = \int_0^{\pm\infty} g(s)\, ds = \infty$$

If $f(0, 0) < 0$, but for some $y_0 > 0$ $f(y, y') \geq 0$ wherever $|y| \geq y_0$. Further, for $y_1 > y_0$, let

$$\int_{y_0}^{y_1} f(s, y')ds \geq 10My_0$$

where for $|y| \leq y_0$, $f(y, y') \geq -M$, and $y' < 0$ is an arbitrary decreasing positive function of y. Then

$$y'' + f(y, y')y' + g(y) = 0$$

has at least one periodic solution.

5.2.4 Dragilev–Ivanov's Theorem

Consider the equation

$$y'' + f(y)y' + g(y) = 0$$

and let $y' = v$, then the equation is transformed into the system

$$y' = v$$

$$v' = -f(y)v - g(y)$$

Now let $w = v + F(y)$, where

$$F(y) = \int_0^y f(s)\, ds$$

and we obtain

$$y' = w - F(y)$$

$$w' = -g(y)$$

In such a system we can state the theorem:

Let

(1) $g(y)$ satisfy $yg(y) > 0$ for $y \neq 0$

$$\int_0^\infty g(y)dy = \infty$$

(2) *F(y) be single-valued for all y* $(-\infty < y < \infty)$, *satisfying a Lipschitz condition in every finite interval*
(3) $yF(y) < 0$, $y \neq 0$ *and* $|y|$ *sufficiently small*
(4) *Constants N, k,* k_l *exist such that*

$$F(y) \geq k, \qquad y > N$$
$$F(y) \leq k_l, \qquad y < -N$$
$$k_l < k$$

then

$$y' = w - F(y)$$
$$w' = -g(y)$$

has at least one limit cycle.

5.3 LIMIT CYCLES

By defining the characteristics of linear systems, and by extending these concepts to nonlinear systems, we were able to examine the nature of the periodic solution of Van der Pol's equation. By applying Lienard's theorem, we learned that there is a unique periodic solution to the equation.

The definition of a periodic solution,

$$y(x) = y(x + \omega)$$

implies a closed path, as the endpoint at $x + \omega$ has the same value as the beginning point x, and this is true for all x.

Therefore, it is reasonable to define a limit cycle as a periodic solution of the differential equation, which is depicted as a closed trajectory on the y–v plane ($v = dy/dx$). Thus, the linear oscillator and the elliptical trajectories for the simple pendulum of Section 5.4.2 are limit cycles. Van der Pol's equation is another example of limit-cycle behavior.

Definition

A closed trajectory $y(t)$ is a limit cycle if there exists an open trajectory $y_1(t)$ which approaches $y(t)$ as $t \to \pm\infty$. If $p(t)$ is a point on $y(t)$

$$|y_1(t) - p(t)| \to 0 \qquad \text{as } t \to \pm\infty$$

Now, however, it is convenient to digress in order to introduce the nature of the solution paths and thereby to allow further insights.

In the search for characteristics of the equations defining our system, it is useful to know whether or not a closed trajectory exists. So far, the theorems we've included all deal with special cases when solutions are periodic or exhibit limit-cycle behavior. A theorem developed by Bendixson provides a means of determining when a closed trajectory does *not* exist. This theorem is stated as follows.

5.3.1 Bendixson's Theorem

In the simply connected domain in the real plane x–y the system equation

$$dy(x)/dx = f[x, y(x)]$$

admits to no closed trajectories if

$$\nabla f(y) = \partial f/\partial x + \partial f/\partial y$$

does not change sign nor is it identically zero in any portion of the domain.

Proof

The proof of this theorem follows from application of Green's theorem:

Let R be a closed bounded region in the x–y plane whose boundary C consists of a finite number of smooth curves. Let $P(x, y)$ and $Q(x, y)$ be continuous functions with continuous partial derivatives $\partial P/\partial x$ and $\partial Q/\partial y$ everywhere in some domain containing R. Then

$$\iint_R (\partial P/\partial x + \partial Q/\partial y)\, dxdy = \int_c (Pdx - Qdy)$$

where the integration continues along the entire boundary of C of R, such that R is on the left.

Using this result, we assume that C is a closed curve representing the solution trajectory of

$$dx/dt = P(x, y)$$

$$dy/dt = Q(x, y)$$

Since C is a closed curve, $\int_c = 0$, and thus $\iint = 0$. But this is only true if the integrand is zero or changes sign. As this is not admissible by the statement of the theorem, C cannot be a closed curve, thereby proving Bendixson's theorem.

Example

Van der Pol's equation is known to have a limit cycle; thus, we would expect that at least one of the two conditions imposed by Bendixson's theorem to not hold. Forming

$$dx_1/dt = f_1 = x_2$$

$$dx_2/dt = f_2 = (1 - x_1^2)x_2 - x_1$$

and computing ∇f or

$$\partial f_1/\partial x_1 + \partial f_1/\partial x_2 = 1$$

$$\partial f_2/\partial x_1 + \partial f_2/\partial x_2 = -x_1(x_1 + 2x_2)$$

we see that ∇f_1 never changes sign and never is zero, while ∇f_2 can be zero at $x_1 = x_2 = 0$ or when $x_1 = -2x_2$. Thus, the conditions are not satisfied and a closed trajectory *does* exist—as we knew beforehand.

5.3.2 Almost Periodic Functions

However, what do we do with a function that is not periodic—that is, where the definition $f(t + T) \equiv f(t)$ is not satisfied? As an example,

$$f(t) = \sin 2\pi t + \sin 2\sqrt{2}\pi t$$

is not periodic as no T can be found to satisfy the definition.

Numerical analysis and/or computerized solutions have a way to respond to this issue by saying, for some degree of accuracy, we get close enough. Thus, we can say that π is 22/7, 3.14, 3.141592, or carry the concept of accurately depicting π to as many decimal places as we wish to suit our needs. That is, by the time we reach the fifth or hundredth or . . . decimal place, the computational technique cannot distinguish between the values of two functions and the functions are considered to be the same.

To allow us to mathematically express this concept, we introduce *almost periodic functions*, where the degree of accuracy is arbitrarily small. We state that for any $\epsilon > 0$, where ϵ is as small as we wish, we can find another number T, such that $(2)^{1/2}T$ differs from another integer by less than $\epsilon/2\pi T$.

$$\begin{aligned} f(t + T) &= \sin 2\pi(t + T) + \sin 2\sqrt{2}\pi(t + T) \\ &= \sin 2\pi t + \sin (2\sqrt{2}\pi t + \theta e), \qquad |\theta| \leq 1 \\ &= f(t) + \theta_0 e, \qquad |\theta_0| \leq 1 \end{aligned}$$

We will therefore adopt the notation that if

$$f(x + T) = f(x) + \theta_0 e$$

then $f(x)$ is an *almost periodic function* with infinitely many values of T available to satisfy this relationship.

Clearly, this has great validity and usefulness in computational problems. A direct correlation to Fourier series exists and it is possible to show that if $\Sigma|A_n|^2$ converges, the almost periodic function $f(x)$ has a Fourier series given by the definitions in Section 5.1.1.

Furthermore, the ability to define solutions by means of trigonometric series, which are almost periodic functions, permits us to approximate solutions to within an arbitrarily small error. Therefore, the concept of developing solutions to differential equations, which are asymptotic to a function, creates a means of developing practical algorithms for computational purposes. (The reader is referred to Besicovitch, Ince, and Erdelyi for further thoughts on these topics.)

5.4 NONLINEAR SYSTEM BEHAVIOR

The behavior characteristics of a nonlinear system can be ascertained by considering the characteristics of the eigenvalues of the linearized system and how they relate to singular points or singularities (to be defined) of the original equation.

By isolating these singularities, we are able to sketch the resultant trajectories corresponding to the solutions without actually solving the equation. Therefore, we will be interested in developing an analogous set of relationships to those of linear systems for nonlinear equations.

5.4.1 Singular Points

We begin by examining the basic second-order equation

$$y'' + f(x, y)y' + g(x, y)y = 0$$

and setting $y' = v$, we obtain

$$dy/dv = P(y, v)/Q(y, v)$$

We seek solutions such that

$$P(y, v) = 0 \qquad \text{and } Q(y, v) = 0$$

to uncover points (y_e, v_e) in the plane corresponding to the system's equilibrium state(s). These solution points are the singularities we seek and are known as the poles and zeros, branch points, equilibrium points, critical points, singular points, attractors, etc., where such terms represent specific special cases, but are often interchangeable. The reader is referred to texts on complex variables (ref. 1) and differential equations (refs. 5, 8, 10, and 13), Chaos theory (ref.

4) and Control theory (refs. 5, 11, and 12) for further discussions on these topics.

As will be seen shortly, the Liapunov Stability Theorem establishes the nature of the trajectories in and around the singular point by examining each such point through the simultaneous solution of the equations

$$P(x, y) = 0$$

$$Q(x, y) = 0$$

Applying this condition to Van der Pol's equation, $P = v$, $Q = ax - by - axy^2$, results in one and only one singularity at $x = 0$, $y = 0$.

The behavior of the solutions to differential equations of second order can be displayed on the x–y plane, typically referred to as the phase plane when x–x' are plotted. These solution curves are termed trajectories, and four types of trajectories associated with singular points have emerged in the classical theory: spirals (foci), saddlepoints, centers, and nodes. In the study of Chaos theory that, in essence, deals with the issue of bounded solutions (therefore, stable solutions), which inexplicably become either unbounded or result in a jump (see Chapters 7 and 9), bifurcations arise. These are shown pictorially in Figure 5-5 in the x–v plane, where we have arbitrarily assumed the origin to be the singular point.

Moon's 1987 text, an excellent annotated bibliography of Chaos theory, shows that Duffing's equation, Lorenz's equations (which result from the Navier–Stokes partial differential equations for fluid flow), Volterra's population

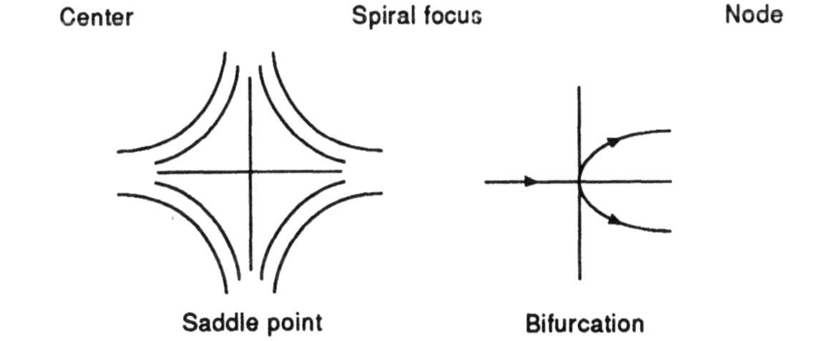

Figure 5-5. Types of singular points.

growth equations, and the use of the phase plane and difference equations all lead to the concepts of bifurcations. We briefly discuss these concepts in Chapters 7 and 9.

By examining the characteristic equation of the differential equation, we can establish conditions under which the singular points can be described. In general, for equations with constant coefficients, the use of Laplace transforms allows the conversion of the linear n^{th}-order differential equation into an algebraic equation which has n roots $s_1, s_2, \cdots, s_n$. A similar approach is to use matrix algebra, where we diagonalize the matrix to establish the n eigenvalues $\lambda_1, \lambda_2, \cdots, \lambda_n$.

The phase plane is concerned with second-order systems and, thus, either only s_1, s_2 or λ_1, λ_2. In this simplified case, it becomes easy to investigate the nature of the singularities and the behavior of the solution of the differential equations in the neighborhood of the singularities. The eigenvalues are very helpful. Conceptually, phase space extends these analyses to n^{th}-order systems.

5.4.2 The Phase Plane

A second-order differential equation is defined by

$$F(x, y, y', y'') = 0, \qquad (\ )' = d(\ )/dx$$

which can usually be manipulated to yield $y'' = f(x, y, y')$, and typically we seek solutions to this equation in the form $y = g(x)$. This can obviously be extended to n^{th}-order systems.

We shall be analyzing the trajectory and/or stability characteristics of these equations for the case in which the independent variable does not explicitly appear. Such systems are termed autonomous.

$$y'' = f(y, y') \qquad \text{or} \qquad F(y, y', y'') = 0 \qquad (\ )' = d(\ )/dx$$

By introducing the transformation $y' = v$, we obtain

$$v' = f(y, v) \qquad \text{or } F(y, v, v') = 0$$

and it becomes possible to examine the graph of v varying with respect to y throughout the y–v plane, with the slope of v being defined at each point (y, v) along the trajectory, which allow us to map the vector field which defines the direction of the trajectory. Furthermore, with the definition of initial conditions, each point along the trajectory corresponds to a particular value of the independent variable.

Let us illustrate these concepts with two simple examples. First, consider the linear harmonic oscillator

$$x'' + \omega^2 x = 0 \qquad (\ )' = d(\ )/dt$$

and then the simple undamped pendulum

$$\theta'' + \alpha \sin \theta = 0 \qquad (\ )' = d(\ )/dt$$

Considering the oscillator first, we can obviously solve the equation to yield $x = c \sin (\omega t + \phi)$, where ϕ and c are constants of integration determined by the initial conditions assigned. By a suitable choice of initial conditions (or a change in variables so that $\psi = \omega t + \phi$, the equation's solution becomes

$$x = c \sin \psi \qquad \text{and, thus, } x' = c\omega \cos \psi$$

Since each point on the x-x' plane is related by the distance formula $r^2 = x'^2 + x^2$, we obtain an ellipse (or if $\omega = 1$, a circle of radius c) on the phase plane. If we assume the initial condition $x(0) = 0$ and $x'(0) = c\omega$, then the graphs in the t-x and x-x' planes appear as shown in Figure 5-6.

The simple pendulum's equation does not have a solution in elementary functions as was discussed in Chapter 2, but can be "solved" graphically by this technique. Let

$$v = \theta'$$

then

$$v' = \theta'' = (dv/d\theta)(d\theta/dt) = (dv/d\theta)v$$

and thus we can write

$$v(dv/d\theta) = -\alpha \sin \theta$$

Separating variables and integrating yields

$$v^2 = 2\alpha \cos \theta + c$$

and it becomes possible to obtain the graph of v versus θ, even though the graph of θ versus t is quite difficult to define (c = constant of integration).

We note first that dependent upon the relative values of α and c, different

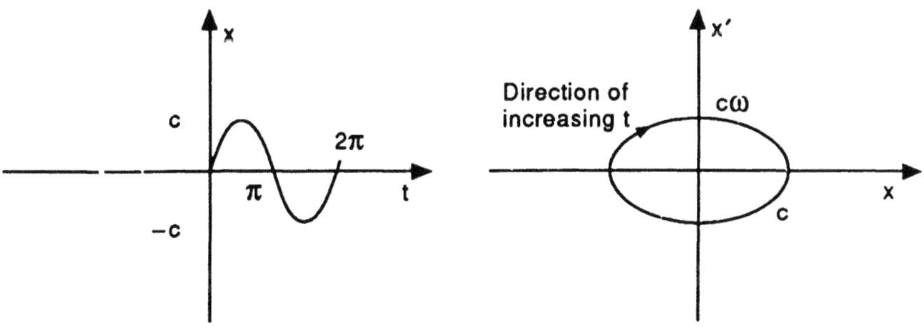

Figure 5-6. Oscillator characteristics.

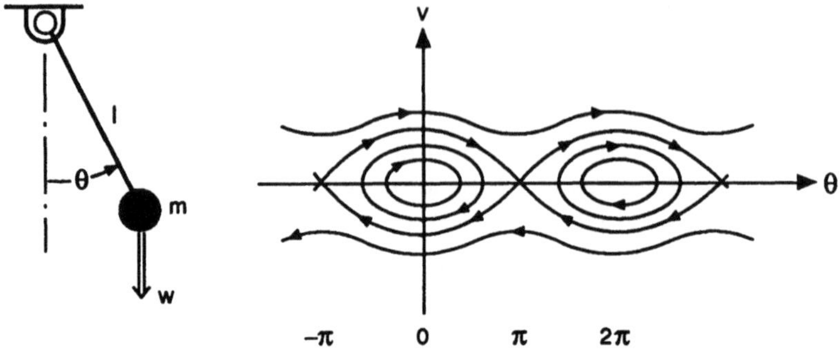

Figure 5-7. Simple pendulum.

results are obtained. In particular, if $c = 0$, then a parametric family of ellipses results (which is accurate for a first-order approximation of $\theta^2 \approx \cos\theta$, and inaccurate otherwise), each family centered along the θ axis at every 2π interval. Second, if $c > 2\alpha$, a continuous wave results oscillating about a dc level and is either positive or negative, depending upon whether we examine the positive or negative root of the right-hand side of the equation. Lastly, if $c = 2\alpha$, a sinusoidal wave results. These are shown Figure 5-7.

These curves in the v–θ plane can be discussed in terms of the potential and kinetic energy of the pendulum, and the motion related to the physical system when $v = 0$, or otherwise. The arrows depict the direction of increasing t.

We also note that the independent variable (time is the usual choice) can be described point-by-point along these curves or trajectories. For example, in the oscillator, when $\psi = 0$, $x = 0$, and $x' = 1$, when $\psi = \pi/2$, $x = 1$, and $x' = 0$, etc. Thus, the trajectories are directional and have a "starting point." In the pendulum's case, if $t = 0$ and we define $v = 0$ and θ represents a specified value, we can identify where we begin the swing. In the case when $c = 2\alpha$, the stability points are $\theta = 0°$ and $180°$, as was suggested in the introduction to this chapter.

To determine the direction of the resultant motion, we recognize that $dv/d\theta$ represents the slope of the solution trajectory at any point. Since

$$dv/d\theta = -(\alpha/v)\sin\theta$$

by restricting v to either the upper or lower half plane, and since $\alpha > 0$, the direction of the trajectory can be determined by the relative magnitude of the amplitude of the solution and the opposite of the sign of $\sin\theta$.

Consider, however, the definition

$$v = d\theta/dt$$

Solving for dt and converting to finite (i.e., delta) increments yields

$$\Delta t = (1/v)\Delta\theta$$

For the upper half plane, $v > 0$ and $v < 0$ in the lower half plane. Thus, by observation of the location of the trajectory on the phase plane and by requiring $\Delta t > 0$, the advance of the trajectory requires θ to increase in the upper half plane and decrease in the lower half plane. This, then, yields the arrows on the trajectories as shown in Figure 5-7.

Of interest is the fact that the harmonic oscillator equation is also an approximation to the pendulum equation for small θ. Since the series expansion of sin θ involves the sum of odd-powered terms of θ, we can approximate sin θ by θ. Thus, the linearized pendulum equation becomes

$$\theta'' + \alpha\theta = 0$$

which is solvable. Note, however, the difference between the y–v graphs for the pendulum and the oscillator (approximated pendulum). This comparison points out the erroneous conclusions one may encounter in "approximating" nonlinear functions or equations and moving out of an "acceptable" neighborhood.

5.4.2.1 Isoclines. The method of construction of the curves on the phase plane involves the method of isoclines, or can be accomplished by the method of analytic continuation (see next section). The isocline method creates a "mapping" of the plane by means of the slope of the trajectories and develops relationships between the variables of the equation and the instantaneous slope. In other words, the direction of the tangents to the solution trajectories creates a vector field which is used to sketch the actual trajectories.

As a simple example, consider the system

$$dx_2/dt = -x_1 - x_2, \qquad dx_1/dt = x_2$$

Forming dx_2/dx_1 (the slope in the x_2–x_1 plane) and setting the slope of the tangent (dx_2/dx_1) equal to m yields $dx_2/dx_1 = m$ or

$$x_1 = -(m + 1)x_2$$

which, when plotted on the x_1–x_2 plane for specified values of m yield loci for each value of m selected. This is shown in Figure 5-8(a).

Each curve/line represents the loci of points of the vector field with the same slope m. Then, to assist in sketching a solution trajectory, short line segments are drawn on the individual loci with the slope m, as shown in Figure 5-8(b).

Lastly, having established the field of isoclines (loci of lines/curves with slopes representing the constant and instantaneous tangents), starting at any

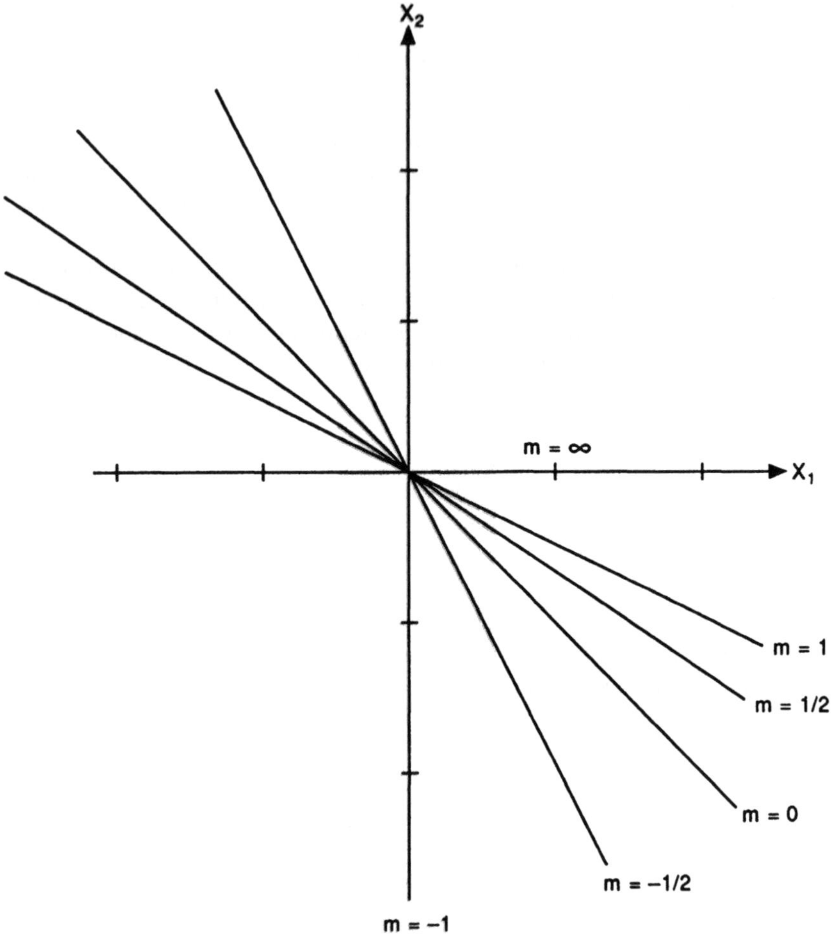

(a)

Figure 5-8. Constructing phase plane trajectories.

initial condition, $x_1(0)$, $x_2(0)$, the trajectory can be sketched using the indicated slopes to "fair-in" the solution trajectory. The illustration of Figure 5-8(c) selected first $x_1(0) = 2\text{-}3/4$, $x_2(0) = 0$, and also $x_1(0) = -3$, $x_2(0) = +1$, which both result in spirals terminating at the origin.

In the Van der Pol equation, as another example, using the parametric form of v and v', we form

$$v'/y' = dv/dy = (av - by - avy^2)/v$$

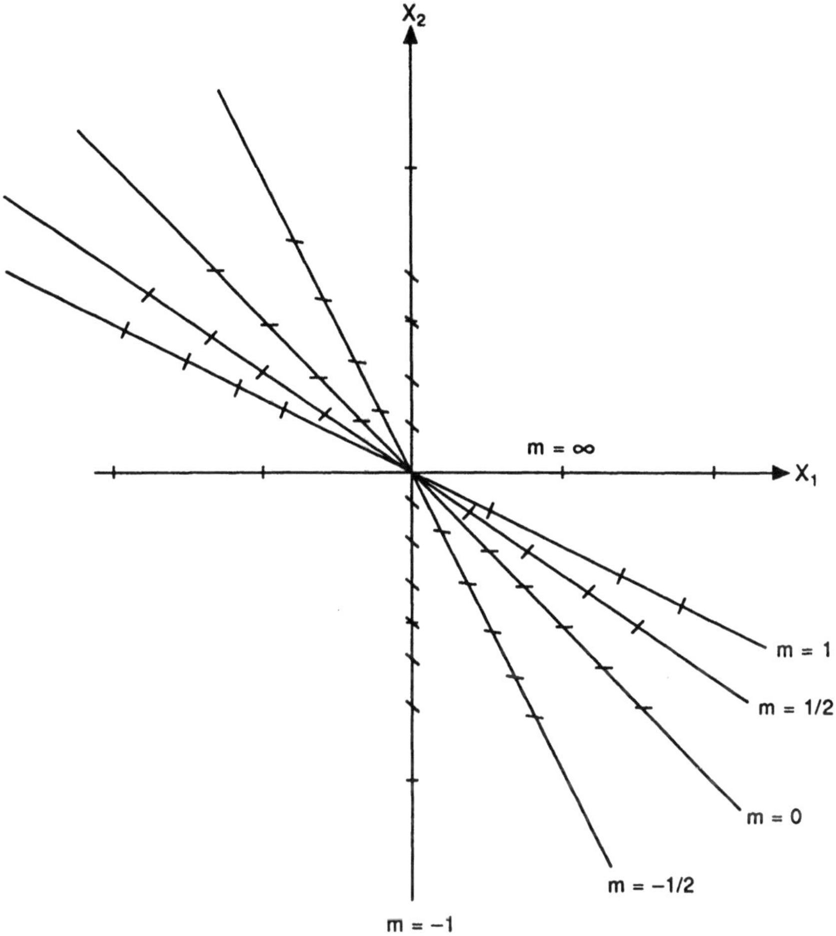

(b)

Figure 5-8. (Continued)

and set the slope = $m = dv/dy$. Thus,

$$mv + by + avy^2 - av = 0$$

which can be solved for constant m everywhere in the y–v plane for a particular set of system constants a, b; the locus of lines of constant m are the isoclines.

The trajectory representing Van der Pol's equation is shown in Figure 5-9 on the y–v plane, where to simplify the analysis we have included the scale factor av for the abscissa.

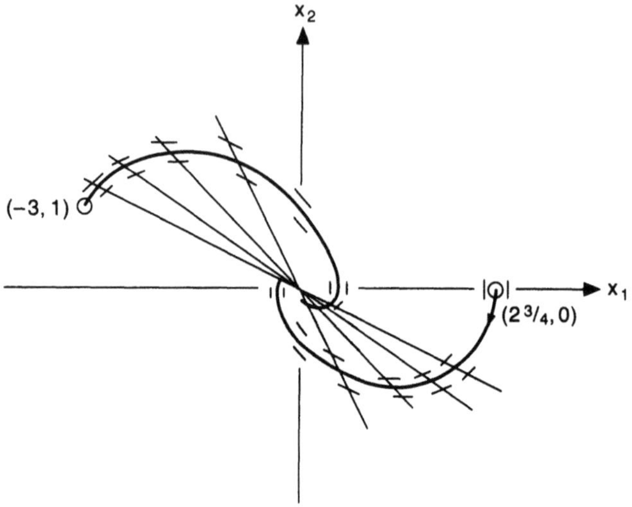

(c)

Figure 5-8. *(Continued)*

In the Van der Pol equation, the relationship between y and v has a geometric/physical relationship to the velocity and position (or, as the original equation was developed for a triode's performance, in terms of the current and its first derivative). However, the phase plane need not have this connotation at all. Consider the Volterra predator–prey equations

$$\dot{x}_1 = -x_1 + x_1x_2$$

$$\dot{x}_2 = x_2 - x_1x_2$$

By defining the phase plane as the variation of one variable (x_2) with the other (x_1), we form

$$\frac{dx_2}{dx_1} = \frac{(1 - x_1)x_2}{(x_2 - 1)x_1}$$

and plot the isoclines for constant $dx_2/dx_1 = m$. Such a sketch is shown in Figure 5-10, recognizing the existence of two singularities at $(x_1, x_2) = (0, 0)$ and $(1, 1)$.

In essence, the method involves plotting the locus of lines of constant slope, identifying that slope for each line and then starting at a particular initial or boundary condition, sketching the curve on the plane. This approach is ap-

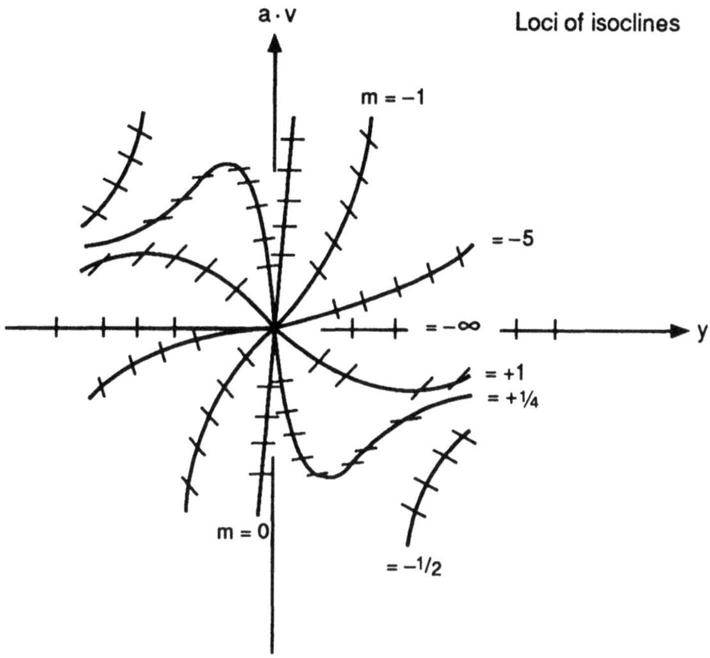

Figure 5-9. Van der Pol's equation.

proximate, of course, but it does define the nature of the trajectory—which is the objective. Use of a computer can refine the process to provide one or more families of trajectories to suit appropriate initial or boundary conditions.

Duffing's equation:

$$y'' + ay + by^3 = k \sin \omega t, \qquad a > 0$$

without a forcing function ($k = 0$), is dependent upon the frequency (related to b) and the amplitude of the motion. Note that this is an approximation to the pendulum equation, where $\sin y \approx c_1 y + c_2 y^3$.

Let us examine the equation with $k = 0$, and selecting $a = 1$ (which corresponds to $t = ct'$, where c is selected to allow $a = 1$ and creating a new constant for the coefficient of y^3 which we choose $= 1$)

$$x_1' = x_2$$

$$x_2' = -x_1(1 + x_1^2)$$

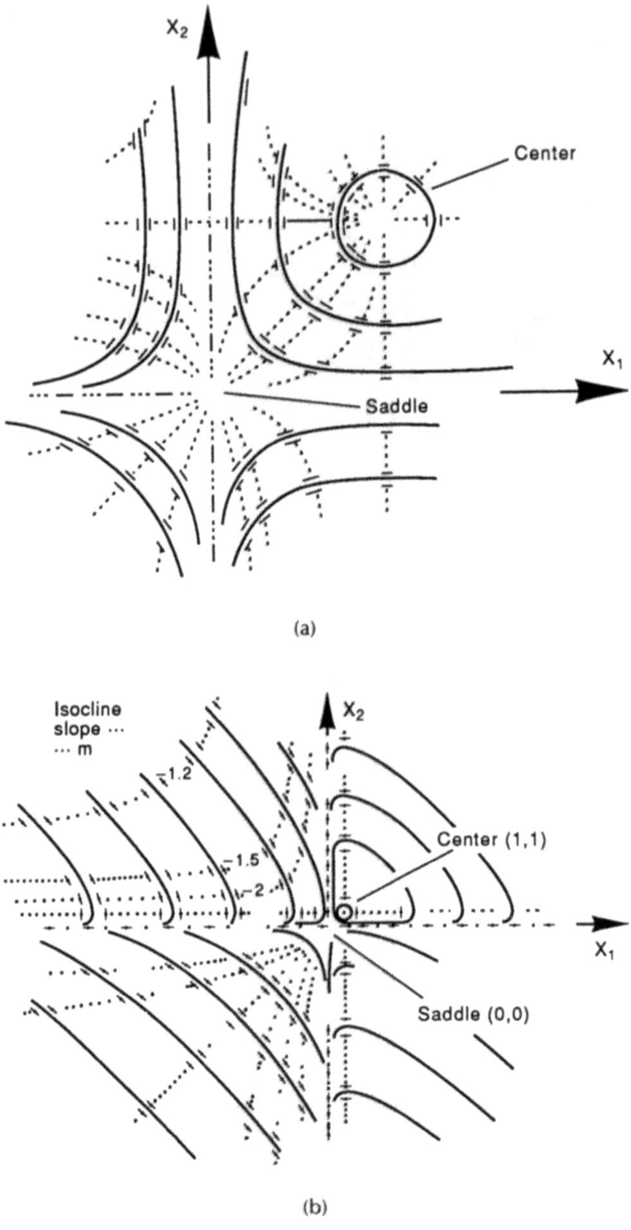

Figure 5-10. Predator prey equation. (a) Expanded view of the origin, (b) the phase plane, (c) the first quadrant.

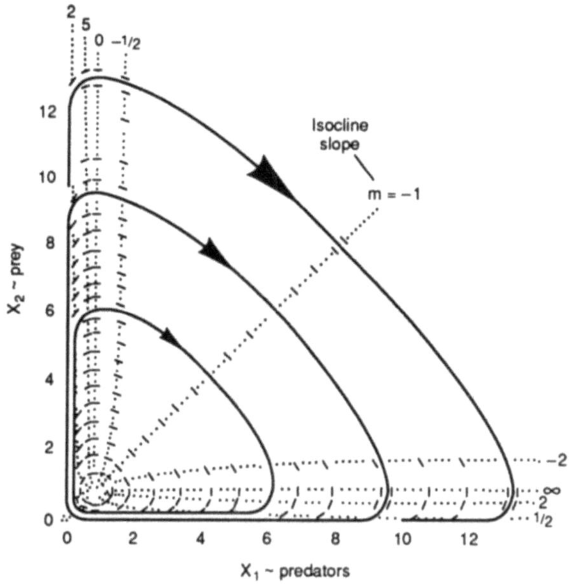

(c)

Figure 5-10. *(Continued)*

The singularities are (0, 0) and $(\pm i, 0)$. However, since we can only admit to real solutions for x_1, the imaginary roots are discarded and the singularities are reduced to the one at the origin (0, 0). The isoclines are found from

$$dx_2/dx_1 = -x_1(1 + x_1^2)/x_2 = m$$

and the loci of lines of constant slope m are found

$$x_1^3 + x_1 + mx_2 = 0$$

The isocline map for Duffing's equation, therefore, is a field of cubics all passing through the origin. For the special case $m = 0$, the x_2 axis is the isocline.

5.4.2.2 Analytic Continuation. Analytic continuation is a technique wherein the circle of convergence of a Taylor series expansion of an analytic function is extended. A function $f(z)$ is analytic within a region in the complex plane which excludes its singularities. Thus, starting at some point z_0, the circle of convergence is

$$|z - z_0| \leq R_0$$

where R_0 is the radial distance from z_0 to the singularity $z = z_{s0}$.

If we now move along the solution trajectory to a new point z_1, where $\Delta z = z_1 - z_0$, then a new circle of convergence with radius R_1 is created. This procedure is repeated along the entire trajectory, provided, however, the trajectory must not pass through a singularity. The R_i are of various lengths and conceivably R_i and R_j are such that they are formed by the same singularity; in the case where only one singularity exists, then all R_i are determined by the one singularity (Figure 5-11).

Thus, as the Δz_i advance, the Taylor series can be written point by point, allowing an algorithm to be developed for computational purposes. Hence, starting at some initial condition z_0, and attempting to reach some final condition z_f, the computations can be established by use of the Taylor series approximation based on a truncation of higher-order terms.

We assume we have a solution $y = y(x)$, which can be expanded in a Taylor series of the form

$$y = y_0 + y_0'\Delta x + (y_0''/2!)(\Delta x)^2 + (y'''/3!)(\Delta x)^3 + \cdots + (y_0^n/n!)(\Delta x)^n + R_n$$

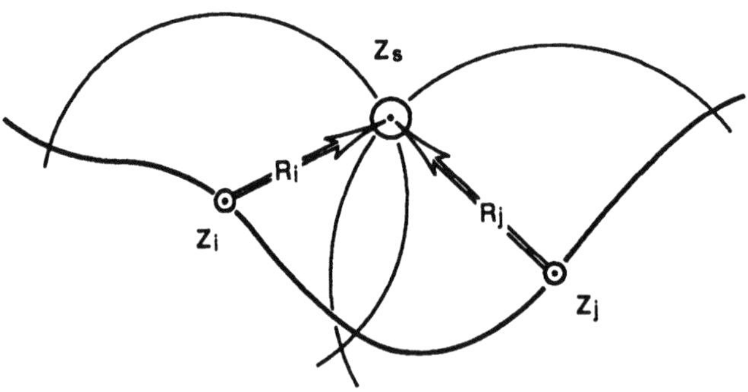

Figure 5-11. Analytic continuation—region of validity.

where the initial point corresponds to $y = y_0$ and $\Delta x = x - x_0$. Now forming y' from this equation, where the only variable is y_0 and its derivatives, since Δx is a constant,

$$y' = y_0' + y_0''\Delta x + (y_0'''/2!)(\Delta x)^2 + \cdots$$

and continuing this process so that we form y'', y''', etc. These results are then evaluated at the values of the initial condition for x_0, y_0, and y_0', substituted and solved for y_1; repeating this procedure yields an iterative formula which allows us to reach y_f.

For example, if we begin with a differential equation representing the simple pendulum, written in terms of $y'' = f(x, y, y')$

$$y'' = -\sin y$$

then $y''' = -\cos y$, $y^{iv} = \sin y = y''$, $y^{v} = y'''$, etc. And, assuming we have some starting values for x_0, y_0, and y_0' (which may be arbitrarily assumed)

$$y_1 = y_0 + y_0'\Delta x + [(-\sin y_0)/2!](\Delta x)^2 + [(-\cos y_0/3!](\Delta x)^3 + \cdots$$

$$y_1' = y_0' + (-\sin y_0)\Delta x + [(-\cos y_0)/2!](\Delta x)^2 + \cdots$$

since we are moving along the solution trajectory, the point corresponding to the increment $\Delta x = x_1 - x_0$. Now, we truncate these series to any degree of accuracy desired, where we recognize that the remainder R and R' are given by the respective n^{th} term of the Taylor series and its derivative.

$$R_n = [y^{(n+1)}/(n + 1)!](\Delta x)^{n+1}$$

$$R_n' = [y^{(n+1)}/n!](\Delta x)^n$$

It is important to recognize that the truncation introduces an error (the remainder) for each iteration. Thus, the total error is NR and NR', where N equals the number of iterations.

In general, we begin with the equation

$$y'' = f(x, y, y')$$

and form y''' and y^{iv} (and higher-order derivatives if desired)

$$y''' = f_x + f_y y' + f f_{y'}$$

$$y^{iv} = f_{xx} + f(f_y + f_{xy'} + f_{y'x}) + y'f(f_{y'y} + f_{yy'}) + y'(f_{yx} + f_{xy} + f_{y'}f_y) + f_{y'}f_x + f^2 f_{y'y'} + (y')^2 f_{yy} + f(f_{y'})^2$$

where we have substituted for y'' and y''' as appropriate to allow us to express these equations for higher-order derivatives in terms of only x, y, y'.

On the assumption that we will accept the accuracy generated by a truncation of the fourth order, we can then develop the Taylor series iterative formula for y and y'. The expression for y_1 and y_1' are presented above. Recognizing that $i = 0$ for the first increment, $i = 1$ for the second, etc., these formulae *are* the iterative expressions we seek. Thus,

$$y_{i+1} = y_i + y_i' \Delta x_i + y_i''(\Delta x_i)^2/2! + y_i'''(\Delta x_i)^3/3! + y_i^{iv}(\Delta x_i)^4/4!$$

$$y_{i+1}' = y_i' + y_i'' \Delta x_i + y_i'''(\Delta x_i)^2/2! + y_i^{iv}(\Delta x_i)^3/4!$$

where $\Delta x_i = x_{i+1} - x_i$ and y_i, y_i', etc., are evaluated at the point (x_i, y_i, y_i') by using their calculated values.

For the second-order equations of Chapter 2, the iterative formulae are obtained from the above where:

Theorem 2

$$y'' + p(x)y' + q(x)G(y) + r(x)Z(y) = (\dot{Z}/Z)y'^2,$$

$$(\ )' = d(\ )/dx \qquad (\ )^{\cdot} = d(\ )/dy$$

$$f = -p(x)y' - q(x)G - rZ + (\dot{Z}/Z)y'^2$$

$$f_x = -p_x y' - q_x G - r_x Z$$

$$f_{xy} = f_{yx} = -q_x G_y - r_x Z_y$$

$$f_y = -qG_y - rZ_y + y'^2 d/dy(\dot{Z}/Z)$$

$$f_{y'} = -p + 2(\dot{Z}/Z)y'$$

$$f_{y'x} = -p_x$$

$$f_{y'y} = f_{yy'} = 2y' d/dy(\dot{Z}/Z)$$

$$f_{y'y'} = 2(\dot{Z}/Z)$$

where

$$d/dy(\dot{Z}/Z) = (Z\ddot{Z} - \dot{Z}^2)/Z^2$$

Theorem 5

$$y'' = (\dot{G}/G)y'^2 + aG(y) + bZ(y) + [c/G(y)]Z^2(y) + [d/G^2]Z^3(y)$$

$$f = (\dot{G}/G)y'^2 + aG + bZ + [c/G]Z^2 + [d/G^2]Z^3$$

$$f_x = f_{xy} = f_{xy'} = f_{yx} = f_{y'x} = f_{xx} = 0$$

$$f_y = y'^2 d/dy(\dot{G}/G) + aG_y + bZ_y + c\,d/dy(Z^2/G) + d\,d/dy(Z^3/G^2)$$

$$f_{y'} = 2(\dot{G}/G)y'$$

$$f_{y'y'} = 2y' d/dy(\dot{G}/G) = f_{yy'}$$

$$f_{yy} = y'^2 d^2/dy^2(\dot{G}/G) + aG_{yy} + bZ_{yy} + c\,d^2/dy^2(Z^2/G) + d\,d^2/dy^2(Z^3/G^2)$$

Theorem 6

$$y'' + a(y)\,G(y') + b(y)\,H(y') = 0$$

$$f = -a(y)\,G(y') - b(y)\,H(y')$$

$$f_x = f_{xx} = f_{xy} = f_{xy'} = f_{y'x} = f_{yx} = 0$$

$$f_y = -a_y G - b_y H$$

$$f_{yy'} = f_{y'y} = -a_y G_{y'} - b_y H_{y'}$$

$$f_{yy} = -a_{yy} G - b_{yy} H$$

$$f_{y'y'} = -aG_{y'y'} - bH_{y'y'}$$

Obviously, this can also be applied to the first-order equations as well. We offer the results for the case which includes the solution of the Ricatti equation. A first-order equation is usually of the form

$$y' = f(x, y)$$

and, thus,

$$y'' = f_x + ff_y$$

$$y''' = f_{xx} + f_y f_x + f[f_{xy} + f_{yx} + f_y^2] + f^2 f_{yy}$$

and for the equation of Chapter 2's Theorem 1,

$$y' + a(x)\, G(y) + b(x)\, H(y) = 0$$

$$f = -(aG + bH)$$

$$f_x = -(a_x G + b_x H)$$

$$f_y = -(aG_y + bH_y)$$

$$f_{xy} = f_{yx} = -(a_x G_y + b_x H_y)$$

$$f_{xx} = -(a_{xx} G + b_{xx} H)$$

$$f_{yy} = -(aG_{yy} + bH_{yy})$$

The iterative equations then become (for a third-order approximation)

$$\begin{aligned} y_{i+1} = y_i &- (aG + bH)_i \Delta x_i - [(a_x G + b_x H)_i \\ &- (aG + bH)_i (aG_y + bH_y)_i] \Delta x_i^2/2! \\ &- \{(a_{xx}G + b_{xx}H) - (a_x G + b_x H)\,(aG_y + bH_y) \\ &- (aG + bH)[2a_x G_y + 2b_x H_y + (aG_y + bH_y)^2] \\ &- (aG + bH)^2 (aG_{yy} + bH_y)\}_i \Delta x_i^3/3! \end{aligned}$$

$$\begin{aligned} y'_{i+1} = &-(aG + bH)_i - [a_x G + b_x H - (aG + bH)(aG_y + bH_y)]_i \Delta x_i \\ &- \{a_{xx} G + b_{xx} H + a_x G_y + b_x H_y - (aG + bH)[2a_x G_y + 2b_x H_y \\ &- (aG + bH)^2] + (aG + bH)^2 (aG_{yy} + bH_{yy})\}_i \Delta x_i^2/2! \end{aligned}$$

In the case of Van der Pol's equation and Volterra's equations, Davis (Chapter 9) develops the derivatives to the fourth order; and in Davis's Chapter 12, Duffing's equation is similarly studied. For specific values of constants in these equations, the technique of analytic continuation provides a means of computing the values of the trajectory and thereby allowing us to examine the properties of the solution.

5.4.3 Poincare–Bendixson

The behavior of the trajectories admits a simple characterization of an index. This index is an integer which allows the rapid determination of the trajectory characteristics about the singularity in question.

Conceptually, the technique is developed in topological terms and involves the encircling of the singularity similar to that of the Nyquist encirclements. By use of vector fields, Green's theorem, and related issues, each singularity is classified and "valued" by an index.

In the x–x' plane, we shall examine the general equation

$$y' = P(x, y), \qquad x' = Q(x, y), \qquad ' = d/dt$$

At each point in the plane there is a vector field which can be defined in terms of the basic equations of the system being analyzed. Clearly, the manner of definition of the field can be quite helpful; thus, we select the direction of the field as the direction tangent to the trajectories of our equation's (system's) solution curve. We next accept the fact that at least one singularity exists, and if there are more than one, we allow only those which are isolated (i.e., are in a distinct neighborhood that can be encircled by a closed curve—no matter how small—and that that closed curve contains no other singularity). The vector field's direction defines the subsequent infinitesimal advance of the solution trajectory. This direction makes an angle Θ with respect to an arbitrary datum. Now we move along the trajectory and reach a point that intersects with a closed curve C, which encircles the singularity we wish to examine. (The curve C is a Jordan curve[1] which we construct for our purposes.) At the intersection, we move along the Jordan curve C in a counterclockwise (i.e., positive) direction until we return to the starting point (the point of initial intersection).

By observing the change in the angle Θ as it varies around the curve C, it will be seen to rotate through a net number of encirclements of the singularity. The number of encirclements is the index I of the curve C. As there are only a few types of singularities, it turns out that there are only a few values of I. They are:

Singularity	Poincare index
Node	+1
Center	+1
Spiral	+1
Saddle point	−1

It is instructive to recognize that if the curve C does *not* contain a singularity, a rotation around C results in no change in angle for Θ and thus corresponds to a zero value.

The concept is easily expanded to allow the simply connected closed Jordan curve to contain as many isolated singularities as may exist for the system's equations. In this instance, the index of the closed curve is the algebraic sum of the indices of the individual singularities. Furthermore, any closed trajectory, which is a center, must contain at least one singularity; thus, it has index +1. As a third point, any closed trajectory contains $2n + 1$ singular points, and the

[1]If $x(t)$ and $y(t)$ are continuous real functions of t in $\alpha \le t \le \beta$ and $z(t) = x(t) + iy(t)$, then $z(t)$ is a continuous curve. If two distinct points (x_1, y_1) and (x_2, y_2) correspond to two distinct values of t, t_1 and t_2, then no multiple points exist. The curve $z(t)$ is then called a Jordan arc. A closed Jordan curve is a continuous curve having $x(\alpha) = x(\beta)$ and $y(\alpha) = y(\beta)$, but otherwise no multiple points. See K. Knopp, *Theory of Functions*, Part I (Dover, New York, 1945).

quantity of these points of index +1 must exceed by one the number of saddle points.

To analytically express these concepts, we recognize that the equations

$$y' = P(x, y), \qquad x' = Q(x, y)$$

describe a vector field over the x–y plane, and we seek to describe the variation of the direction of that field with respect to a datum and the motion defined by these equations. It is convenient to arbitrarily select the tangent to the field at each point (x, y) and to associate with each point an angle Θ measured relative to the datum. Then we impose a closed Jordan curve C on the vector field, where C either does or does not encircle a singularity (Figure 5-12).

At the intersection of the tangent to the system solution and the curve C, we move counterclockwise and "add up" all Θ's. Analytically, this is equivalent to $\int_C d\Theta$ and equal to 2π radians times the index I. By selecting the datum to be parallel to the x axis, the angle Θ has the property

$$\tan \Theta = dy/dx = P/Q$$

and, thus,

$$2\pi I = \int_C d[\arctan P/C] = \int_C (QdP - PdQ)/(P^2 + Q^2)$$

This is the equation of a line integral around a curve C. Its value is zero, provided that P, Q and their derivatives are continuous (see Cauchy's theorem for an analytic function).

Thus far, we have been discussing the concept of an index of a closed

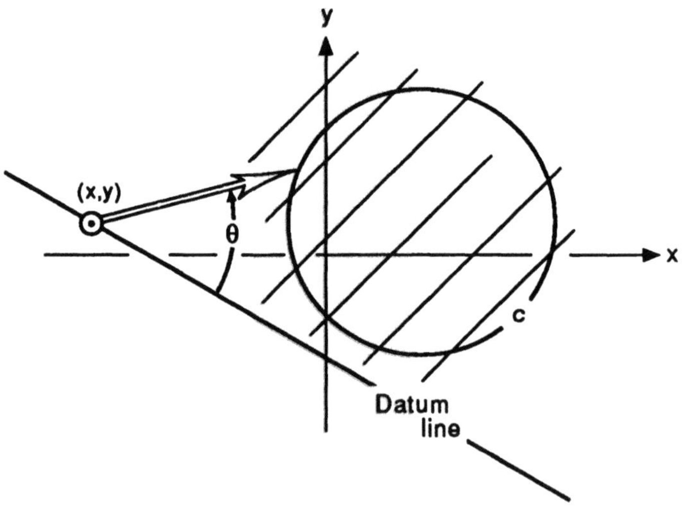

Figure 5-12. A vector field.

Jordan curve. We now will consider the concept of an index of the singular point itself. Simply said, the index of a singular point is the index of a simply connected Jordan curve which surrounds only one singularity.

The evaluation of I proceeds as follows. Consider the system

$$dx_1/dt = x_2 = P$$

$$dx_2/dt = -x_1 = Q$$

The eigenvalues for this system are $\pm i$; thus, a center is the resultant trajectory. Select the circle

$$x_1^2 + x_2^2 = R^2$$

as the Jordan curve C and substitute. This results in

$$2\pi I = (1/R^2) \int_C [x_1 dx_2 - x_2 dx_1]$$

Change variables so that

$$x_1 = R\cos\Theta, \qquad x_2 = R\sin\Theta, \qquad R = \text{constant}$$

which results in

$$2\pi I = \int_C d\Theta = \Theta$$

and since Θ varies between 0 and 2π, $I = 1$.

For the case where we change the system described above to $P = x_2$, $Q = x_1$, the same analysis yields eigenvalues ± 1, which is the characteristic of a saddle point. The same approach yields $I = -1$.

Consider next the system where $P = x_2$ and $Q = -ax_2 - bx_1$, then if $a^2 > 4b$, the eigenvalues are both real corresponding to a node. If $a^2 < 4b$, the eigenvalues are complex conjugates and the solution trajectory becomes a spiral or focus.

Selecting the Jordan curve to be the ellipse

$$x_2^2 + (ax_2 + bx_1)^2 = 1$$

Substitution into the integral yields

$$2\pi I = b \int_C [x_1 dx_2 - x_2 dx_1]$$

Differentiating the equation for the ellipse and substituting results in $I = 1$.

5.4.3.1 Poincare–Bendixson Theorem. An obvious interpretation of a solution to a system of equations relates to its interval of existence. Usually, the mathematics deals with the entire real line $(-\infty, \infty)$, yet physically we like to envision things happening from some existing state of time into the future. In other words, we seek to predict system performance.

The Poincare–Bendixson theorem uses this concept by defining half trajectories. That is, either $(-\infty, t_0]$ or $[t_0, \infty)$ is the region of validity for the solution trajectory. With this we can state the following.

Poincare–Bendixson Theorem

If a half trajectory remains in a closed bounded region, and does not approach a singular point, then the half trajectory is either a closed trajectory (a limit cycle) or approaches a closed trajectory.

Geometrically, the application of this theorem implies the existence of an annular region where all trajectories enter the annulus (i.e., none leave) and the region contains no singular points.

Example

Van der Pol's equation is known to have a limit cycle (see Figure 5-13) about its singular origin; thus, we define the closed bounded region as shown in Figure 5-13.

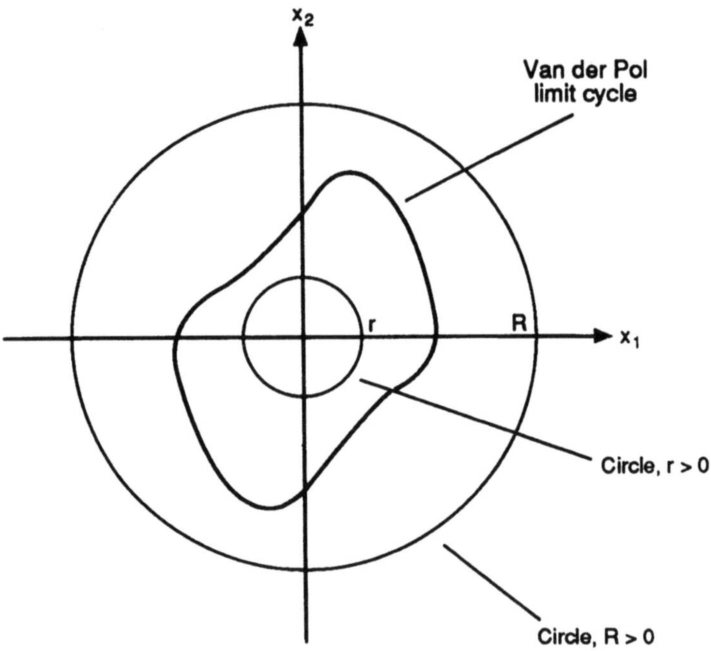

Figure 5-13. Van der Pol's limit cycle within a closed bounded region.

5.5 SOME PHYSICALLY REALIZABLE NONLINEARITIES

As will be discussed in the chapter on the describing function, the introduction of nonlinear elements into a system block diagram in the Laplace domain has its analog in the time domain. We begin by identifying a group of nonlinearities and the equations defining the properties. In all cases, we assume a system input signal $x(t)$. See Figures 5-14 to 5-18.

In each of these cases, as well as for others in which we can piece together segments which represent our nonlinearities, we can analyze the system on the

$$\text{Output} = \begin{cases} +M, & x > 0 \\ -M, & x < 0 \end{cases}$$

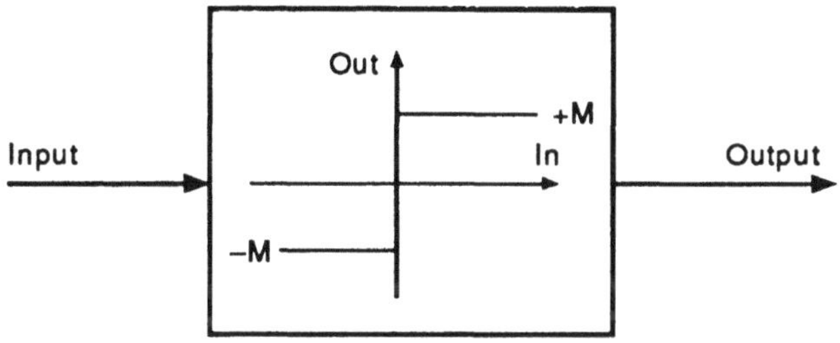

Figure 5-14. Ideal relay.

$$\text{Output} = \begin{cases} +K\cdot x, & x > +D \\ 0, & -D < x < +D \\ -K\cdot x, & x < -D \end{cases}$$

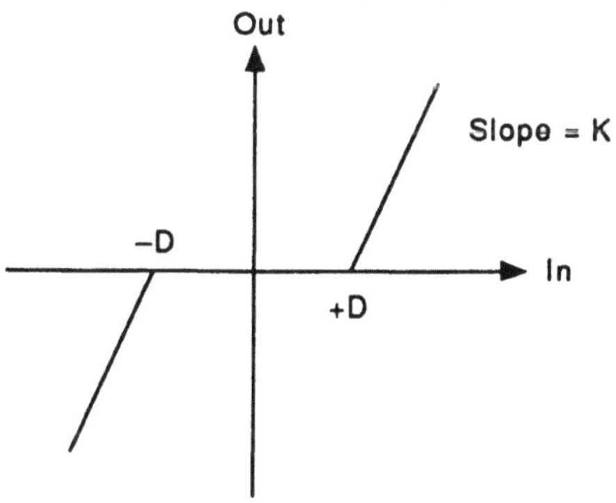

Figure 5-15. Dead band.

$$\text{Output} = \begin{cases} M & x > A \\ Kx & -A < x < A \\ -M & x < -A \end{cases}$$

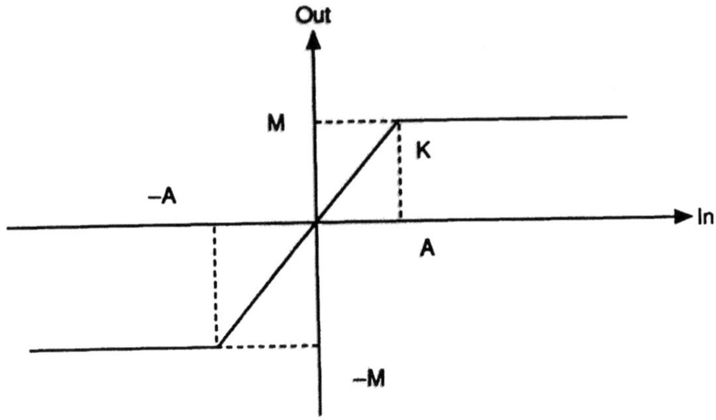

Figure 5-16. Saturation with a linear gain.

$$\text{Output} = \begin{cases} M & x > A \\ -M & x < -A \\ & -A < x < A, \text{ see Figure 5-18} \end{cases}$$

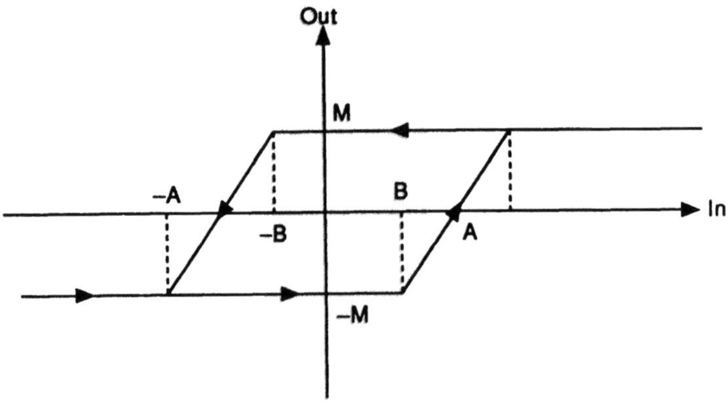

Figure 5-17. Hysteresis.

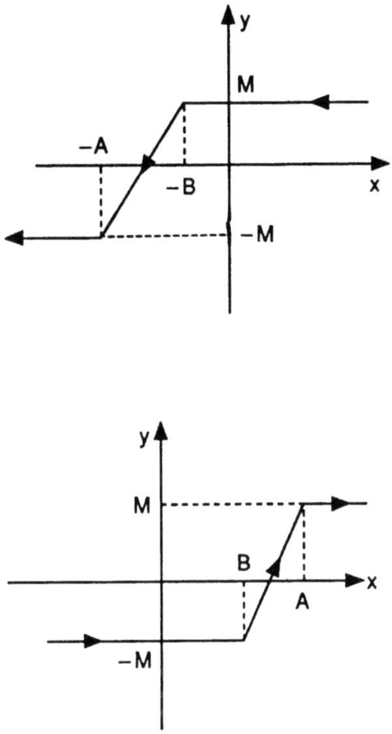

For x decreasing

$x \geq -B$	$y = M$
$-A \leq x \leq -B$	$y = \frac{2M}{A-B}\left(x + \frac{A+B}{2}\right)$
$x \leq -A$	$y = -M$

For x increasing

$x \leq B$	$y = -M$
$B \leq x \leq A$	$y = \frac{2M}{A-B}\left(x - \frac{A+B}{2}\right)$
$x \geq -B$	$y = M$

Figure 5-18. Increasing and decreasing paths of the hysteresis characteristic.

phase plane. To accomplish this requires the introduction of switching lines, i.e., curves at which system behavior changes.

5.5.1 Example of a Non-Linear System

Consider a second-order system represented by the equation

$$y'' + ay' + by = 0$$

and let us assume that the system is a mechanical apparatus where we also assume $a = b = 1$ for simplicity. Further, let

$$y = x_1 \text{ and } x_1' = x_2 = y'; \text{ then } x_2' = y'' = -y' - y$$

and let us interpret the states x_1, x_2 as the position (x_1) of the system we wish to analyze and its velocity (x_2).

These transformations result in the state variable equations and the system block diagram of Figure 5-19.

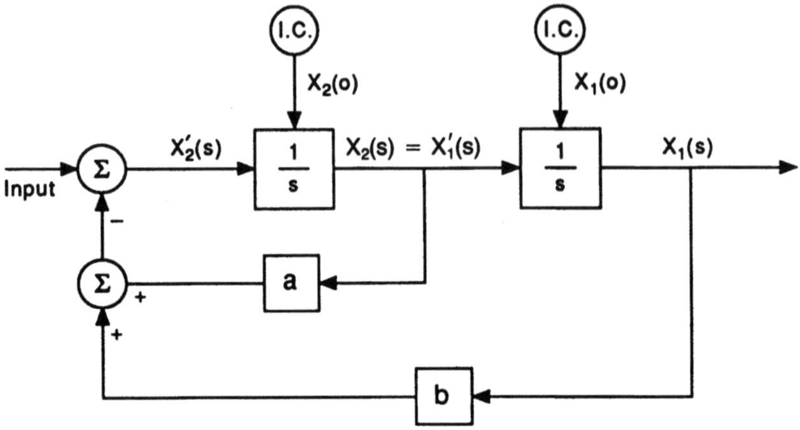

Figure 5-19. A linear second-order system.

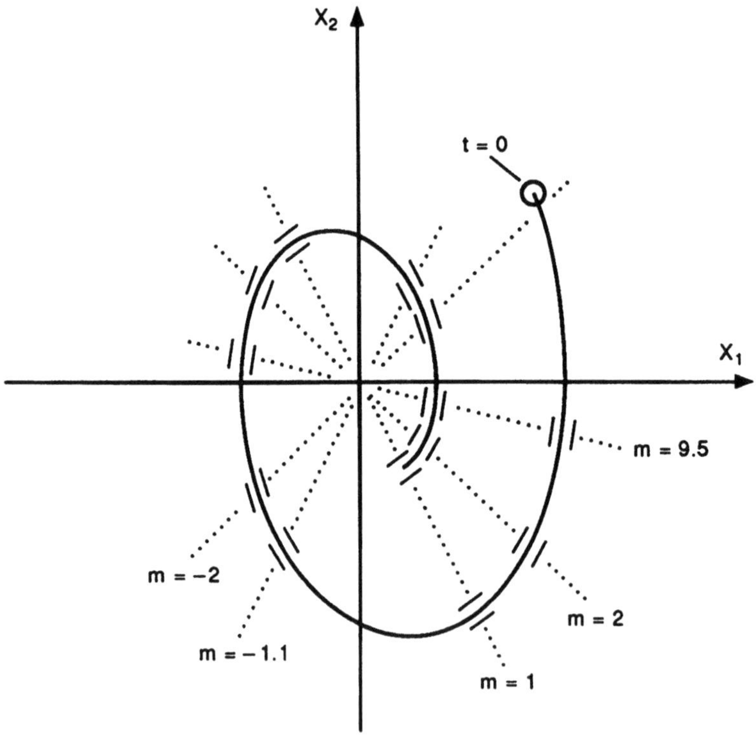

Figure 5-20. Isoclines for a linear second-order system.

$$x_2' = -x_1 - x_2$$
$$x_1' = x_2$$

Following the phase plane's isocline method leads to the equation for m, the slope of the trajectory at any point in the x_1-x_2 plane.

$$dx_2/dx_1 = -1 - x_1/x_2 = m$$

As the system is linear, the phase plane (x_2 vs x_1) plot of points having constant trajectory slope m are given by the equation

$$x_1 = -(1 + m)x_2$$

which are straight lines all passing through the origin, as shown in Figure 5-20.

Let us now introduce a nonlinear element so that velocity is limited. (Clearly we can extend this concept to limit position as well, and further, to limit both.) We introduce this nonlinear element by imposing a bound on the integrator 1/s so that the system operates linearly so long as $-M < x_2 < M$, and nonlinearly otherwise, as shown in Figure 5-21. Depicting the saturating

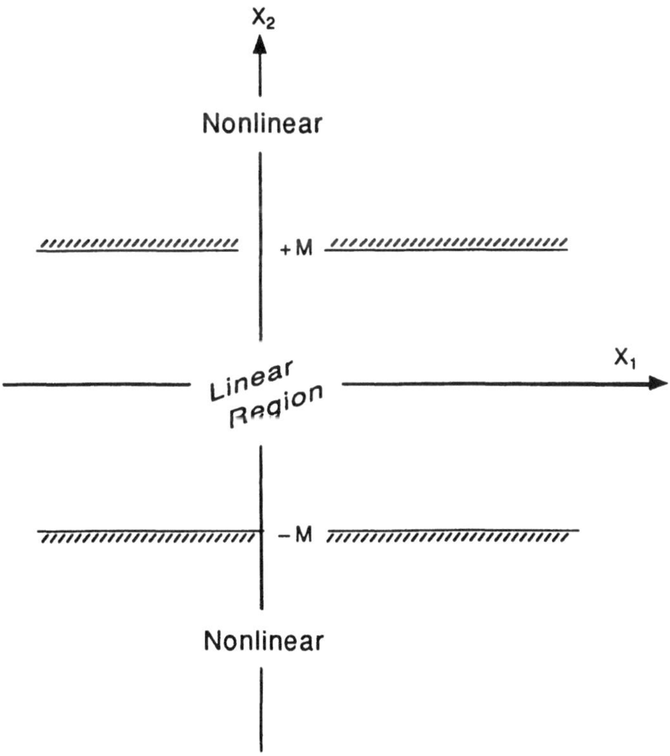

Figure 5-21. Boundaries for a velocity limited system.

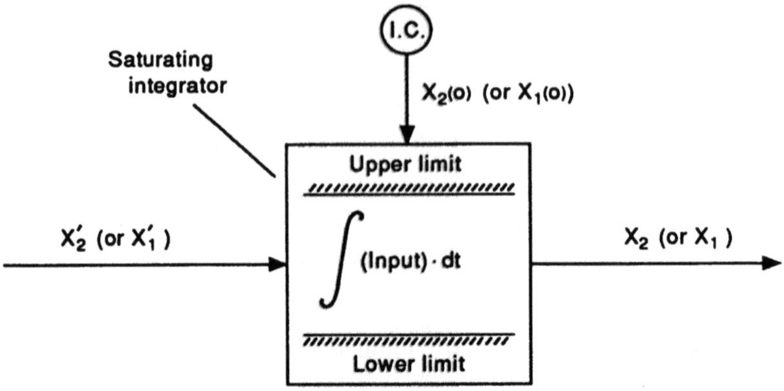

Figure 5-22. Velocity (position) saturation limiter.

integrators requires some care in order not to convey the notion that the integrator is in series with a saturation element. Thus, they are depicted so as to specifically show that the saturation is a part of, and not a modifier of, their integration process as in Figure 5-22.

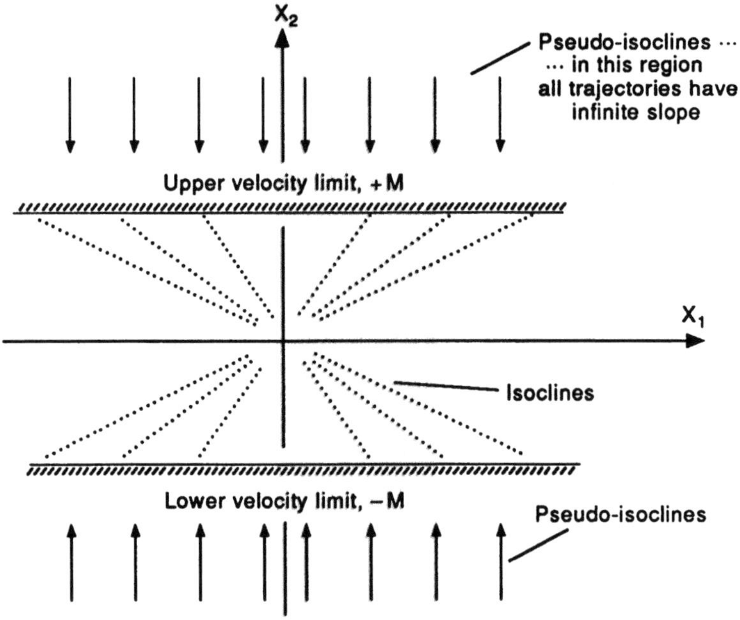

Figure 5-23. Isoclines for a velocity limited second-order system.

If velocity is limited (for example, a governor on an engine restricts fuel-flow once a preset speed is reached), then, assuming the saturation levels are $\pm M$ we obtain either

$$x_1 = -(m + 1)M \text{ or } x_1 = M(m + 1)$$

These equations represent pseudo-isoclines lying in a region denied by the velocity saturation (i.e., outside the band of linearity). We shall shortly revisit the concept of "pseudo" isoclines, but for now let us continue to refer to the isoclines without that connotation. These isoclines are vertical lines at each value of x_1 terminating at the horizontal lines $x_2 = \pm M$ as shown in Figure 5-23. Between them lies the linear region.

If, on the other hand, position is limited to $\pm N$ (for example, a pendulum whose swing is restricted by mechanical stops to an angle of, say, $\pm 30°$ or a

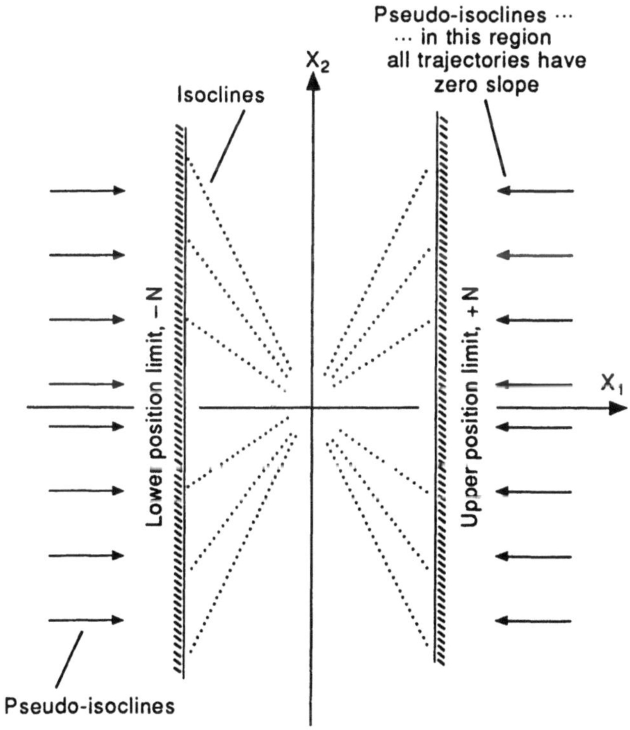

Figure 5-24. Isoclines for a position limited system.

piston which can only move within a specified band), the isocline equations become,

$$x_2 = -N/(m + 1) \text{ or } x_2 = N/(m + 1)$$

which are again pseudo-isoclines consisting of horizontal lines terminating at $x_1 = \pm N$ as shown in Figure 5-24. The linear region lies in between, and is criss-crossed by radiating isoclines, as before.

We have therefore defined the cases where either the velocity or position is limited and the regions in which isoclines are defined.

> *Note*: The concept of a pseudo-isocline is one which must be examined more closely. Each isocline is a locus of points which admit to solution trajectories having a constant slope, m. The equations just defined represent the isoclines in the nonlinear region. They are fictitious! The physical restraints on our system do not permit a solution trajectory to exist

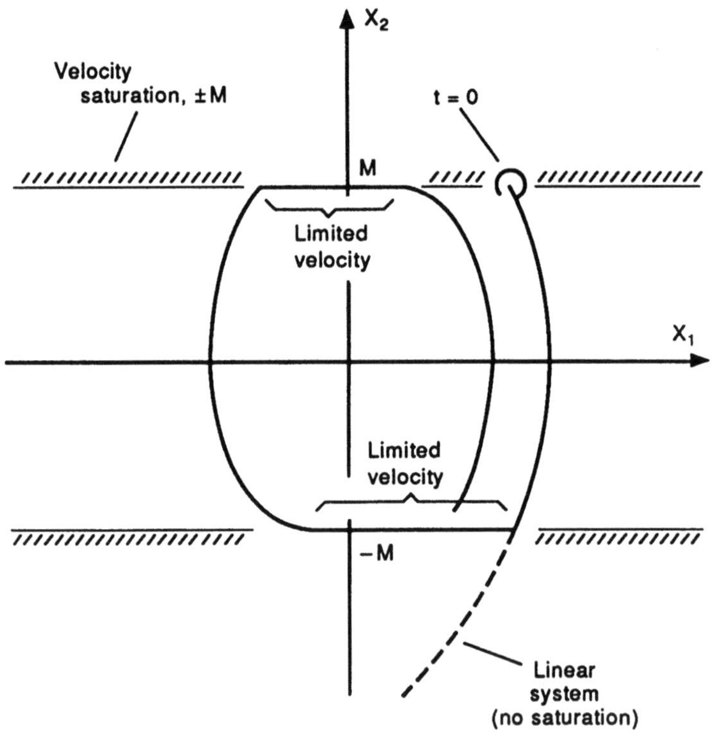

Figure 5-25. Velocity limited system.

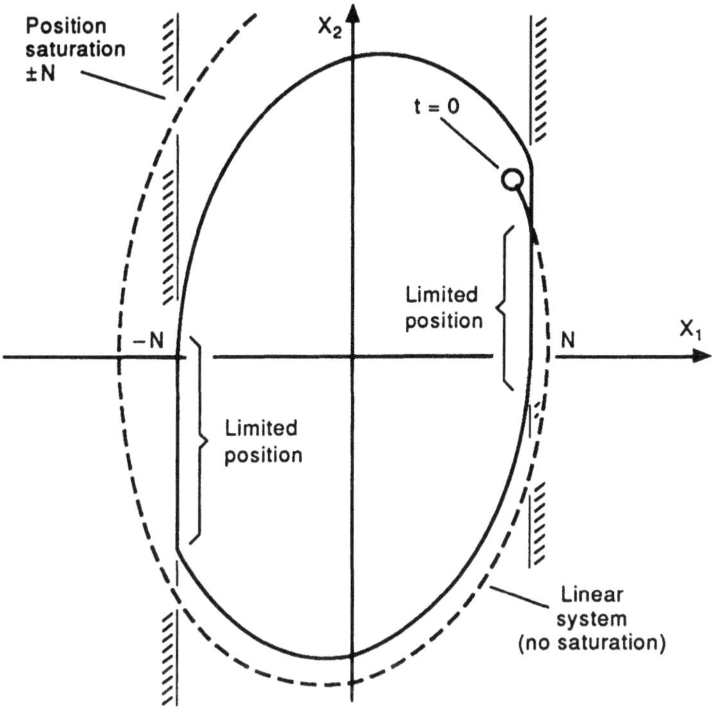

Figure 5-26. Position limited system.

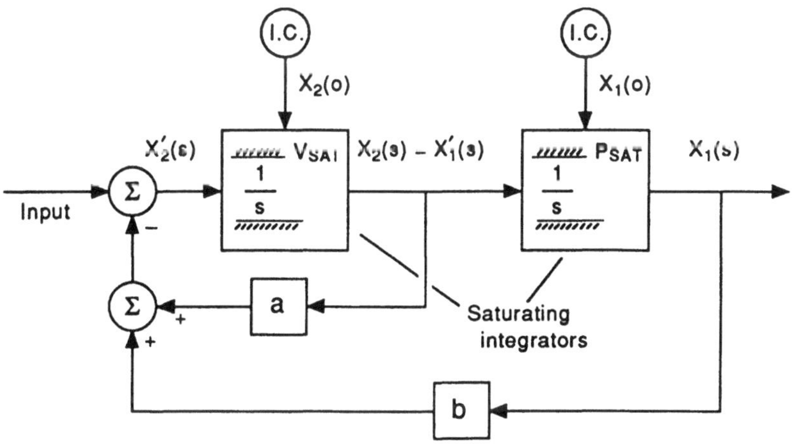

Figure 5-27. Second-order system with velocity and position limiting.

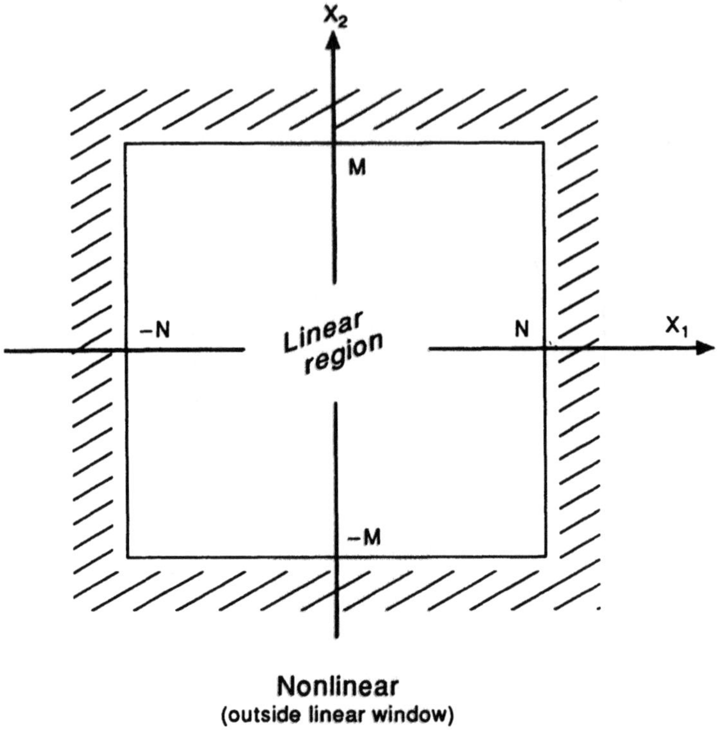

Figure 5-28. Region of admissible solutions.

beyond the $\pm M$ or $\pm N$ boundaries. Hence we must reject the potential for solutions defined by the horizontal or vertical isoclines of Figures 5-23 and 5-24 in this system configuration.

Now, with a system which cannot admit solutions which extend into the regions defined as "nonlinear" in Figure 5-21, then the solutions must lie on the boundary of the linear-nonlinear region, creating an envelope for the system solutions. The only admissable solutions lie in the linear range or along the saturation boundaries. Typical solutions for velocity saturation or position saturation are shown in Figures 5-25 and 5-26.

In the case of a system having both velocity-saturation and position-saturation, its block diagram appears as in Figure 5-27 where care must be exercised to avoid modeling the saturation as an integrator whose output goes into a limiter. The isoclines lie inside the linear rectangle (Figure 5-28), with solutions confined therein, and a typical trajectory is shown in Figure 5-29.

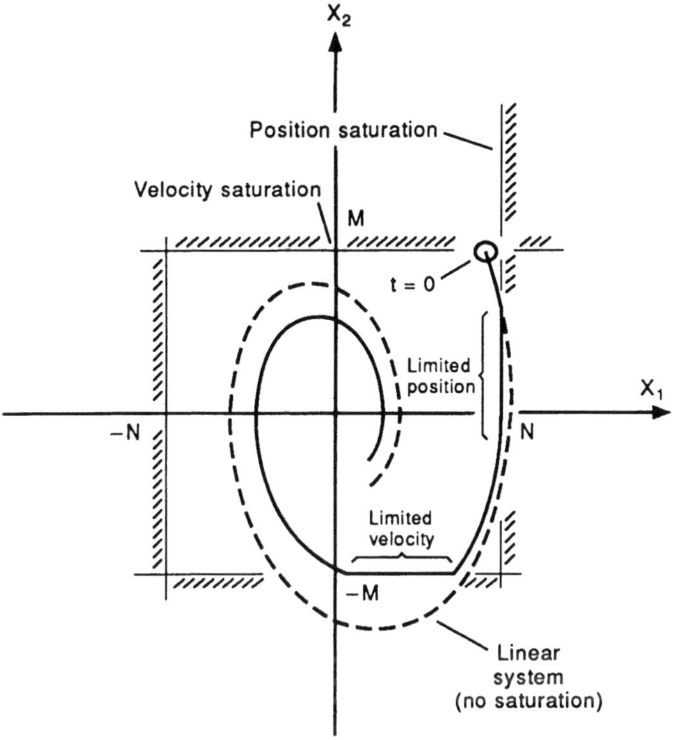

Figure 5-29. Typical solution trajectories.

5.6 PROBLEMS

1. Transform Hill's equation $y'' + Q(x)y = 0$ into Mathieu's equation

$$y'' + (a - 2b \cos 2t)y = 0$$

where $' = d/dt$.

2. (a) Show that the phase plane yields circular trajectories for the system

$$x'' + x = 0$$

(b) Evaluate the damped pendulum $x'' + cx' + k \sin x = 0$ using the phase plane. Compare the result with the linear equation

$$x'' + cx' + kx = 0$$

3. Plot the phase plane trajectory for:

(a) $x'' + x = 2 \cos t$

Is this system stable?

(b) $x'' + 2x' + 2x = 0$
Is this system stable?
(c) If the forcing function in (a) had a frequency other than $\omega = 1$, would the system be stable?
(d) In (a)–(c), discuss the concept of resonance.

4. Show that if $y(x) = y(x + w)$, then the solution to the linear differential equation $ay'' + by' + cy = 0$ (a, b, c-constants) is of the form $y = c_1e^{Ax} + c_2e^{Bx}$, where $r = A, B$ satisfies the characteristic equation $ar^2 + br + c = 0$.
5. Show that the system

$$\dot{x}_1 = ax_1 + bx_2$$
$$\dot{x}_2 = cx_1 + ex_2$$

is equivalent to the system

$$\dot{x}_1 = \lambda_1 x_1$$
$$\dot{x}_2 = \lambda_2 x_2$$

in that they have the same characteristic equations leading to λ_1, λ_2 by use of the transformation

$$x = \alpha x_1 + \beta x_2$$
$$y = \gamma x_1 + \delta x_2$$

6. Show the conditions under which the equations

$$x'' + f(x)x' + g(x) = 0, \qquad ' = d/dt$$
$$x'' + f(x, x')x' + g(x) = 0$$

do not contain any closed trajectories using Bendixson's theorem.
7. Consider the equation of a pendulum with damping and a constant torque applied.

$$A\theta'' + B\theta' + C \sin \theta = T, \qquad A, B, C, T = \text{constants}$$

Show there are no periodic solutions outside of the range

$$\frac{T - C}{B} < u < \frac{T + C}{B}$$

using Poincare's method of contact curves (see p. 134 of Nemytskii and Stepanov).
8. Using Bendixson's theorem:
(a) Demonstrate that Duffing's equation

$$y'' + ay' + by + cy^3 = 0$$

has no closed trajectories if $a \neq 0$.

(b) Find the singular points for the Volterra predator–prey equations

$$\dot{x}_1 = -x_1 + x_1x_2$$
$$\dot{x}_2 = x_2 - x_1x_2$$

and show that closed trajectories do exist.

9. Find the singular points for the pendulum equation with damping

$$x'' + cx' + k \sin x = 0$$

10. What are the singularities for

$$y'' + f(y)y' + g(y) = 0$$

and for

$$y'' + f(y, y')y' + g(y) = 0?$$

11. Find the singular points for each of the nine theorems of Chapter 2.
12. For the second-order system

$$dx_1/dt = +x_2 \text{ and } dx_2/dt = -x_1$$

(a) Define the second-order differential equation.
(b) Establish the isocline map on the x_1–x_2 plane.
(c) Sketch the resulting curve of system performance using the initial condition

$$x_1(0) = 2, \qquad x_2(0) = 1$$

13. Repeat Problem 12 for the system

$$dx_1/dt = +x_2 \text{ and } dx_2/dt = +x_1$$

14. (a) Using the concept of the phase plane, determine the direction of increasing time (i.e., positive motion along the trajectory) for the general system:

$$x_1' = x_2$$
$$x_2' = f(x_1, x_2)$$

(b) For the linear system:

$$x' = x, \qquad y' = -y$$

with the solutions

$$x = Ae^t, \qquad y = Be^{-t}$$

find the singularity, sketch the trajectories, and show the direction of increasing t.

15. For the system

$$x_1' = x_2 + \alpha x_1(\beta^2 - x_1^2 - x_2^2)$$
$$x_2' = -x_1 + \alpha x_2(\beta^2 - x_1^2 - x_2^2)$$

define an annulus which establishes the region of validity for the Poincare–Bendixson theorem.

Hint: Use polar coordinates where

$$r^2 = x_1^2 + x_2^2, \qquad \tan\phi = x_2/x_1$$

5.7 REFERENCES AND RELATED LITERATURE

5.7.1 References

1. Churchill, R. V., *Fourier Series and Boundary Value Problems*, McGraw-Hill, New York (1941).
2. Besicovitch, A. S., *Almost Periodic Functions*, Dover, New York (1954).
3. Arscott, F. M., *Periodic Differential Equations*, Macmillan, New York (1964).
4. Moon, F. C., *Chaotic Vibrations*, John Wiley & Sons, New York (1987).
5. Vidyasagar, M., *Nonlinear Systems Analysis*, Prentice-Hall, Englewood Cliffs, New Jersey (1978).
6. Whitaker, E. T., *A Treatise on the Analytical Dynamics of Particles and Rigid Bodies*, Cambridge University Press, London (1964).
7. Stern, Thomas E., *Theory of Nonlinear Networks and Systems—An Introduction*, Addison-Wesley, Reading, Massachusetts (1965).
8. Davis, H. T., *Introduction to Nonlinear Differential and Integral Equations*, Dover, New York (1962).
9. Moulton, F. R., *An Introduction to Celestial Mechanics*, 2nd Rev. Ed., Macmillan, New York (1914).
10. Ince, E. L., *Ordinary Differential Equations*, Dover, New York (1956).
11. Langill, A. W., Jr., *Automatic Control System Engineering*, Vol. II, Prentice-Hall, Englewood Cliffs, New Jersey (1965).
12. Ku, Y. H., *Analysis and Control of Nonlinear Systems*, Ronald Press, New York (1958).
13. Nemytskii V. V., and Stepanov, V. V., *Qualitative Theory of Differential Equations*, Dover, New York (1989).
14. Erdelyi, A., *Asymptotic Expansions*, Dover, New York (1956).

5.7.2 Related Literature

1. Truxal, John G., *Automatic Feedback Control System Synthesis*, McGraw-Hill, New York (1955).
2. DeRusso, Paul M., Roy, Rob J., and Close, Charles M., *State Variables for Engineers*, John Wiley & Sons, New York (1965).

Chapter 6
Liapunov Stability

6.0 INTRODUCTION

In previous chapters we alluded to the concept of stability without a rigorous mathematical definition, being content to accept the concept of boundedness to guarantee stability. In the linear case this is sufficient; in the nonlinear case we need to be more specific. The concepts of local, global, absolute, uniform, and asymptotic stability will be defined in Section 6.2.2, and examples given of each. Each depends upon regions of validity of the analysis about singular points (See Chapter 5). In the case of linear equations, the issue is moot, since the linear system exhibits its behavior characteristics throughout the real space in which it is defined—typically the entire x–y plane or n-dimensional space. In nonlinear systems, we must restrict our attention to neighborhoods around the operating point.

The concept of stability of a physical system is often intuitively obvious. However, the differential equations which describe a system can be complex, so that the a priori determination of stability is very difficult. In fact, even simple equations can involve difficult analyses to determine if they yield stable solutions.

We must also recognize that stability is not always the designer's goal; for example, the detonation of a bomb corresponds to an unstable physical state—which is the desired result.

Clearly, if a system comes to rest, it appears to be stable; however, if it is perturbed from the rest state, certain systems return to rest while others do not. It is instructive to recognize at the outset that not all solutions which reach a state of rest remain "stable." For example, the pendulum has a natural point at which it will come to rest. That point is aligned with the gravity vector, such that the weight of the mass is closest to the attracting gravity force, and is referred to as "down." However, there is also another stable location 180° in the opposite direction, normally referred to as "up," which is also stable. Reference is made to the Figure 5-7, which shows that the two stable points are at 0° and 180° (i.e., 0 and π) and multiples thereof. The slightest disturbance, such as a breeze or background vibration will, of course, dislodge the mass from the "up" position and the pendulum will swing to the "down" stability orientation. Although the "up" position is not of practical importance and only a scientific curiosity, our methods of analysis, to be both complete and useful in other situations where the dynamics are not so obvious, must include a means of identifying such points. Usually, it is possible to do so.

Often, it is possible to determine a system's stability without solving its dynamical equations. This chapter is concerned with several of the techniques available to establish the likelihood of a system's stability.

6.1 LIAPUNOV STABILITY: AN OVERVIEW

Liapunov's direct method generalizes the proposition that a system seeks its lowest energy state. This is as true of water "seeking its own level" by flowing downhill as it is of a football rolling end-over-end (about its axis of maximum inertia); thus, in the face of such everyday experience, the notion that a system is attracted to its lowest energy state is readily accepted. In fact, the converse is also appealing because things in an agitated state eventually self-destruct if the agitation escalates enough. Bridges fall down, tires burst, and, so, we are tempted to relate stability to energy.

Historically, energy has served as a basis for stability studies. Early attempts at stability criterion for feedback amplifiers were largely based on energy arguments. That they were unsuccessful is a tribute to the level of abstraction found necessary by Harry Nyquist, which stands even today as a beautiful expository of complex variables, Cauchy's winding number, and contour mapping.

Liapunov generalizes a functional V, the Liapunov functional, being a measure of energy using coordinates consisting of the system's state variables $\mathbf{x}(t)$, $\mathbf{x}(t) = [x_1(t), x_2(t), \cdots, x_n(t)]$ and its derivatives. He asserts that the system will seek the lowest level of generalized energy if it is stable. In so doing, a sufficient (if not necessary) condition is that his candidate for the generalized energy evermore decrease, suggesting its derivative with time be negative. That the system be exclusively nonlinear is of no consequence, so Liapunov's approach applies equally well to linear systems.

It should be noted that Liapunov's direct method does not suggest the precise form the measure of generalized energy should take, leaving it to be discovered and to be specialized for each system being analyzed. Nor did he specify that there is only one such candidate from which we must admit the possibility of many. And finally, he never even guaranteed that there is a candidate! Thus, he asks us to go forward much as Diogenes seeking the truthful man, but doesn't underwrite its existence (or uniqueness). Happily for us, every so often we can find appropriate candidates to establish stability characteristics conclusively.

6.1.1 Liapunov Stability Theory

The solution of a linear second-order differential equation as well as for stability considerations involves examination of the characteristic or eigenvalue equation. Analogous considerations exist in nonlinear equations.

In general, we seek solutions of the nonlinear autonomous system,

$$dy/dt = F(x, y)$$

$$dx/dt = G(x, y)$$

We shall first show that it is permissible to examine the linearized system to determine behavior characteristics (including stability) of the nonlinear system.

Let $x = w + p$ and $y = u + q$, where the general function point (x, y) is expanded about the operating point (p, q). Then

$$du/dt = F(w + p, u + q)$$

and

$$dw/dt = G(w + p, u + q)$$

and if F and G are analytic (i.e., continuous and differentiable) in a particular region of the x–y plane (which could include the entire plane), then we can expand F and G in a Taylor series to obtain

$$\begin{aligned} F(w + p, u + q) &= F(p, q) + wF_x(p, q) + uF_y(p, q) \\ &+ (\tfrac{1}{2!})[p^2F_{xx}(p, q) + 2pqF_{xy}(p, q) \\ &+ q^2F_{yy}(p, q)] + \cdots \end{aligned}$$

and

$$\begin{aligned} G(w + p, u + q) &= G(p, q) + wG_x(p, q) + uG_y(p, q) \\ &+ (\tfrac{1}{2!})[p^2G_{xx}(p, q) + 2pqG_{xy}(p, q) \\ &+ q^2G_{yy}(p, q)] + \cdots \end{aligned}$$

where subscripts x, y denote partial differentiation with respect to x or to y. The circle of convergence is defined by the neighborhood in which the series is valid and is related to the truncation level selected.

The singular points of the nonlinear differential equations are the intersections of the curves $F = 0$ and $G = 0$, that is,

$$F(p, q) = 0, \qquad \text{and } G(p, q) = 0$$

at the point (p, q).

Linearizing these equations around the point $P_0 = P(p_0, q_0)$ results in discarding all terms of second order and higher. Therefore, we set

$$wF_x(p_0, q_0) + uF_y(p_0, q_0) = 0$$

$$wG_x(p_0, q_0) + uG_y(p_0, q_0) = 0$$

and solving this system for its eigenvalues is equivalent to

$$\begin{vmatrix} F_x - \lambda & F_y \\ G_x & G_y - \lambda \end{vmatrix} = \lambda^2 - \lambda(F_x + G_y) + F_xG_y - F_yG_x = 0$$

where we have used F_x in place of $F_x(p_0, q_0)$ for ease of notation. We can therefore state the following.

LIAPUNOV'S STABILITY THEOREM

In the neighborhood of the point $P(p_0, q_0)$ of the solution to the equation

$$dy/dt = F(x, y)$$

$$dx/dt = G(x, y)$$

the eigenvalues (roots) of the characteristic equation

$$\lambda^2 - \lambda(F_x + G_y) + F_xG_y - F_yG_x = 0$$

establish the stability characteristics of the solution in the same sense as they do in linear systems.

Thus, this theorem permits the use of the concepts of linear system stability and the phase plane for nonlinear systems to yield conclusions about trajectories in the neighborhood of singular points.

Example: Van der Pol's Equation Revisited

Applying this concept to Van der Pol's equation

$$y'' - a(1 - y^2)y' + by = 0, \qquad a, b > 0$$

Let $y' = v$ and, thus, $v' = av - by - avy^2$. Set

$y = u + q$, $v = w + p$, where p, q are constants.

Then

$$y' = u' = w + p$$

$$v' = w' = ap - bq + aw - bu - a[wu^2 + 2puw + q^2w + pu^2 + 2p^3u + pq^2]$$

Following Liapunov's Stability theorem, we expand the functions about a point p_0, q_0 using a Taylor series, discard terms higher than the first order, and obtain

$$wv_x + uv_y = 0$$

$$w[av_x - av_x q_0^2] + u[av_y - b - av_y y^2 - 2avq_0] = 0$$

Clearly, $w = u = 0$ is a solution. Furthermore, as p, q is an arbitrary translation from the origin, we can set $p = q = 0$ to see that one singularity is the origin. From the Taylor series expansion, we see that if $p = 0$, $q = 0$ is the *only* possible solution (by setting all constant terms = 0); thus, only one singularity exists, and it is at the origin of the y–v plane. Thus, the equations yield

$$u' = w$$

$$w' = aw - bu$$

and the characteristic equation yields its eigenvalues as

$$\lambda = a/2 \pm 0.5(a^2 - 4b)^{1/2}$$

But we know that a periodic solution exists (Lienard's theorem, Chapter 5) and thus we need to examine the eigenvalues to determine the behavior of the trajectories. As there is only one singular point (the origin in a translated system), we know that the periodic trajectory must move toward a closed curve encircling the origin. Thus, we examine regions within a certain domain around and including the origin, having a boundary which is the limit cycle (i.e., closed trajectory), and then the region exterior to the limit cycle.

If $a^2 < 4b$, the nature of the roots yields an unstable focus. If $a^2 > 4b$, then an unstable node results. Thus, in both cases, trajectories move away from the origin to the limit cycle.

We recognize that the Taylor series expansion is valid only in a neighborhood of the point p_0, q_0 (the origin in this example), and thus the analysis breaks down when we move much beyond that location.

If, however, we move to another arbitrary point in the phase plane (p_1, q_1), external to the limit cycle, expand the functions about that point in a Taylor series, we obtain the same expression for the functions as before, except we must substitute p_1, q_1 for p_0, q_0, and then linearize with respect to u, w to yield

$$u' = w + p_1$$

$$w' = ap_1 - bq_1 + aw - bu - aq_1^2 w + 2p_1^3 u + p_1 q_1^2$$

which has the form

$$u' = Aw + B$$

$$w' = Cw + Du + E$$

where $A = 1$, $B = p_1$, $C = a(1 - q_1^2)$, $D = 2p_1^3 - b$, $E = ap_1 - bq_1 + p_1q_1^2$.

By a translation of axes to the point p_1, q_1, we force the constant terms to vanish and the characteristic equations becomes

$$\begin{vmatrix} A - \lambda & 0 \\ C & D - \lambda \end{vmatrix} = \lambda^2 - (A + D)\lambda + AD = 0$$

and, therefore,

$$\lambda = A + D \pm 0.5[(A + D)^2 - 4AD]^{1/2}$$

A similar argument as before leads to the conclusion that all points move toward the limit cycle.

6.1.2 Liapunov Functions: The Mathematics

In the search for methods of determining if a system of equations is stable, Liapunov introduced what has become known as a Liapunov function. Unfortunately, these functions admit to having specific properties which are easily defined, yet their identity often remains elusive as there is, as yet, no universal method which allows their construction. That is, neither existence nor uniqueness is guaranteed.

We begin by defining a Liapunov function $V(\mathbf{x})$, where $\mathbf{x}$ is an n-vector $(x_1, \cdots, x_n)$ with the following properties:

1. $V(\mathbf{x})$ is continuous as are its first derivatives.
2. $V(\mathbf{0}) = 0$.
3. $V(\mathbf{x}) > 0$ for all $|\mathbf{x}| > 0$.
4. $\lim_{|x| \to \infty} V(\mathbf{x}) = \infty$ for all $\mathbf{x}$.

The function V is termed positive-definite whenever conditions 2 and 3 exist. Substituting the $<$ sign changes the definition to negative-definite; and if condition 3 uses the $\geq$ or $\leq$ sign the function is termed positive- or negative-semidefinite.

Typical Liapunov functions include:

$$V(\mathbf{x}) = x_1^2 + x_2^2$$

or

$$V(\mathbf{x}) = (x_1^2 + x_2^2)^2$$

and often can be identified as the energy of the system, although such a physical connotation is not necessary, nor is it always physically valid.

Other selections typically involve quadratic forms of the state variables.

A quadratic form is

$$V = \sum_i \sum_j a_{ij} x_i x_j$$

$$a_{ij} = a_{ji}$$

By Sylvester's theorem (see ref. 5) a quadratic form is positive-definite if and only if the determinants of *all* of its minors are positive. These determinants are

$$|a_{11}| \quad \begin{vmatrix} a_{11} & a_{12} \\ a_{21} & a_{22} \end{vmatrix} \quad \begin{vmatrix} a_{11} & a_{12} & a_{13} \\ a_{21} & a_{22} & a_{23} \\ a_{31} & a_{32} & a_{33} \end{vmatrix}, \text{ etc.}$$

The direct method of Liapunov generally follows the procedure of defining a Liapunov function *candidate* $V(\mathbf{x})$, forming the derivative $V'(\mathbf{x}) = W(\mathbf{x})$, substituting the derivatives of the state variables x_i into the expression for W, and applying the tests of the definitions to determine if we have selected a form of V wisely. Then by applying the tests for stability, conclusions are reached.

Examples

(a) $V(\mathbf{x}) = (x_1 + x_2)^2, \qquad n = 3$
positive-semidefinite as x_3 is arbitrary and x_1, x_2 can be zero
(b) $V(\mathbf{x}) = x_1^2 + x_2^2 + \cdot \cdot \cdot + x_n^2$
positive-definite for all n
(c) $V(\mathbf{x}) = a_{11}x_1^2 + a_{12}x_1x_2 + a_{21}x_1x_2 + a_{22}x_2^2$, where $a_{12} = a_{21}$
positive-definite by Sylvester's theorem.

$$|a_{11}| > 0, \quad \begin{vmatrix} a_{11} & a_{12} \\ a_{12} & a_{22} \end{vmatrix} > 0$$

implies $a_{11}a_{22} - a_{12}^2 > 0$

The main difficulty with the stability theorems of Liapunov is that they require the identity of an appropriate V, yet if the V selected does not satisfy the theorem's requirements, that does not mean that the system's stability or instability has been determined. All that has been accomplished is that one *candidate* for V has been eliminated. However, if the conditions of the theorem have been satisfied, the search is over and a determination is made.

We shall continue our discussion with the definition of stability classifications—namely absolute, asymptotic, and uniform stability on a local and a global basis. With these definitions, we will then examine examples of each type to illustrate the Liapunov method.

6.1.3 Types of Stability

The concept of stability is one of boundedness. That is, if a trajectory (solution) of the differential equation remained bounded, in Chapter 5 we accepted that definition as one which satisfied stability requirements. Now, however, we shall be more specific.

We shall concern ourselves with a system of differential equations which are expressible in terms of the first-order vector equation

$$\mathbf{dy}/dx = \mathbf{f}[x, \mathbf{y}(x)], \qquad x \geq 0, \qquad \mathbf{y} = (y_1, \cdots, y_n)$$

We shall also define our equilibrium point (singularity) as one at which the system satisfies

$$\mathbf{f}[x, \mathbf{y}(x_0)] = 0, \qquad \text{for all } x \geq x_0$$

Shifting coordinates by the quantity x_0, the equilibrium point can be made to coincide with the origin of the coordinate system.

Of interest to Liapunov was the question of the behavior of the trajectory $\mathbf{y}(x)$ if the initial state of the system does not exist at, or is perturbed from, an equilibrium point. Does the system return to the equilibrium point or not? To answer this question, a set of definitions of different types of stability was developed based on $x = 0$ representing the equilibrium point (through a translation of coordinates, if necessary).

The equilibrium point is *stable* at state $\mathbf{x}_0$ if, for each $\epsilon > 0$, there is a $\delta > 0$, $\delta = \delta(\epsilon, x_0)$, such that for all $x \geq x_0$ the trajectory $\mathbf{y}(x)$ remains within a circle (in two dimensions) or a ball/sphere (in n dimensions) of radius δ, that is

$$\|\mathbf{y}(x_0)\| < \delta \text{ whenever } \|\mathbf{y}(x)\| < \epsilon, \qquad \text{for all } x \geq x_0$$

where $\|y\|$ denotes the norm of y (the Euclidean norm is the distance between two points $\|y_1 - y_0\|$, such as the radius from the origin $\|y - 0\|$) (Figure 6-1).

In the definition, $\mathbf{y}(x)$ represents the reaction (response) of the system at state x (often, the independent variable x represents time), to a perturbation $\mathbf{y}_0 = \mathbf{y}(x_0)$. The interpretation of the mathematics involves establishing a bound on how far from the equilibrium point we are permitted to initially move (δ) due to the perturbation, so that for all subsequent motions the system response does not exceed the outer bound (ϵ). Clearly, this definition is consistent with our concepts previously studied.

The equilibrium point $\mathbf{y}(x_0)$ is termed *unstable* if it is not stable at x_0.

The equilibrium point is *uniformly stable* (Figure 6-2) in the interval $[x_0, \infty)$, if for every $\epsilon > 0$ there exists a $\delta(\epsilon) > 0$ such that the trajectory $\mathbf{y}(x)$, after

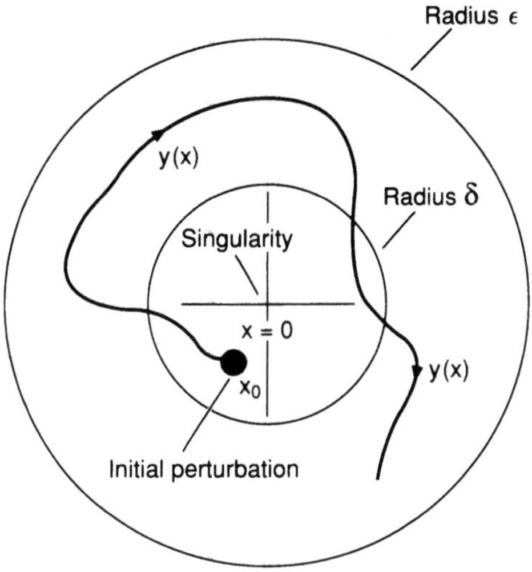

Figure 6-1. A stable trajectory.

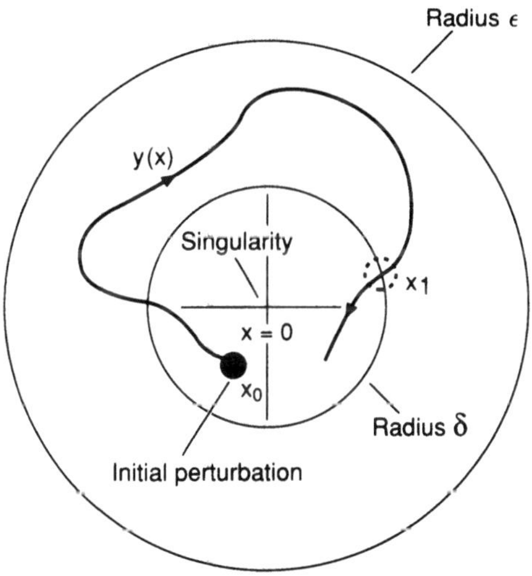

Figure 6-2. Uniformly stable trajectory.

state x_1 remains inside a ball for all $x_1 > x_0$. That is,

$$\|\mathbf{y}(x_1)\| < \delta(\epsilon)$$

provided that for $x_1 \geq x_0$

$$\|\mathbf{y}(x)\| < \epsilon$$

for all $x > x_1$.

Note that uniform stability does not require δ to be dependent upon the initial perturbation, as was required for the definition of stability.

These definitions allow for conditions under which the system is perturbed (i.e., moved away from an equilibrium point), yet still the system behavior remains stable as long as the trajectory does not stray too far from the equilibrium point. In a sense, we have described boundedness. Stability thus involves the need to restrict the ϵ, δ process to be dependent upon the initial state; whereas uniform stability involves all states being dependent upon a specific state other than the initial perturbed state.

Having studied the Van der Pol equation before, consideration of these definitions should be consistent with our prior findings. Since the definition of stability requires us to locate at least one ϵ, δ pair which is valid at the equilibrium point (the origin). Furthermore, since we know that a periodic solution (a limit cycle) will always result (except for the trivial solution $x_1 = x_2 = 0$) for any perturbation, if we select our δ too small, the trajectory will move outside the δ limit; therefore, the search for a suitable δ will be unsuccessful, and we will conclude the trajectory is unstable. However, if δ is large enough, encompassing at least the outer reaches of the Van der Pol limit cycle, it will be possible to find an ϵ satisfying the definition. The origin then is considered a stable equilibrium point.

Is this a contradiction? No, because the ϵ, δ process requires defining *suitable* values of these radii. Furthermore, the origin is not stable, but rather is unstable; the *system* is stable in the sense that all solutions end in a limit cycle around the origin and after some time t (or after some initial state $\mathbf{x}_0$) remain bounded. Thus, for the proper values of ϵ, δ, the Van der Pol equation is seen to be uniformly stable.

We further define *asymptotic stability* (Figure 6-3) of the equilibrium point $\mathbf{y}_0$ (the origin) at the initial state x_0 if it is both stable and the trajectory begins within a ball of radius δ_1 and tends toward zero (the origin) as the independent variable increases to infinity. That is, if

(a) $\mathbf{y}(x_0)$ is stable; and
(b) $\|\mathbf{y}(x_0)\| < \delta_1(x_0)$ while $\|\mathbf{y}(x)\| \to 0$ as $x \to \infty$.

We observe that a system with a limit cycle is excluded from the concept of asymptotic stability.

$\mathbf{y}_0$ is *uniformly asymptotically stable* (Figure 6-4) during the interval $[x_0, \infty)$ if it is uniformly stable during that same interval, and at state x_1 a number δ_1 can be found such that the trajectory's state at x_1 is contained within a ball of radius less than δ_1 and for all subsequent states $x_1 \geq x_0$, and the trajectory tends toward zero as the independent variable tends toward infinity. That is,

$$\|\mathbf{y}(x_1)\| < \delta_1, \qquad \text{for all } x_1 \geq x_0$$

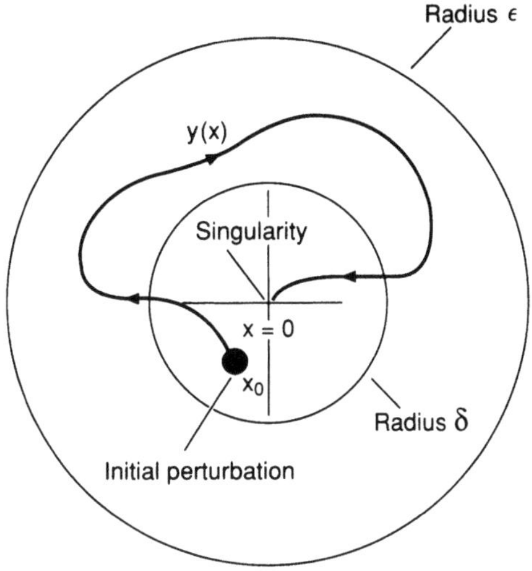

Figure 6-3. Asymptotically stable trajectory.

and

$$\|\mathbf{y}(x)\| \to 0, \qquad \text{as } x \to \infty$$

Note that an asymptotically stable equilibrium point allows a trajectory to wander outside the ball of radius δ_1 which exists at state x_0, but at some future state it will return to the ball and remain within the ball thereafter, eventually

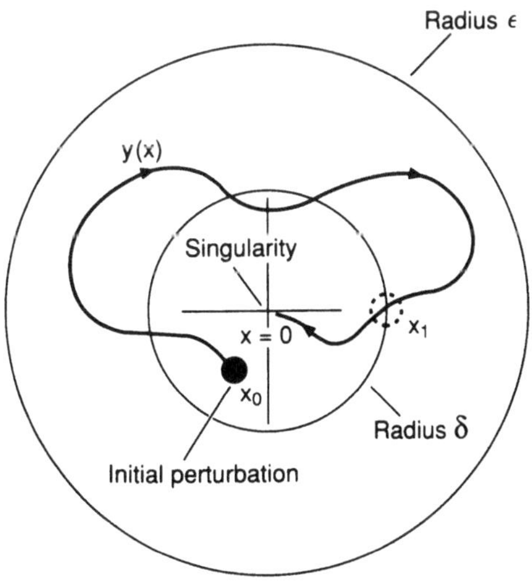

Figure 6-4. Uniformly asymptotically stable trajectory.

approaching the equilibrium point. This concept is often referred to in the literature as approaching arbitrarily close to the equilibrium point, since the mathematical definition involves reaching the equilibrium state at infinity. The concept of uniform asymptotically stable trajectories eliminates the dependence on the initial state of perturbation but rather involves establishing behavior characteristics of the system state at some point x_1 later on ($x_1 > x_0$).

Finally, we define the equilibrium point to be *globally asymptotically stable* at state $\mathbf{x}_0$ if after the initial state the trajectory $\mathbf{y}(x)$ approaches the equilibrium point as the independent variable x increases without limit.

These definitions refer to the behavior of the particular trajectory (solution) in a local neighborhood of the equilibrium point; whereas global definitions involve the entire state space of all trajectories as they tend toward or away from the equilibrium point.

6.1.4 Liapunov's Stability Criteria

Having defined a variety of stability types, we now examine a means of determining how to uncover which, if any, of these stability measures exist within our system. Returning to the definition of Liapunov's functions, defined in Section 6.1.2, we say:

> *The system is stable throughout the region where dV/dt is negative-semidefinite for all nonzero x and $t > 0$.*
>
> *The system is asymptotically stable in the region if dV/dt is negative-definite.*
>
> *The system is globally asymptotically stable if dV/dt is negative-definite in the entire state space governed by the variable x.*

It is often convenient to define $W(\mathbf{x}) \equiv dV(\mathbf{x})/dt$. Then we need to examine the behavior/definition of W and V. The use of the system energy for V is a good first start since this satisfies all conditions for V under the definition.

It is also important to recognize that if the conditions for V, W are valid in a region, the stability conclusions are also valid in that region.

Examples

1.

$$x_1' = -x_2 - x_1^3$$

$$x_2' = +x_1 - x_2^3$$

$$V(x_1, x_2) = x_1^2 + x_2^2$$

$$W(x) = dV/dx = (\partial V/\partial x_1)dx_1 + (\partial V/\partial x_2)dx_2$$

$$W = 2x_1x_1' + 2x_2x_2' = -2(x_1^4 + x_2^4)$$

Since $V > 0$, $V(0) = 0$, V is positive-definitive

Since $W < 0$, $W(0) = 0$, W is negative-definitive

Thus, the equilibrium point (0, 0) is asymptotically stable.

2. Linear oscillator:

$$y'' + y = 0$$

$$x_1' = x_2$$

$$x_2' = -x_1$$

$$V(x_1, x_2) = x_1^2 + x_2^2$$

$$W(x_1, x_2) = 0$$

Hence, the system is globally stable.

3.

$$x_1' = x_2 - ax_1(x_1^2 + x_2^2)$$

$$x_2' = -x_1 - ax_2(x_1^2 + x_2^2)$$

$$V(x_1, x_2) = x_1^2 + x_2^2$$

$$W(x_1, x_2) = -2aV^2$$

Since $V(0) = 0$, $V > 0$, $W < 0$, $\lim_{x \to \infty} V = \infty$

The system is globally asymptotically stable.

6.1.5 Liapunov's Instability Theorems

Theorem 1

If $W(x) < 0$ for all nonzero x, $W(0) = 0$ and $\text{Lim}\ W(x) \to -\infty$ for all x, then the system is unstable in the region where these conditions apply provided that $V(x)$ is not positive-semidefinite.

Theorem 2

The system is globally unstable if $V(x)$ is not positive-semidefinite everywhere.

6.2 CONSTRUCTION OF LIAPUNOV FUNCTIONALS

The Liapunov function is often considered as a generalized energy. Consequently, in mechanical systems the use of the position variable x_1 is tied to potential energy and the velocity x_2 to kinetic energy. Therefore, it is reasonable to construct a function $V(x_1, \cdots, x_n)$ in n-dimensional systems, and to seek means of assuring the minimum of these functions so as to bring the system V represents to an equilibrium state.

As we are willing to accept a quasiphysical interpretation of the Liapunov functionals, we can therefore be imprecise about the formation of the functionals in terms of energy. Thus, we are comfortable in using the *analogy* of energy and seeking functionals which are useful.

Consider the two-dimensional equations

$$x'' + f(x') + g(x) = 0$$

$$x'' + f(x)x' + g(x) = 0$$

$$x'' + f(x, x')x' + g(x) = 0$$

In each of these equations, we can use the state variable notation where $x_1 = x$ and $x_2 = x'$ and construct an energy functional of the type

$$V(x_1, x_2) = x_2^2/2 + U(x_1)$$

and examine its properties in keeping with Liapunov's theorems. Obviously, we need to assume various properties for U. It is not unreasonable to assume U is related to $g(x)$ and has the following properties (usually called a nonlinear spring)

$g(x)$ is continuously differentiable for $|x| < k, \quad k > 0$

$g(0) = 0$

$xg(x) > 0, \quad x \neq 0$

This then leads to selecting

$$U = \int_0^x g(s)ds$$

and, thus,

$$V = x_2^2/2 + \int_0^x g(s)ds$$

and

$$W = x_2x_2' + g(x_1)x_1'$$

The condition $xg(x) > 0$ requires that the functional relation behave as shown in Figure 6-5 and the integral is

$$\int_0^{x_1} g(s)ds > 0, \qquad 0 < |x_1| < k$$

and $V(x_1, x_2)$ is positive-definite.

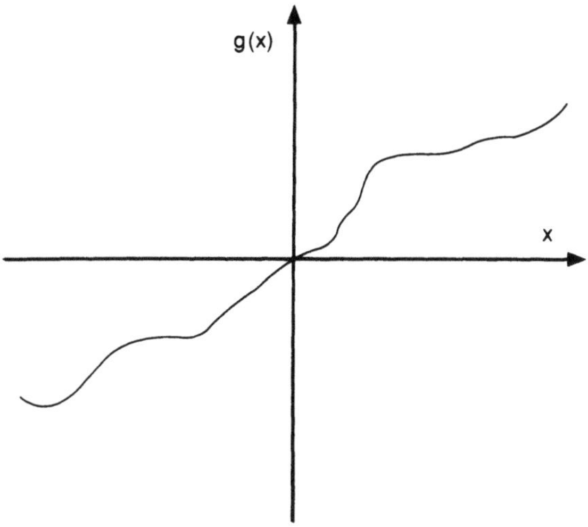

Figure 6-5. Characteristics of a nonlinear spring.

Other variations relate to using a modified $V(x_1, x_2)$, such as

$$V(x_1, x_2) = ag(x_1)x_2 + x_2^2/2 + \int_0^{x_1} g(s)ds$$

Example

Lienard's equation

$$x'' + x' + f(x) = 0$$

can be rewritten as

$$x_1' = x_2$$

$$x_2' = -f(x_1) - x_2$$

with the assumptions on the form of $f(x_1)$ being

$$|x| < k, \qquad k > 0$$

$$xf(x) > 0, \qquad x \neq 0$$

Then we select

$$V(x_1, x_2) = x_2^2/2 + \int_0^{x_1} f(s)ds$$

which is positive-definite for $|x_1| < k$ and all x_2, and

$$W(x_1, x_2) = x_2x_2' + f(x_1)x_1' = -x_2^2$$

Thus, W is negative-semidefinite since $W(x_1, 0)$ is zero on the x_1 axis; the system is stable, but not asymptotically stable.

Example

Consider the equation

$$x'' + g(x, x')x' + h(x) = 0$$

and let $h(x)$ have the properties of a nonlinear spring, while $g(x, x')$ represents a nonlinear damping term that is continuously differentiable and

$$g(x, x') > 0 \qquad \text{for all } (x, x') \neq (0, 0)$$

where $|x|, |x'| \leq k, < k$, so that we may assure that the damping term is positive-definite.

Then, we set $x = x_1$, $x_1' = x_2$ and obtain

$$x_1' = x_2$$

$$x_2' = -g(x_1, x_2)x_2 - h(x_1)$$

Construct the Liapunov function as

$$V(x_1, x_2) = x_2^2/2 + \int_0^{x_1} h(s)ds$$

and obtain

$$W(x_1, x_2) = x_2x_2' + h(x_1)x_1' = -g(x_1, x_2)x_2^2$$

Since $g(x_1, x_2)$ is positive-definite, $W(x_1, x_2)$ is negative-definite; thus, we conclude that the system is stable if x_1 is not explicitly present in $g(x_1, x_2)$, or asymptotically stable if it is.

6.3 THE LUR'E PROBLEM

The Lur'e problem, also known as the regulator problem, is concerned with establishing solutions to a class of systems rather than to an individual system. Further, the formulation of the problem includes control functions, and it is the inclusion of these direct or indirect controls which differs from many of the other techniques included in this text.

The problem statement is formulated in a general form and is quickly transformed into a combination of a linearized system on which nonlinear controls are applied. The analysis then proceeds by investigating the nature of the system stability characteristics under the influence of the control functions.

We begin by recognizing that ordinary differential equations, in their most

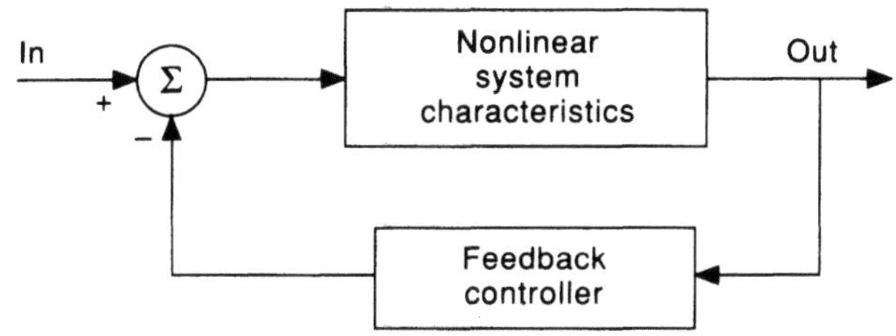

Figure 6-6a. Direct control.

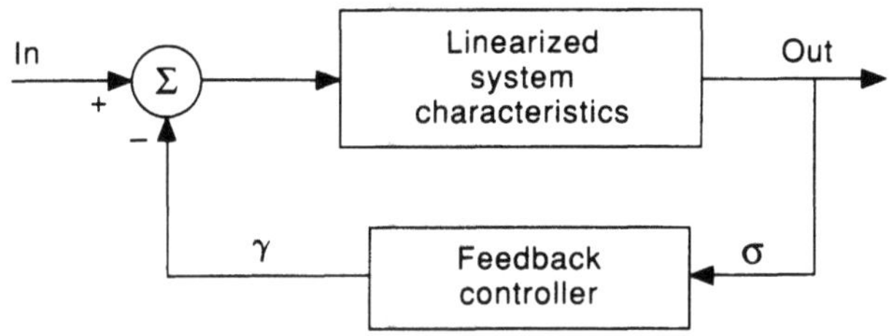

Figure 6-6b. Indirect control.

general form, are nonlinear. We observe that linear equations allow the stability characteristics of the system to be determined throughout the space of admissible solutions; whereas nonlinear equations (when solutions do exist) are usually localized. That is, we seek equilibrium or singular points and investigate the behavior of the system when subjected to a perturbation in the vicinity of the singularity. Such studies, when solutions in an analytic sense are readily obtainable even if they exist, involve the determination of asymptotic stability criteria.

Lur'e also concerned himself with the criteria which allowed a linear system with nonlinear controls to achieve asymptotic stability (Figure 6-6). He classified the control systems into two types: direct or indirect control. By direct control is meant the introduction of a feedback control signal σ, which operates directly on the system; whereas the indirect feedback signal σ operates on one or more system derivatives.

6.3.1 The Lur'e Problem in One Dimension

It is useful to consider a first-order system $x' = kx$ to illustrate the basics, where $k \neq 0$ and is a constant. Such a system with a direct control, takes the form

$$x' = kx + \phi(cx)$$

or in state variable form

$$x' = kx + \gamma$$

$$\gamma = \phi(\sigma)$$

$$\sigma = cx, \qquad c \neq 0$$

Thus, if we can express the characteristic nonlinear function ϕ mathematically, we can integrate the equation yielding

$$t - t_0 = \int_{x_0}^{x} ds/[ks + \phi(cs)]$$

In this context, it is necessary to establish certain conditions on the control function to assure that solutions to the system equation exist and are unique. The assumptions we make are:

$\phi(\sigma)$ is continuous and defined for all σ

$$\phi(0) = 0$$

$$\sigma\phi(\sigma) > 0, \qquad \sigma \neq 0$$

$$\int_0^{\pm\infty} \phi(\sigma)d\sigma = \infty$$

We note that the third property requires first and third quadrant characteristics as shown in Figure 6-7, and the second that the origin is a point on the characteristic curve of the nonlinear function. Note also the similarity with the nonlinear spring.

As the intent of the control problem is to achieve a return of the system to the singular or equilibrium point, we first observe that the singularity exists for $\gamma = 0$ (i.e., no control) and $x = 0$. This corresponds to $\gamma = \phi(0)$, which is obtained from

$$\gamma = \phi(\sigma) = 0 \qquad \text{and } \sigma = cx = 0$$

Since an equilibrium point corresponds to the system at rest (i.e., the velocity decreases until it reaches the equilibrium state) we recognize that a velocity x' always of opposite sign to the position vector x is adequate. Thus, we require

$$x' = f(x) = -x[kx + \phi(cx)] > 0$$

We note that if $k < 0$, the linear system with solution $x = c_1e^{kt}$ is already stable and, to make it more so, we continue to increase $f(x)$. As $-kx^2 > 0$,

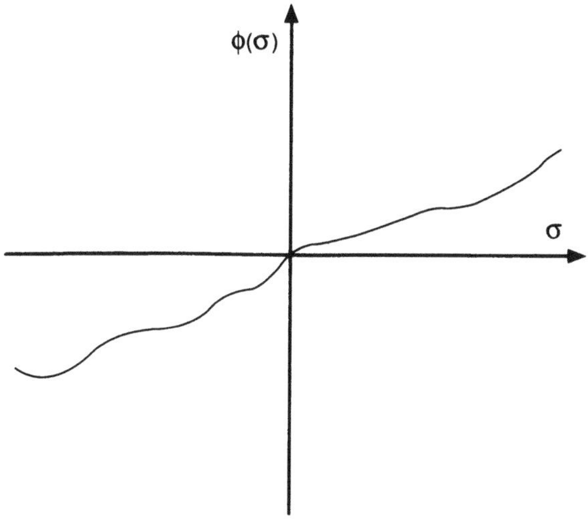

Figure 6-7. A nonlinear control function.

then, since

$$-[(cx)/c]\phi(cx) > 0$$

the only requirement we need impose is that $c < 0$.

Now, should $k > 0$, c is selected < 0, so that the graph $\phi(-cx)$ is below the line $y = kx$ for $x < 0$, and above the line for $x > 0$. With this convention, the control function will force the unstable system into stability as all solutions will tend toward the origin ($x = 0$) as $t \to \infty$.

We now turn to the issue of indirect control and examine the linear system

$$x' = -kx, \qquad k > 0$$

which is asymptotically stable. Using derivative control, we increase the order of the system by one and introduce the equivalent state variable equations

$$\begin{aligned} x' &= -kx - b\gamma \\ \gamma' &= \phi(\sigma) \\ \sigma &= cx - \rho\gamma \end{aligned} \tag{1}$$

which have the control parameters b, c, and ρ. Lur'e then posed the following problem:

Find the necessary and sufficient conditions to assure the asymptotic stability of this system independent of initial conditions and without constraining the choice of admissible control functions $\phi(\sigma)$.

Posed in this way, Lur'e sought conditions for the system's absolute stability. In practice, however, there are usually bounds on the system as well as the control variables. Thus, if one need not be so mathematically precise, the Lur'e problem actually provides more than one needs.

Let us redefine our system to allow for certain simplifications in concept. First, we recognize that b is a scale factor and we can normalize variables. Thus, our system becomes

$$x' = -kx + \gamma$$

$$\gamma' = \phi(\sigma)$$

$$\sigma = cx - \rho\gamma$$

with a singular point corresponding to $x = 0$, $\gamma = 0$.

Second, let us make the linear transformation from x, γ coordinates to y, σ using

$$y = -kx + \gamma, \qquad \sigma = cx - \rho\gamma$$

to obtain

$$y' = -ky + \phi(\sigma)$$

$$\sigma' = cy - \rho\phi(\sigma)$$

The linear transformation has a unique solution provided its determinant is nonsingular. That is,

$$\begin{vmatrix} -k & 1 \\ c & -\rho \end{vmatrix} = k\rho - c \neq 0$$

Further, as we have assumed that only one singular point exists and is located at the origin in the x, γ coordinate system, this must be the case also in the y, σ system

$$-ky + \phi(\sigma) = 0$$

$$cy - \rho\phi(\sigma) = 0$$

We recall that $\phi(\sigma) = 0$ when $\sigma = 0$, which corresponds to $y = 0$. Thus, $(y, \sigma) = (0, 0)$ is the only singularity.

Thus, we have established conditions under which we have a linear system with nonlinear direct or indirect controls and are seeking means of assuring absolute stability. It is possible to modify the equations further to seek asymptotic stability. This can be accomplished by using the geometrical concepts of Liapunov.

We digress for a moment to introduce this geometrical concept. In essence, the y, σ system is a second-order nonlinear (later it will be expanded to

n^{th} order) system, and we seek contours on the y–σ plane such that we continuously approach the origin. In the one-dimensional examples of Chapter 1, we can construct such contours where we recognize that we seek $x \to 0$ as $t \to \infty$ for asymptotic stability.

Using the equation

$$x' + x = 0$$

which has the solution $x = ce^{-t}$, we select the Liapunov candidate $V = x^2$, which corresponds to $V' = -x^2/2$. This clearly satisfies Liapunov's conditions for asymptotic stability. Now with reference to Figure 6-8, we have drawn solution curves and lines of constant V which depict the solution trajectories moving from higher to lower values of V.

Considering the equation

$$x' + |x| = 0$$

which is obviously nonlinear, by separating it into two parts ($x > 0$, $x < 0$), we can again use $V = x^2$, where $V' = -2x|x|$. Constraining solution curves to the upper-half plane ($x > 0$), corresponds to solution trajectories $x = ce^{-t}$ and for the lower-half plane ($x < 0$), to $x = ce^{+t}$. These yield unstable solutions for $x < 0$ and stable solutions for $x > 0$ (Figure 6-9).

Extending this concept to two dimensions, we construct Liapunov functions, such as

$$V = x_1^2 + x_2^2$$

which correspond to circles on the x_1–x_2 plane. For asymptotic stability, we again seek conditions under which the solution trajectories move toward the

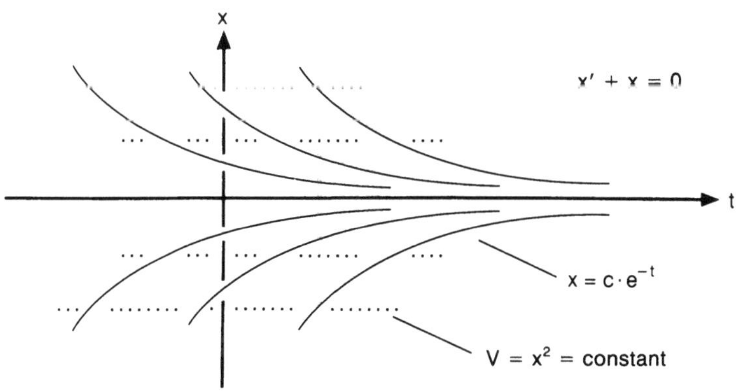

Figure 6-8. One-dimensional linear system and a Liapunov function.

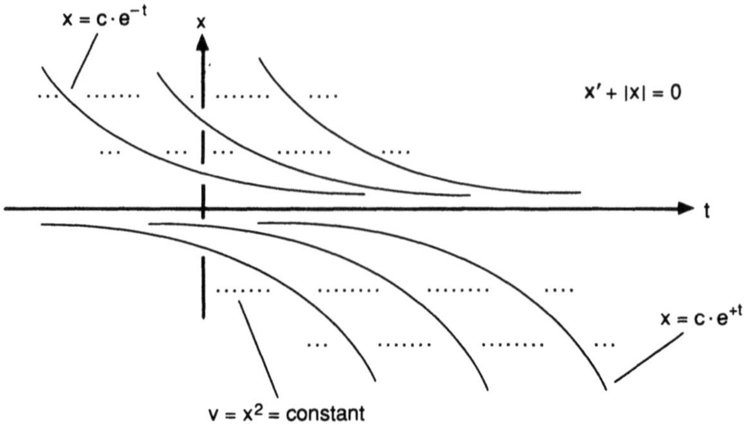

Figure 6-9. One-dimensional nonlinear system Liapunov function.

origin (i.e., always inward or at a constant radius, but never away from the origin). This concept can obviously be extended to n-dimensions with surfaces of constant V (Figure 6-10).

Returning now to the one-dimensional Lur'e problem, we seek Liapunov

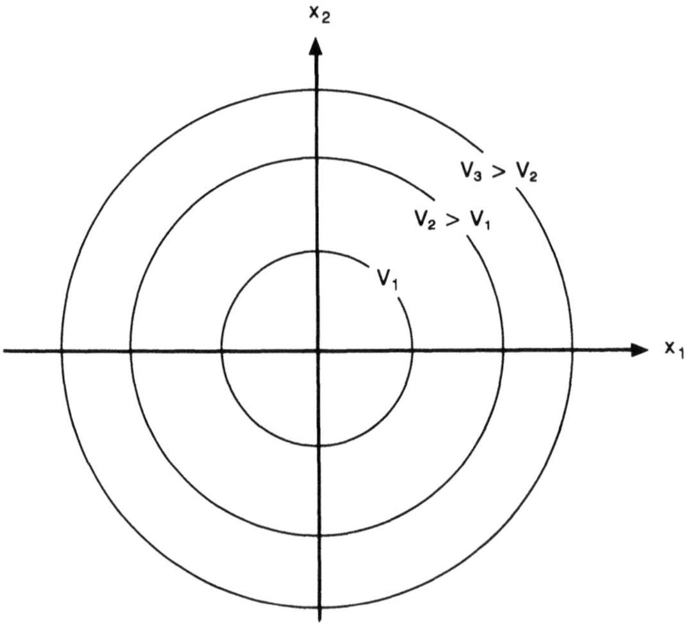

Figure 6-10. Two-dimensional Liapunov function.

functionals which also have these properties. As the states are y, σ, we select

$$V = my^2 + \Phi(\sigma), \qquad m > 0, \qquad m = \text{constant}$$

$$\Phi(\sigma) = \int_0^{\sigma} \phi(s)ds$$

and seek solutions such that $V = h = \text{constant} > 0$. These correspond to

$$y = \pm\{[h - \Phi(\sigma)]/m\}^{1/2}$$

and have contours symmetric with the σ axis as shown in Figure 6-11.

For asymptotic stability $V' < 0$

$$V' = 2myy' + \phi(\sigma)\sigma'$$
$$= -2mky^2 + (2m + c)y\phi(\sigma) - \rho\phi^2(\sigma)$$

which is negative-definite since $mk > 0$, provided that

$$\phi(\sigma)[(2m + c)y - \rho\phi(\sigma)] < 0$$

and asymptotic stability is assured.

Example

For the case of direct control

$$x_i' = -k_i x_i - b_i\phi(\sigma), \qquad i = 1, 2$$

$$\sigma = c_1x_1 + c_2x_2$$

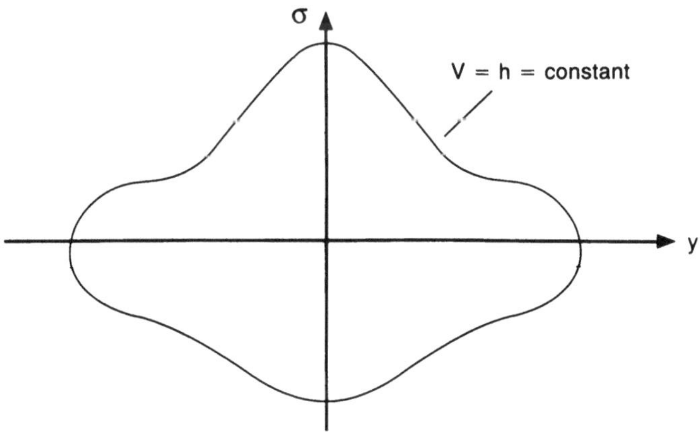

Figure 6-11. Lur'e system with nonlinear controller.

Select

$$V(x_1, x_2) = x_1^2 + x_2^2 + \Phi(\sigma)$$

and establish conditions under which absolute stability is assured.

Answers

The origin is the only singular point if

$$-k_i x_i + b_i\phi(c_1x_1 + c_2x_2) = 0, \qquad i = 1, 2$$

and $V' < 0$ requires

$$b_1c_1 + b_2c_2 = (b_1 - c_1/2)^2/2k_1 + (b_2 - c_2/2)^2/2k_2$$

6.3.2 The General Lur'e Problem

These concepts are usually included in the literature with vector/matrix notation, creating a means of analyzing n^{th} order solutions. In such a case, the system again offers a feedback system of the type shown in Figure 6-12.

The linearized system is represented by $\mathbf{x}' = \mathbf{A}\mathbf{x}$, and the direct controlled system is then represented by

$$\mathbf{x}'(t) = \mathbf{A}\mathbf{x}(t) + \mathbf{b}\gamma(t), \qquad t \geq 0$$

$$\sigma(t) = \mathbf{c}^*\mathbf{x}(t) + d\gamma(t)$$

$$\gamma(t) = -\phi[t, \sigma(t)]$$

where $\mathbf{A}$ is an $n \times n$ matrix; $\mathbf{b}$, $\mathbf{c}$, and $\mathbf{x}$ are n-vectors; d, σ, and γ are scalars; and the superscript * denotes the transpose operation. It is also assumed that ϕ is continuous in both arguments. We also assume that the solution is unique for each initial condition $\mathbf{x}(0)$, and that the solution depends continuously on $\mathbf{x}(0)$.

We make the following further assumptions:

1. $\mathbf{A}$ is a stability (or Hurwitz) matrix (i.e., all eigenvalues have negative real parts).
2. The nonlinearity $\phi(t, \sigma)$ is zero when $\phi(t, 0)$ and $\sigma\phi(t, \sigma) \geq 0$.

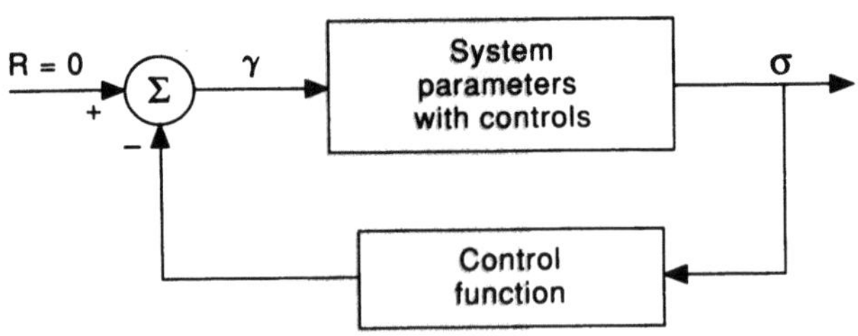

Figure 6-12. The Lur'e problem.

The generalized Lur'e problem is then posed as

Given a system formulated as above, find the necessary and sufficient conditions on the system parameters which ensure that the equilibrium point $\mathbf{x} = \mathbf{0}$ *is globally asymptotically stable for any admissable nonlinearity* ϕ.

Associated with Lur'e's problem are two conjectures, both of which have been proven false by counterexamples and are included here because of their prominence in the early literature. They are the Aizerman and Kalman conjectures.

We begin by defining the concept of a sector. A function $Z(t)$, satisfying a condition such as

$$k_1\sigma^2 \leq \sigma Z(t) \leq k_2\sigma^2$$

for all σ, where k_1 and k_2 are real numbers, is said to lie in a sector $[k_1, k_2]$ (Figure 6-13).

Suppose $d = 0$ and ϕ is within the sector $[k_1, k_2]$, where

$$k_1\sigma^2 \leq \sigma\phi(t, \sigma) \leq k_2\sigma^2$$

$$k_2 \geq k_1 \geq 0$$

and for each k_1, k_2 the matrix $\mathbf{A} - \mathbf{bc}^*k$ is Hurwitz. Then Aizerman proposed that the system equilibrium point ($\mathbf{x} = \mathbf{0}$) is globally asymptotically stable.

Had Aizerman's conjecture been true, then the study of nonlinear systems, under his hypothesis, would have been reduced to the study of linear systems—a very tempting result.

Kalman proposed that if

$$k_3 \leq \partial\phi(t, \sigma)/\partial\sigma \leq k_4$$

and $\mathbf{A} - \mathbf{bc}^*k$ is Hurwitz for all $k \in [k_3, k_4]$, then the nonlincar system is globally asymptotically stable. Unfortunately, this thought was also shown to be invalid.

In order to proceed, we require the statement of several theorems which

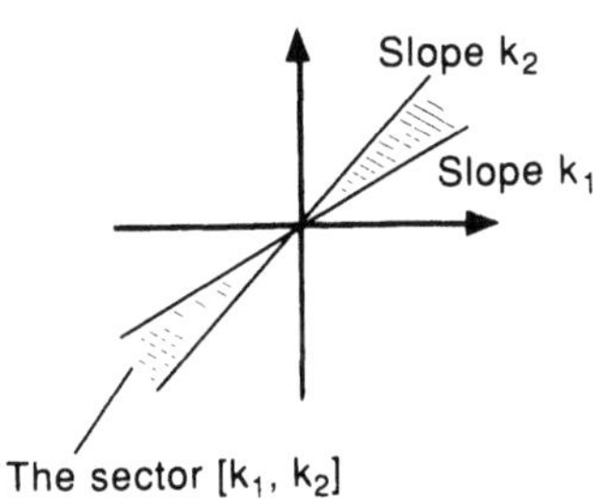

Figure 6-13. A sector.

set forth the hypotheses and conclusions under which the Lur'e problem can be solved. In essence, the proofs of these theorems use Liapunov functions. For proofs and the matrix algebra, the reader is directed to refs. 1 and 2, and Atherton in Related Literature. We will present the theorems and an example of their individual application.

6.3.2.1 Stability of the General Linear System. The analysis of this general linear system begins by assigning a control function ϕ, which operates on the feedback signal σ to yield a control signal $\phi(\sigma)$. Then by assuming different forms of Liapunov functions V to assure asymptotic stability (i.e., $V > 0$, $V' < 0$), conditions are imposed upon the system parameters. We note first that the linear system $\mathbf{x}' = \mathbf{Ax}$ is absolutely stable (and hence asymptotically stable as well) if $\mathbf{A}$ is Hurwitz. Thus, a Liapunov function of the form

$$\mathbf{V} = \mathbf{x^*Px}$$

will be common to all such systems, and modifications to the system by means of the controller will require appropriate modifications to the additive terms to the Liapunov function.

Let us examine the system $\mathbf{x}' = \mathbf{Ax}$, where $\mathbf{A}$ is Hurwitz, and show that the Liapunov function $V(\mathbf{x}) = \mathbf{x^*Px}$ establishes asymptotic stability for the system, where $\mathbf{P}$ is a real symmetric matrix (i.e., satisfies Sylvester's theorem).

Let

$$\mathbf{V} = \mathbf{x^*Px}$$

Then

$$\begin{aligned}\mathbf{V}' &= \mathbf{x^{*\prime}Px} + \mathbf{x^*Px'}\\ &= \mathbf{x^*(A^*P + PA)x}\end{aligned}$$

we define

$$-\mathbf{Q} = \mathbf{A^*P} + \mathbf{PA}$$

to be the Liapunov matrix equation and obtain

$$\mathbf{V}' = -\mathbf{x^*Qx}$$

(We refer the reader to ref. 2 for the proof of the following theorem.)

Theorem

Given an $n \times n$ matrix $\mathbf{A}$ *satisfying the Liapunov matrix equation, then the following statements are all equivalent:*

(a) All eigenvalues of $\mathbf{A}$ *have negative real parts.*

(b) *Some positive-definite $n \times n$ matrix exists such that*

$$-Q = A^*P + PA$$

*has a unique solution for **P**, and that **P** is then positive-definite.*

(c) *For every $n \times n$ positive-definite matrix Q, $-Q = A^*P + PA$ has a unique solution for **P**, and that **P** is positive-definite.*

Example

Select

$$\mathbf{A} = \begin{vmatrix} -2 & 1 \\ -1 & 1 \end{vmatrix} \quad \mathbf{Q} = \begin{vmatrix} q_1 & 0 \\ 0 & q_2 \end{vmatrix} \quad \mathbf{P} = \begin{vmatrix} p_1 & p_2 \\ p_1 & p_3 \end{vmatrix}$$

and substitute into $-\mathbf{Q} = \mathbf{A^*P} + \mathbf{PA}$ yielding

$$\begin{vmatrix} -q_1 & 0 \\ 0 & -q_2 \end{vmatrix} = \begin{vmatrix} -2 & -1 \\ 1 & 1 \end{vmatrix} \begin{vmatrix} p_1 & p_2 \\ p_2 & p_3 \end{vmatrix} + \begin{vmatrix} p_1 & p_2 \\ p_2 & p_3 \end{vmatrix} \begin{vmatrix} -2 & 1 \\ -1 & 1 \end{vmatrix}$$

$$= \begin{vmatrix} -4p_1 - 2p_2 & p_1 - p_2 - p_3 \\ p_1 - p_2 - p_3 & 2p_2 + 2p_3 \end{vmatrix}$$

which yields the following simultaneous equations

$$-q_1 = -4p_1 - 2p_2$$

$$0 = p_1 - p_2 - p_3$$

$$-q_2 = 2p_2 + 2p_3$$

Now, by selecting values of q_1, q_2 we obtain results which yield conclusions about the eigenvalues of **A** and whether or not **P** is positive-definite.

(a) If $q_1 = q_2 = 1$ (i.e., **Q** is selected to be the identity matrix **I**), then we obtain

$$p_1 = -\tfrac{1}{2}, \quad p_2 = \tfrac{3}{2}, \quad p_3 = -2$$

which yields a matrix for **P** which is not positive-definite. Hence, not all eigenvalues of **A** have negative real parts and **A** is not Hurwitz.

(b) Suppose

$$\mathbf{A} = \begin{vmatrix} -4 & 2 \\ 2 & 1 \end{vmatrix}$$

then the simultaneous equations become

$$-q_1 = -8p_1 - 4p_2$$

$$0 = 2p_1 + p_2 - 2p_3$$

$$-q_2 = 4p_2 + 2p_3$$

which reduce to the condition

$$q_1 + 4q_2 = 0$$

Thus, if q_1, q_2 do not satisfy this requirement, there can be no solution and hence not all eigenvalues of **A** have negative real parts.

(c) If $\mathbf{A} = \begin{vmatrix} 2 & -2 \\ 3 & -5 \end{vmatrix}$ and $\mathbf{Q} = \begin{vmatrix} -1 & 0 \\ 0 & -1 \end{vmatrix} = -\mathbf{I}$

then we conclude that

$$\mathbf{P} = \begin{vmatrix} +3\frac{5}{8} & -1\frac{3}{8} \\ -1\frac{3}{8} & +\frac{3}{4} \end{vmatrix}$$

and since **P** is positive-definite, all eigenvalues of **A** have negative real parts. Hence, the linear system is both absolutely and asymptotically stable.

Now we examine the introduction of a controller and study the system's required characteristics through the use of Liapunov functions. That is, we return to the Lur'e problem and construct appropriate Liapunov functions, so that we can assure asymptotic stability. We begin again with the linearized, dynamical system without control and consider first

$$\mathbf{V} = \mathbf{x}^*\mathbf{P}\mathbf{x} + \alpha\Phi(\sigma)$$

and then

$$\mathbf{V} = \mathbf{x}^*\mathbf{P}\mathbf{x} + \alpha\Phi(\sigma) + \beta\sigma^2$$

where

$$\Phi(\sigma) = \int_0^\sigma \varphi(s)ds$$

The derivation of $\mathbf{V}'$ in conjunction with the Lur'e system formulation using $\mathbf{V} = \mathbf{x}^*\mathbf{P}\mathbf{x}$ yields a circle criterion analogous to the Nyquist criterion. Following with the added term $\alpha\Phi(\sigma)$ generates a more limited theorem (Lur'e–Postnikov) with the requirement that $d = 0$ and φ be time-invariant; as the circle criterion is the more useful and less constraining, we shall not pursue the Lur'e–Postnikov result further, but merely repeat the theorem below. Lastly, we use the Liapunov function of the form $\mathbf{V} = \mathbf{x}^*\mathbf{P}\mathbf{x} + \alpha\Phi(\sigma) + \beta\sigma^2$ and obtain the results attributable to Popov.

Theorem (Lur'e–Postnikov)

Let φ be time-invariant in the sector $[0, \infty)$ and in the Lur'e formulation, $d = 0$. Using the Liapunov function $\mathbf{V} = \mathbf{x}^*\mathbf{P}\mathbf{x} + \alpha\Phi(\sigma)$, let Q be the positive-definite Liapunov matrix and $\mathbf{P}$ the solution of

$$-Q = A^*P + PA$$

Let

$$v = Pb - 0.5\alpha A^*c$$

then

$$ac^*b - v^*Q^{-1}v > 0$$

is a sufficient condition for the absolute stability of the Lur'e system.

6.3.3 The Circle Criterion

Using the Lur'e formulation for direct control (Figure 6-14)

$$\mathbf{x}' = \mathbf{A}\mathbf{x} + \mathbf{b}\gamma$$
$$\gamma = -\phi(\sigma)$$
$$\sigma = \mathbf{c}^*\mathbf{x} + d\gamma$$

and the Liapunov function $V = \mathbf{x}^*\mathbf{P}\mathbf{x}$, we seek conditions under which $V' < 0$ throughout the real space, so that the system is absolutely stable. The derivation follows that presented by Vidyasagar and Lefschetz (refs. 1 and 2) in

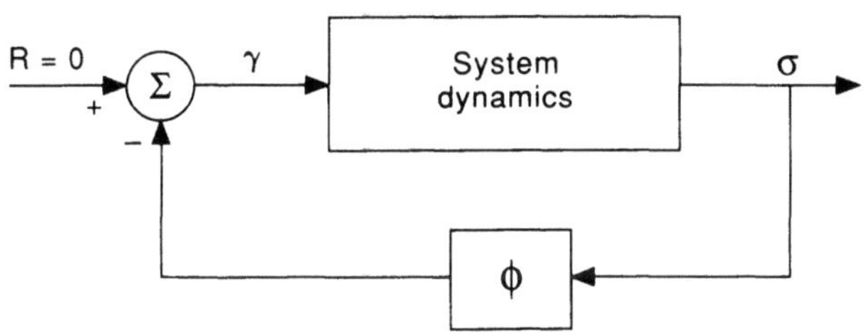

Figure 6-14. Lur'e system, direct control.

which it is shown that:

> If the matrix **A** is Hurwitz, the vector **b** is such that (**A**, **b**) is controllable (i.e., Rank $[\mathbf{b}|\mathbf{Ab}| \cdots |\mathbf{A}^{n-1}\mathbf{b}|] = n$), **Q** and **P** exist as positive-definite matricies, then
>
> $$Re\ h(j\omega) > 0$$
>
> where the system open-loop transfer function is $g(s)$ and
>
> $$h(s) = 1 + kg(s)$$

Using the relationship $Re\ h(j\omega) > 0$, and the fact that the nonlinearity lies in the sector $[0, k]$, then it can be shown that a sufficient condition for the absolute stability of the Lur'e system is

$$Re[1 + kg(j\omega)] > 0$$

provided that

$$1 + kd > 0$$

We note that if the nonlinearity is in the sector $[0, \infty)$, then

$$\lim_{k \to \infty} (1/k)Re[1 + kg(j\omega)] = Re\ g(j\omega) > 0$$

By continuing to examine properties of the system transfer function, the forgoing theorem's requirement for absolute stability, the concept of a sector, and the linear system's Nyquist criteria, it is possible to reduce the analysis into the circle criterion, which can be stated as the following theorem.

Theorem

Given the Lur'e problem formulation for direct control where **A** has no eigenvalues on the imaginary axis, η eigenvalues with positive real parts, and ϕ lies in the sector $[\alpha, \beta]$. Then any of the following four statements is a sufficient condition for absolute stability:

(a) If the Nyquist plot of $G(j\omega)$ does not enter the disk $D(\alpha, \beta)$ and encircles the disk η times in the counterclockwise direction, where $0 < \alpha < \beta$.
[Note: a disk $D(\alpha, \beta)$ is a circle in the complex plane which passes through the points $-1/\alpha + j0$ and $-1/\beta + j0$ and is centered on the real axis. See Figure 6-15.]

(b) If the Nyquist plot of $G(j\omega)$ lies in the half plane, $Re\ s > -1/\beta$, where $0 = \alpha < \beta$.

(c) If $\alpha < 0 < \beta$, the Nyquist plot lies in the interior of the disk $D(\alpha, \beta)$.

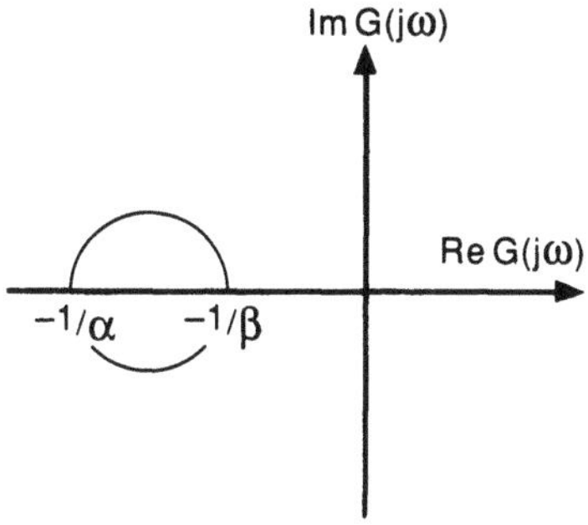

Figure 6-15. The disc, $D(\alpha,\beta)$.

(d) If $\alpha < \beta < 0$, then by substituting $G = -G$, α with $-\beta$ and β with $-\beta$, condition (a) applies.

Lastly, we note that if $\alpha = \beta$, the disk becomes the $-1/\alpha + j0$ point and the circle criterion becomes the sufficiency portion of the Nyquist criterion.

Example

Consider the linear oscillator system (which can represent the nonlinear pendulum in linearized form),

$$y'' + y = 0$$

which has the solution $y = c \sin (t + \phi)$ and is known to have a limit cycle which is depicted on the phase plane by the circle of radius c as shown in Figure 6-16.

Using state variables, where $x_1 = y$

$$\begin{vmatrix} x_1' \\ x_2' \end{vmatrix} = \begin{vmatrix} 0 & 1 \\ 1 & 0 \end{vmatrix} \begin{vmatrix} x_1 \\ x_2 \end{vmatrix}$$

which does not yield a Hurwitz matrix, as expected.

Consider, however, the same system with damping

$$y'' + y' + y = 0$$

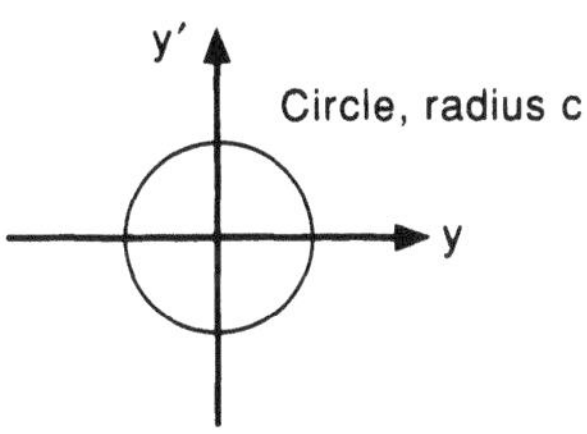

Figure 6-16. The limit cycle for a linear oscillator.

which has the matrix

$$A = \begin{vmatrix} 0 & 1 \\ -1 & -1 \end{vmatrix}$$

Both eigenvalues have a negative $\frac{1}{2}$ real part; thus, the matrix is Hurwitz and the system is stable. Suppose we introduce a forcing function proportional to y of the form $y \sin \omega t$. Then our system equation has the form of a Mathieu equation

$$y'' + y' + (1 - a \sin \omega t)y = 0$$

and the Lur'e problem formulation can be introduced where

$$b = \begin{vmatrix} 0 \\ 1 \end{vmatrix} \quad c = \begin{vmatrix} 1 \\ 0 \end{vmatrix} \quad d = 0$$

$$\phi = \sigma a \sin \omega t$$

The system transfer function is

$$G(s) = \frac{1}{s^2 + s + 1}$$

and ϕ lies in the sector $[-a, a]$, which is defined by the coefficients of the nonlinear term.

Hence, the disk is centered at the origin and has radius $= 1/a$. Since the requirement of the circle criterion is that the Nyquist plot remain interior to the circle in this case, all we need establish is the maximum value of $G(j\omega)$ as ω varies between $0, \infty$ to assure that the graph of $G(j\omega)$ does not extend beyond the circle perimeter.

$$G(j\omega) = \frac{1}{1 - \omega^2 + j\omega} = \frac{(1 - \omega^2) - j\omega}{(1 - \omega^2)^2 + \omega^2}$$

The modulus of G is $(1 - \omega^2 + \omega^4)^{-1/2}$ and by forming $dG/d\omega = 0$ we obtain minimum/maximum values of the modulus at $\omega = 0, \pm(0.5)^{1/2}, \infty$ from which we conclude that the values of the modulus are

ω	0	$\pm\sqrt{2}$	∞
$\lvert G(j\omega)\rvert$	± 1	$\pm 2\sqrt{3}/3$	0

Thus, the modulus maximum value requires the disk radius $1/a > 2\sqrt{3}/3$ or $a > \sqrt{3}/2$.

6.4 THE POPOV CRITERION

By slightly reformulating the Lur'e problem, Popov was able to obtain additional criteria for analyzing system stability characteristics.

Lur'e formulation	Popov formulation
$x' = Ax - \gamma b$	$x' = Ax - b\phi(\sigma)$
$\gamma' = \phi(\sigma)$	$\gamma' = \phi(\sigma)$
$\sigma = c^*x - \rho\gamma$	$\sigma = c^*x - \delta\gamma$

In the above formulations, $\mathbf{x}$, $\boldsymbol{\varphi}$, $\boldsymbol{\sigma}$, γ are vectors, b, ρ, δ scalars, $\mathbf{A}$, $\mathbf{c}$ matricies, and $\mathbf{A}$ is a Hurwitz matrix. The notation $\mathbf{c}^*$ denotes the transpose of the matrix $\mathbf{c}$, while $\mathbf{x}'$ denotes the time derivative. Furthermore, the nonlinearity $\boldsymbol{\varphi}$ has the properties

$$\boldsymbol{\varphi}(\mathbf{0}) = \mathbf{0}$$

$$\boldsymbol{\sigma\varphi}(\boldsymbol{\sigma}) > \mathbf{0}$$

$$\boldsymbol{\Phi}(\boldsymbol{\sigma}) = \int_0^{\sigma} \boldsymbol{\varphi}(s)ds, \qquad \boldsymbol{\Phi}(\boldsymbol{\sigma}) \to \infty \text{ as } \boldsymbol{\sigma} \to \pm\infty$$

The Lur'e and Popov problems are both concerned with determining absolute stability criteria for their system formulations. Note that in both formulations, the singular point corresponds to $\mathbf{x} = \mathbf{0}$, $\boldsymbol{\sigma} = \mathbf{0}$. Popov's system yields *direct* controls when $\delta = 0$, since the system equation can be rewritten as

$$\mathbf{x}' = \mathbf{Ax} - b\boldsymbol{\varphi}(\mathbf{c}^*\mathbf{x})$$

For indirect control (i.e., $\delta \neq 0$) a necessary condition for absolute stability is $\delta > 0$. See Lefschetz (ref. 2, Chapter 7) for other related results concerning invariance of coordinate systems and equivalence of system formulations.

By defining the linear system's transfer function $G(j\omega)$

$$G(j\omega) = c^*A_{j\omega}^{-1}b + \delta/j\omega$$

and where $A_{j\omega}^{-1}$ denotes the use of $s = j\omega$ in the inverse matrix, Popov was able to prove two important theorems:

Theorem 1

A sufficient condition for the absolute stability of the Popov system formulation is

$$Re[(1 + j\omega q)G(j\omega)] = Re[(1 + j\omega q)(c^*A_{j\omega}^{-1}b)] + q\delta \geq 0$$

where $q > 0$ and ω is real.

Theorem 2

If the Popov system formulation can be investigated using Liapunov function $V(x, \sigma)$ of the form where it is "quadratic in x, $\sigma + \beta \int_0^\sigma \phi(s)ds$," so that the system is absolutely stable, then there exists a $q \geq 0$ such that

$$Re[(1 + j\omega q)G(j\omega)] \geq 0$$

In the first theorem, we find there is an analogy with the Nyquist plot, in which we investigate the variation of the system transfer function with variations in gain and frequency on the *Re–Im* plane. The Popov plane examines the transfer function on the *Re–ωIm* plane.

By establishing the Popov inequality requirement, the straight line passing through the origin places an upper bound on the system transfer function to assure absolute stability. Consider the inequality in the theorem, and let us create two real rational functions $S_1(\omega)$ and $S_2(\omega)$ continuous in ω. Then we can write

$$S_1(\omega) + j\omega S_2(\omega) = c^*A_{j\omega}^{-1}b + \delta/j\omega$$

and the inequality takes the form

$$S_1(\omega) - q\omega S_2(\omega) \geq 0$$

which is the line shown in Figure 6-17 when the equality sign holds. We note that an upper bound for the line is the vertical or imaginary axis (to prevent the

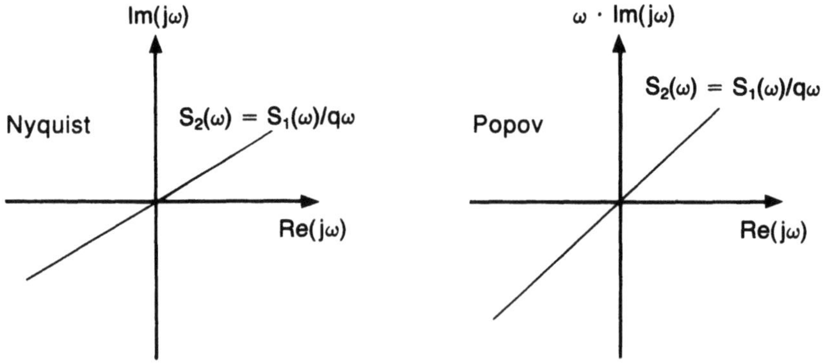

Figure 6-17. Nyquist and Popov boundaries.

inequality from becoming negative), and that all admissable plots will be, at most, tangent to this line or will lie below it.

Thus, the analysis of a *nonlinear* system allows the construction of a Popov plot and/or the evaluation of a class of Liapunov functions. Thus, it becomes necessary to restrict the nonlinearity to a sector defined by the real axis and the line passing through the origin with slope $1/q\omega$.

This can be expressed in terms of the following theorem.

Theorem 3

A sufficient condition for the Popov system formulation to be absolutely stable for all nonlinearities ϕ within the sector [0, k] is that

$$Re[(1 + j\omega q)G(j\omega)] + 1/k > 0$$

for all ω.

Example

For the linear system $G(s) = K/s(s + a)^2$ with the nonlinearity consisting of a relay with deadband of width $2d$ and output signal level at $\pm h$, find conditions for absolute stability (see Figure 6-18).

Since

$$G(j\omega) = K/j\omega(j\omega + a)^2$$

$$ReG(j\omega) = -2a/(a^2 + \omega^2)^2$$

$$ImG(j\omega) = -\omega(a^2 - \omega^2)/(a^2 + \omega^2)^2$$

and the Popov criterion of Theorem 1 requires

$$Re\,[(1 + j\omega q)G(j\omega)] \geq 0$$

or

$$\frac{2a(a^2 + \omega^2)^2}{\omega^2(\omega^2 - a^2)} \geq q$$

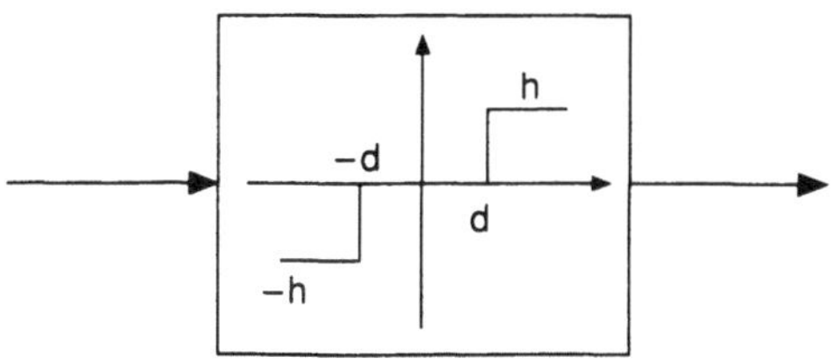

Figure 6-18. Characteristic of a relay with a deadband.

and if we normalize so that $q_1 = q/a$ and $\omega_1 = \omega/a$,

$$\frac{2(1 + \omega_1^2)^2}{\omega_1^2(\omega_1^2 - 1)} \geq q_1 > 0$$

Thus, the linear system is absolutely stable in the Popov sense for all ω_1 satisfying the above relationship.

Now to define the sector, we introduce the relationship from Theorem 3:

$$Re[(1 + j\omega q)G(j\omega)] + 1/k \geq 0$$

which leads to

$$1 + \frac{\omega^2 q(a^2 - \omega^2) - 2a}{(a_2 + \omega^2)^2} + \frac{1}{k} > 0$$

or

$$k > -1 + [2a + \omega^2 q(\omega^2 - a^2)]/(a^2 + \omega^2)^2$$

and ω is that value where the Popov plot crosses the negative real axis. At that point, $ImG(j\omega) = 0$ and, thus, since

$$ImG(j\omega) = \omega(\omega^2 - a^2)/(a^2 + \omega^2)^2$$

the crossing of the real axis occurs when $\omega = \pm a$, 0 and, thus,

$$ReG(ja) = -2a^3/(2a^2)^2 = -\frac{1}{2a}$$

Hence, substituting into the relationship for k,

$$k = -1 + 1/a^3$$

6.5 PROBLEMS

1. Determine the nature of the singularities using Liapunov's Stability Theorem for:
   (a) the predator–prey equations

$$dx_1/dt = -x_1 + x_1x_2$$
$$dx_2/dt = x_2 - x_1x_2$$

   (b) Duffing's equation

$$y'' + y + y^3 = 0$$

2. Using the direct method of Liapunov, establish the stability properties of the equations of the theorems of Chapter 2.

3. For the system

$$x_1' = x_2$$
$$x_2' = -Kx_1 - \omega x_2$$

develop conditions for a_1, a_2, k, ω which assure system stability using $V = a_1x_1^2 + a_2x_2^2$.

4. Repeat Problem 3 using the quadratic form $V = a_{11}x_1^2 + 2a_{12}x_1x_2 + a_{22}x_2^2$, $(a_{21} = a_{12})$.

5. For the system

$$x_1' = x_2$$
$$x_2' = -f(x_1)$$

construct the Liapunov function in terms of the total system energy in the form

$$V = E(x_1, x_2) = x_2^2/2 + \int_0^{x_1} f(u)du$$

and discuss the conclusions you reach regarding system stability.

6. For the equation

$$x'' + f(x)x' + g(x) = 0$$

Using a Liapunov function V in the form of that used in problem 5, determine if V and W are positive or negative definite or semidefinite and whether the system is stable.

Hint: ignore $f(x)$ in the Liapunov function. Why?

7. In the Lur'e problem formulation for direct control, suppose $\phi(\sigma)$ is linear [i.e., $\phi(\sigma) = a\sigma$, a = constant]. Derive the conditions needed to assure asymptotic stability.

8. Compare the result of Problem 7 with that of the Aizerman conjecture.

9. Repeat Problems 7 and 8 for indirect control.

10. Using the circle criterion, develop the requirements on the system parameters where the system equation is

$$y'' + 2\mu y' + (\mu^2 + a^2)y - qf(t, y) = 0$$

and $f(t, y)$ is a nonlinearity restricted to the sector $[-1, 1]$.

11. For the second-order system,

$$x_1' = -k_1x_1, \qquad k_1 > 0$$
$$x_2' = -k_2x_2, \qquad k_2 > 0$$

with the system with controls

$$
\begin{aligned}
x_i' &= -k_i x - \gamma b_i, \qquad i = 1, 2 \\
\gamma' &= \phi(\sigma) \\
\sigma &= c_1 x_1 + c_2 x_2 - \rho\gamma
\end{aligned}
$$

and where $\phi(\sigma)$ has the usual properties:

(a) define the determinant condition which assures the origin is the only singular point

(b) Construct the linear transformation to obtain the new system

$$
\begin{aligned}
y_i' &= -k_i y_i - b_i \phi(\sigma), \qquad i = 1, 2 \\
\sigma' &= c_1 y_1 + c_2 y_2 - \rho\phi(\sigma)
\end{aligned}
$$

(c) Using the Liapunov function

$$V(y, \sigma) = y_1^2 + y_2^2 + \Phi(\sigma)$$

establish conditions for absolute stability.

6.6 REFERENCES AND RELATED LITERATURE

6.6.1 References

1. Vidyasagar, M., *Nonlinear Systems Analysis*, Prentice Hall, Englewood Cliffs, New Jersey (1978).
2. Lefshetz, Solomon, *Stability of Nonlinear Control Systems*, Academic Press, New York (1965).
3. Langill, A. W., Jr., *Automatic Control System Engineering*, Vol. II, Prentice-Hall, Englewood Cliffs, New Jersey (1965).
4. Ku, Y. H., *Analysis and Control of Nonlinear Systems*, The Ronald Press, New York (1958).
5. DeRusso, Paul M., Roy, Rob J., and Close, Charles M., State Variables for Engineers, John Wiley & Sons, New York (1965).
6. Gibson, J. E., *Nonlinear Automatic Control*, McGraw-Hill, New York (1963).
7. Stern, Thomas E., *Theory of Nonlinear Networks and Systems—An Introduction*, Addison-Wesley, Reading, Massachusetts (1965).
8. Brauer, Fred, and Nobel, John, *The Qualitative Theory of Ordinary Differential Equations*, Dover, New York (1969).

6.6.2 Related Literature

1. Atherton, Derek P., *Stability of Nonlinear Systems*, John Wiley & Sons, New York (1981).
2. Sanchez, David A., *Ordinary Differential Equations and Stability Theory: An Introduction*, Dover, New York (1978).
3. Nemytskii, V. V., and Stepanov, V. V., *Qualitative Theory of Differential Equations*, Dover, New York (1989).
4. Saaty, T. L., and Bram, J., *Nonlinear Mathematics*, Dover, New York (1964).

Chapter 7
Recursions and Their Stability

7.1 RECURSIONS

7.1.1 Recursions Formed from Jacobi's Iterations

Recursion formulae are algorithms in which an initial value is used to produce a new value, which in turn is inserted into the algorithm to produce again a new value, the process being repeated as many times as is desired. Jacobi's iteration is just such a recursion, whose algorithm is the basis for solving a set of linear algebraic equations. We consider the simplest case of two such equations in two unknowns,

$$a_{11} \cdot x_1 + a_{12} + x_2 = b_1$$

$$a_{21} \cdot x_1 + a_{22} \cdot x_2 = b_2$$

While Jacobi had in mind a much larger set of equations, this simple example illustrates his method. While the equations can be solved exactly by either Cramer's rule or the Gauss Triangulation and back substitution, such methods are tedious. Jacobi elected to create a recursive method which would start from an initial guess, and produce successively better guesses in each recursion. His approach was to solve each equation for a different unknown, say x_1 in the first and x_2 in the second (or vice versa),

$$x_1 = -\frac{a_{12}}{a_{11}} \cdot x_2 + \frac{b_1}{a_{11}}$$

$$x_2 = -\frac{a_{21}}{a_{22}} \cdot x_1 + \frac{b_2}{a_{22}}$$

He considered the right-side to be the starting point into which an initial value was to be placed. Then the left-side gave the result of the recursion, which would be entered into the right, etc. An index, n, is selected to indicate the sequence for recursing the algorithm,

$$x_1^{(n+1)} = -\frac{a_{12}}{a_{11}} \cdot x_2^{(n)} + \frac{b_1}{a_{11}}$$

$$x_2^{(n+1)} = -\frac{a_{21}}{a_{22}} \cdot x_1^{(n)} + \frac{b_2}{a_{22}}$$

7.1.2 Stability of Jacobi's Iteration

An interesting observation is that ordinarily Jacobi's recursion converges toward the correct solution, and seldom does it diverge. However, the failures are of enough concern that most programs written to solve linear equations contain a preconditioning routine which attempts a linear transformation, yielding "diagonal dominance" (see ref. 1), a relation among the coefficients which renders those on the principal diagonal "largest," and guarantees convergence. While there is no guarantee of achieving diagonal dominance, a system which approaches it stands a better chance of converging than one that does not. However, setting aside the question of diagonal dominance, the convergence (i.e., stability) in linear equations is interesting because it leads to the question of stability in nonlinear equations.

A way to show the stability of Jacobi's iteration is illustrated in the simplest case of two equations in two unknowns, but can be extended to any number (see Problem 1). The equations are made into time-variable equations which can be Laplace-transformed, allowing the eigenvalues to describe stability in the usual way. The z-transform is also used to reexpress the answer, where stability is mapped into a unit circle.

First, the "inputs," b_1 and b_2, are acknowledged to be constants, and can be interpreted as step inputs. Second, the unknowns, x_1 and x_2, are interpreted as functions of time, $x_1(t)$ and $x_2(t)$. In order to distinguish between successive values of the unknowns, a time shift, T, is used, so that $x_1(n + 1)$ is interpreted as $x_1(t + T)$, and likewise for x_2. The equations become

$$x_1^{(n+1)}(t + T) = -\frac{a_{12}}{a_{11}} \cdot x_2^{(n)}(t) + \frac{b_1}{a_{11}} \cdot u_{-1}(t)$$

$$x_2^{(n+1)}(t + T) = -\frac{a_{21}}{a_{22}} \cdot x_1^{(n)}(t) + \frac{b_2}{a_{22}} \cdot u_{-1}(t)$$

where $u_{-1}(t)$ is the unit step function.

Recognizing that the "next" value of x_1 (or x_2), in the algorithm is its "last-value advanced by T," suggests the relations,

$$x_1^{(n+1)}(t + T) = x_1^{(n)}(t), \text{ and } x_2^{(n+1)}(t + T) = x_2^{(n)}(t)$$

The index n is no longer necessary as the advance, T, maintains the order, so it is dropped, and the equations are first Laplace-transformed

$$x_1(s) = -\frac{a_{12}}{a_{11}} \cdot x_2(s)e^{-sT} + \frac{b}{a_{11}} \cdot \frac{1}{s}$$

$$x_2(s) = -\frac{a_{21}}{a_{22}} \cdot x_1(s)e^{-sT} + \frac{b_2}{a_{22}} \cdot \frac{1}{s}$$

and then z-transformed (see Chapter 8)

$$x_{1(z)} = -\frac{a_{12}}{a_{11}} \cdot z^{-1} \cdot x_2(z) + \frac{b_1}{a_{11}} \frac{z}{z-1}$$

$$x_2(z) = -\frac{a_{21}}{a_{22}} \cdot z^{-1} \cdot x_1(z) + \frac{b_2}{a_{22}} \cdot \frac{z}{z-1}$$

The resultant condition of stability, (Figure 7-1), is the region enclosed by a pair of asymmetric hyperbolae.

7.1.3 Diagonal Dominance: Sufficient for Stability

The condition of diagonal dominance, mentioned earlier, asks that the elements on the principal diagonal be larger in magnitude than the sum of its other row element's magnitudes. In our two-dimensional case,

$$|a_{11}| > |a_{12}|, \qquad \text{and } |a_{22}| > |a_{21}|$$

We note that the three-dimensional model would have the equivalent condition,

$$|a_{11}| > |a_{12}| + |a_{13}|, \qquad |a_{22}| > |a_{21}| + |a_{23}|, \qquad |a_{33}| > |a_{33}| + |a_{32}|$$

This condition can be superimposed on the stability boundaries for Figure 7-1, as shown in Figure 7-2, where the region of diagonal dominance lies entirely inside the stable region. Since $1 > |a_{12}|/|a_{11}|$ and $1 > |a_{21}|/|a_{21}|$. Thus, it is clear that diagonal dominance is sufficient but not necessary for stable Jacobi iterations because there are cases lying outside which are stable (see Problem 3). For this reason, there are many systems of equations which recurse

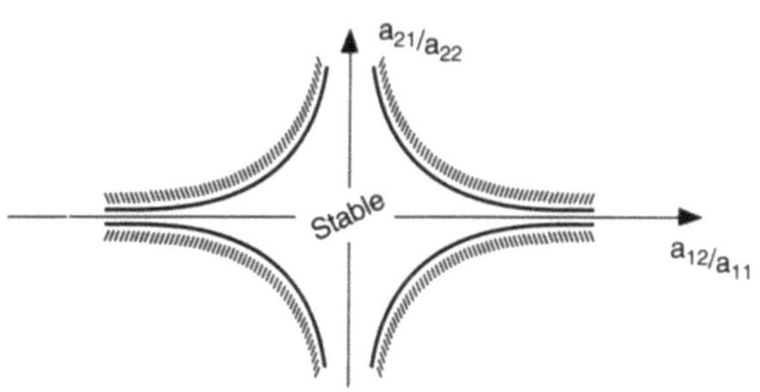

Figure 7-1. Stability boundaries for $N = 2$.

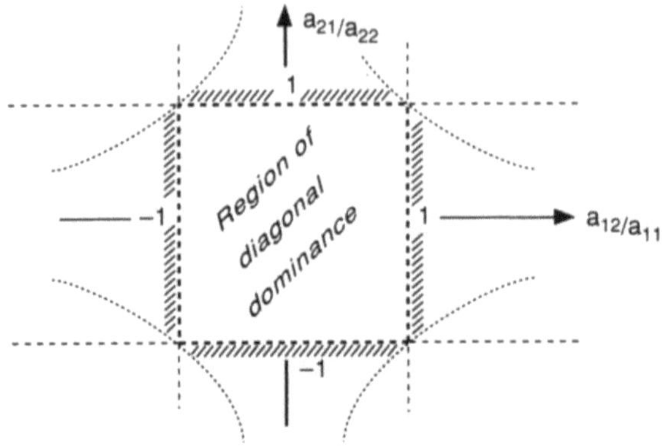

Figure 7-2. Region of diagonal dominance.

stably despite their lack of diagonal dominance, which is a good feature because not all systems can be made diagonally dominant, yet solutions exist uniquely to all algebraic systems.

7.2 THE MECHANIC'S RULE

7.2.1 Discussion of the Rule

The preceding development of stability in Jacobi's iteration is neat, organized, logical, and complete as is any analysis based on the presumption of linearity. The guarantees of existence and uniqueness are the roots which marshall compact analysis, free of contradiction and even ambiguity. Lacking these ingredients, the analysis is more difficult, perhaps more devious, and places more demands on the analyst's mathematical dexterity.

The Mechanic's rule is an example of a nonlinear recursion which not only has utility, but presents an interesting problem in stability which sets the tone for an attack on even more stubborn recursions, such as the arcane Logistics Map, addressed later.

The Mechanic's rule is a recursive method of extracting the square root of a number (usually positive). It consists of an algorithm which refines a guess, producing a better guess. Successive applications get closer to the exact root, so we may say that it "converges." This becomes a statement of stability. Jacobi's iteration does this for linear algebraic equations, and, as has been seen, can either converge to, or diverge from, the solution—the one called stable and the other unstable.

The Mechanic's rule is an old-timer's way to ESTIMATE the square root of a NUMBER, by a GUESS, which is then averaged with the quotient of the NUMBER and the GUESS. Of course, the result can be refined by repeating the process as often as is desired. The recursion formula is,

NUMBER = a, the number whose square root is desired.

GUESS = x_0, a positive number.

Then, the ESTIMATE = $x_1 = [x_0 + a/x_0]/2$

And, the recursion formula is $x(n + 1) = [x(n) + a/x(n)]/2$

and we note that this is a nonlinear relation!

7.2.2 Stability in the Mechanic's Rule

The interesting feature of this rule is that it seems always to converge on the right answer. In support this conjecture, suppose the square root of 4 is desired (and known to be 2). No matter what the starting value, the answer seems to appear, as indicated in Table 7-1.

However, these illustrations, as convincing as they may seem, are not proof of convergence. Were it a *linear* rule, the proof of convergence would be available, but being *nonlinear*, there is no such guarantee and it is precisely this flaw which adds spice to the search for a proof (ref. 2). As a preliminary, we note that if the Mechanic's rule does converge, then the quotient of successive estimates must approach unity, that is,

$$\lim_{n \to \infty} \frac{x_n}{x_{n+1}} = 1$$

And, if this is true, in the limit $x_n = x_{n+1}$. Substitution into the Mechanic's rule formula yields $x_n^2 = a$, which is equivalent to

$$\lim_{n \to \infty} x_n = \pm\sqrt{a}$$

The proof must in some way reflect this observation, and one way to express it is to define "errors" after each recursion. Then the proof can consist of verifying that successive errors get smaller. It is convenient to define a normalized error, in terms of the root, r_2.

TABLE 7-1.

Question: Does each recursion *always* repay with *better* accuracy? An example supports the conjecture (but this is not proof!):

	x_0	x_1	x_2	x_3		
If $a = 4$	1	2.5	2.1	2.0		
	2	2	2	2		
	10	5.2	3.0	2.2	2.0	
	20	10.1	5.2	3.0	2.2	2.0

Define normalized "error,"

$$e_n = \frac{x_n}{r_2} - 1$$

Then, using

$$e_{n+1} = \frac{x_{n+1}}{r_2} - 1, \text{ and } x_{n+1} = \tfrac{1}{2}\left\{x_n + \frac{r_2^2}{x_n}\right\}$$

eliminate x_{n+1}, x_n. Giving,

$$e_{n+1} = \frac{e_n^2}{2 \cdot (1 + e_n)}$$

Now note, $e_0 > -1 \cdots s_0$, $e_1 > 0$ $e_2 > 0$, etc. Conclude

$$|e_n| = e_n \mid n > 0$$

Form the inequality

$$e_{n+1} = \frac{e_n/2}{1 + {}^1/e_n} < e_n/2 \mid n > 0$$

that is,

$$\frac{e_{n+1}}{e_n} < \frac{1}{2}$$

so errors at least halve in each recursion, so, they disappear as $n \to \infty$. Or, more formally,

$$e_2 < e_1/2, \qquad e_3 < e_1/2^2, \qquad e_n < e_1/2^{n-1}$$

so

$$\lim_{n \to \infty} e_n = e_1 \cdot \lim_{n \to \infty} 2^{1-n} = 0 \qquad \text{Q.E.D.}$$

We next consider the anatomy of the Mechanic's rule in order to provide a geometric picture of its sequential operations which will be useful later on in interpreting local stability in recursions, particularly important to an understanding of the arcane behavior of the Logistics Map. We denote the right-hand-side of the Mechanic's rule as a continuous function, f, in the continuous variable, x,

$$f(x) = \frac{1}{2}\left(x + \frac{a}{x}\right)$$

We note that in the limit, $x_{n+1} = y_n$ and, thus, we can say $x = x$. In the graphical approach which compares $f(x) = \frac{1}{2}(x + a/x)$ and $x = x$, we find the solution to be the intersection of the two curves as shown. Analytically, this is seen to be

$$x = \frac{1}{2}(x + a/x)$$

or $x^2 = a$.

We also observe that $df/dx = 0$ yields

$$x^2 = a$$

and we remark that the local minimum of $f(x)$ corresponds to the square root we seek.

Next, plot both f and x against x as in Figure 7-3. Follow the sequence given in Figure 7-3.

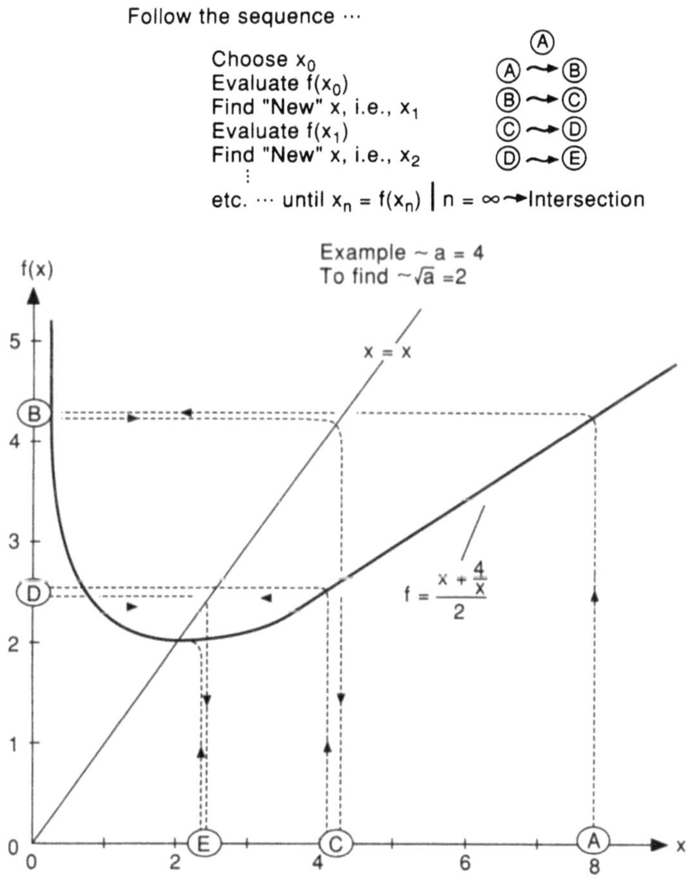

Figure 7-3. Anatomy of the mechanics rule.

From Figure 7-3 the algorithm starts at an initial guess on the x-axis, moves vertically to f, then horizontally to x, vertically to f, and so on (see Problem 5). Two things are apparent: first, no matter what value of x, the iteration approaches the solution from the "right." Second, the minimum value of f occurs at the intersection of f and x (i.e., the solution). The first observation is a cornerstone of the proof of convergence, where it was noted that all errors after the first are positive. The second observation offers an interesting extension to the Mechanic's rule—the possibility of determining the cube root of a number by a recursion formula. Note that the derivative, f-prime is,

$$f'(x) = \frac{df}{dx} = \frac{1}{2} \cdot \left\{1 - \frac{a}{x^2}\right\}$$

Since the intersection of f and x occurs at the minimum, the slope there must be zero, which it is, as the bracketed-term vanishes.

7.2.3 Finding Other Roots

This provokes the prospect of another recursion for the cube root, requiring a new function, f, whose local minimum occurs at the cube root of "a," where it is intersected by "x". Synthesizing this function proceeds as follows,

Define $r_3 \equiv +\sqrt[3]{a}$ i.e., the cube-root of a

Start with $f' = k \cdot \left\{1 - \frac{a}{x^3}\right\}$, and note, $f'(r_3) = 0$

Integrate, constrained to $f(r_3) = r_3$, giving $f = \frac{2}{3}\left\{x + \frac{a}{2x^2}\right\}$

and, the recursion is, $x_{n+1} = \frac{2}{3} \cdot \left\{x_n + \frac{a}{2x_n^2}\right\}$

Following the analogy in the limit, we set $x = x$ and seek intersections with $f(x) = \frac{2}{3}(x + a/2x^2)$ which when solved under the condition $x = f(x)$ yields $x^3 = a$.

7.2.4 Rate of Convergence

An interesting comparison between the Mechanic's rule for extracting the square root, and this extension of the method for extracting the cube root, is illustrated in Table 7-2.

It is evident from Table 7-2 that the square root converges more rapidly than the cube root, implying that it is a more stable recursion. The truth of this

TABLE 7-2. Comparing the Convergence Rates of Square- and Cube-Roots

n	SQRT x_n	Cube-RT x_n	SQRT x_n	Cube-RT x_n
0	.5	.5	100.	100.
1	4.25	5.667	50.020	66.667
2	2.596	3.819	25.050	44.445
3	2.068	2.638	12.605	29.631
4	2.001	1.950	6.461	19.755
5	2.000	1.651	3.540	13.174
6	↓	1.590	2.335	8.790
7		1.587	2.024	5.877
8		↓	2.000	3.957
9			↓	2.732
10				1.995
11				1.665
12				1.591
13				1.587
14				↓
15				

assertion is supported by the graphical comparison of the iteration functions for the two cases, shown in Figure 7-4.

Except for the initial guess, the iterations approach the solution from the "right" and, because the cubic function is everywhere "closer" to the x-curve, more steps are required to approach the intersection. This is more apparent at large values of x, and can be quantified by observing the asymptotic behavior of the recursion functions,

$$\text{as } x \to \text{``Big,''}\ f_{\text{SQRT}} \to \frac{x}{2}, \quad \text{and } f_{\text{CUBE-RT}} \to \frac{2 \cdot x}{3}$$

Each succeeding value of x is half its predecessor in the square root, while it is only $\frac{2}{3}$ in the cube root. In Problem 8, the ratio is $(n - 1)/n$, for the $1/n$th root, showing that the "higher" the root, the "slower" it will converge—a result which forms the basis for a criterion of "local stability," the topic to be addressed next.

Extending this concept to allow the solution of nth roots, one would assume

$$\lim_{n \to \infty} \frac{x_n}{x_{n+1}} = 1 \qquad \text{or } x_{n+1} = x_n$$

and seek the intersection of the iterative formula

$$x_{n+1} = f(x_n, a)$$

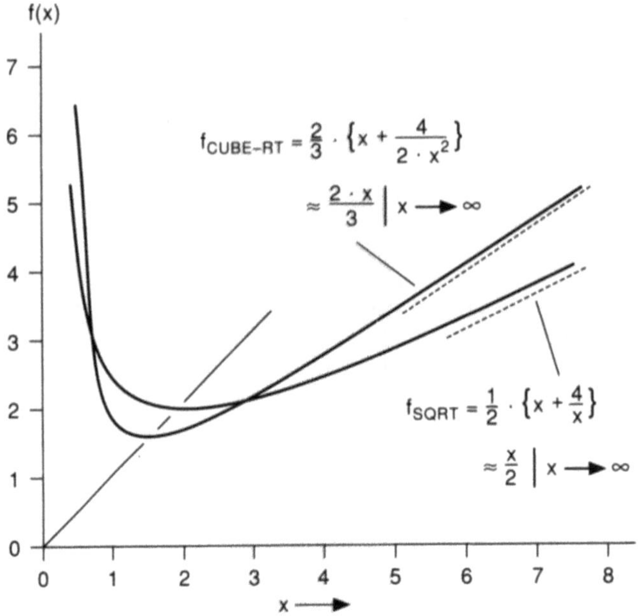

Figure 7-4. Iteration functions for square and cube roots.

and the line $x = x$, so that we are looking for functions satisfying

$$x = f(x, a)$$

where f' assumes the form

$$f' = k(1 - a/x^n)$$

and n is the root of the number a sought, corresponding to a local minimum ($f = 0$). Integration yields

$$f = k\left(x - \frac{a(n-1)}{x^{n-1}}\right)$$

and use of $x = f$ yields

$$x^n(1 - k) = -\frac{a - k}{n - 1}$$

and since $x^n = a$ at the local minimum, which is the intersection we seek to force,

$$k = \frac{n - 1}{n - 2}$$

and the iterative formula becomes

$$x_{n+1} = \frac{n-1}{n-2}\left[x_n - \frac{a(n-1)}{x^{n-1}}\right]$$

7.3 SINGULARITIES AND PECULIARITIES

7.3.1 Nodes, Foci, and Centers

It is apparent that the question of stability of an iterative relationship is analgous to one of convergence (i.e., convergent recursions are stable and divergent recursions are unstable). A criterion for ''local'' stability, in the neighborhood of the intersection of the x-curve and the $f(x)$-curve, is derived from the relationship

$$\lim_{n \to \infty} \frac{x_n}{x_{n+1}} = 1 \qquad \text{or} \qquad f(x) = x$$

The general nonlinear function, $f(x)$, is assumed to be continuous to the extent that it may be approximated by its tangent,

$$f(x) = m_i \cdot x + b \mid m = \frac{\partial f}{\partial x} \text{ at } X_1$$

Five possibilities are illustrated in Figure 7-5, regarding the nature of, and stability at, the intersections.

A summary of local stability is given in Figure 7-6.

7.3.2 Numerical Bifurcations (Limit Cycles)

We are now in a position to examine numerical bifurcations, the iterative analogy of a limit cycle, in which a recursion ''relaxes'' into a cyclic pattern, alternating between two numbers. This is not unlike the result of dividing two rational numbers, typically $\frac{8}{33} = 0.242424$, etc. Since the numbers 2 and 4 repeat, we can say that it is periodic of period 2, and can adopt this analogy. It is clearly a phenomenon which never occurs in a linear recursion, its closest kin being the numerical center; however, unlike the bifurcation, the center alternates between two values uniquely determined by the starting point. The simplest illustration of a bifurcation consists of a nonlinear $f(x)$, shown in Figure 7-7, made up of three straight-line segments, the middle segment having a slope more negative than -1, which makes the intersection an unstable focus. The outer segments have negative slopes more positive than -1, which make the outer system act as if it contained a stable focus. As a result, the recursion

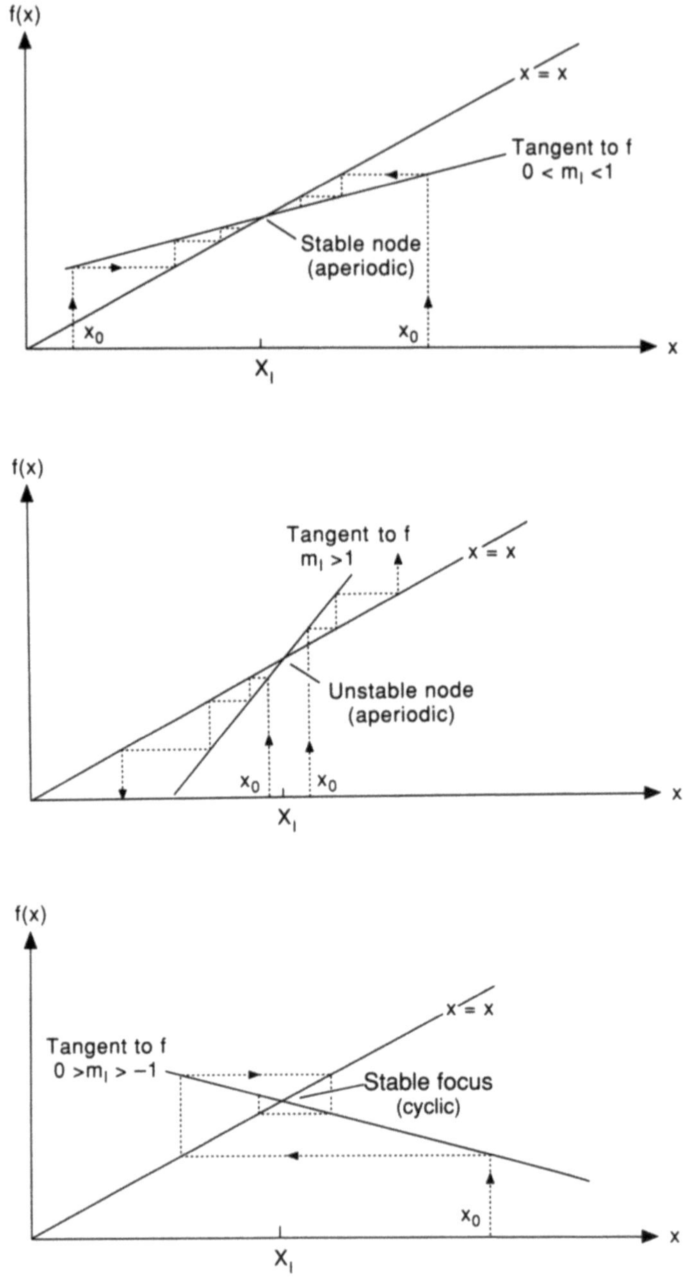

Figure 7-5. Classifying intersections.

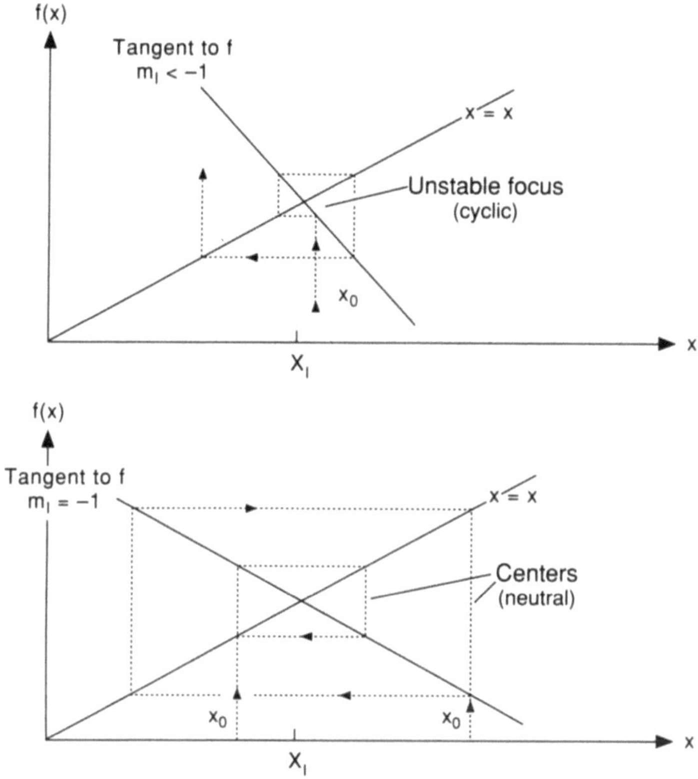

Figure 7-5. *(Continued)*

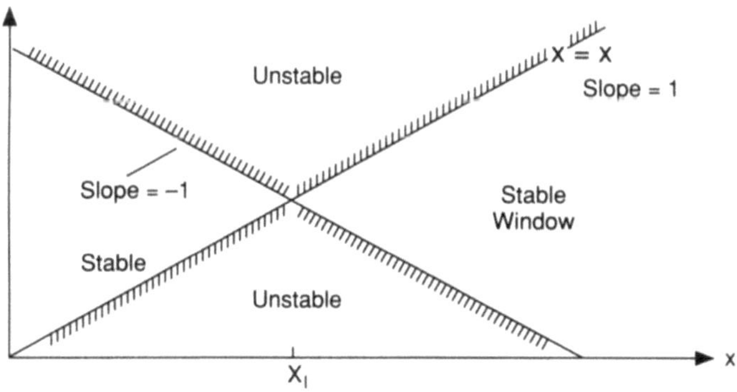

Figure 7-6. Stability summary.

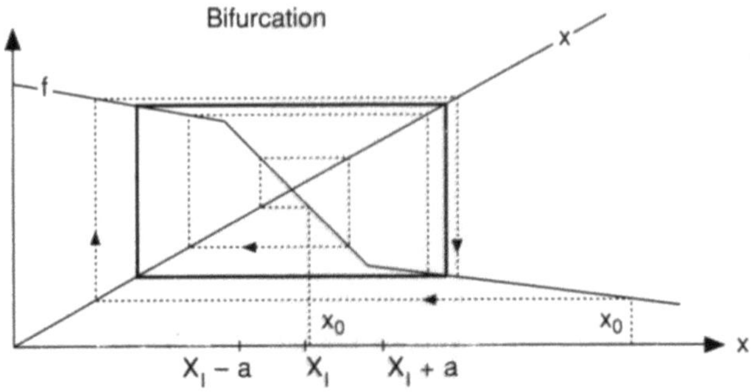

Figure 7-7. Bifurcation.

"relaxes" into a bifurcation indicated by the closed contour in Figure 7-7. While the system must be nonlinear to exhibit a bifurcation, not all nonlinear systems display bifurcations (see Problem 10).

This can be expressed for the case where we have a convergent iterative scheme, for two possibilities as

$$\lim_{n \to \infty} x_n = m_1, \qquad \lim_{n \to \infty} x_{n+1} = m_2$$

then

$$\lim_{n \to \infty} x_n/x_{n+1} = m_1/m_2$$

and

$$\lim_{n \to \infty} x_n/x_{n+2} = 1$$

Further, should we examine trifurcations (period 3) or higher furcations (period k), we can develop equivalent relationships, such as

$$\lim_{n \to \infty} x_n/x_{n+k} = 1$$

7.3.3 Antibifurcations (Unstable Limit Cycles)

Just as there can be unstable limit cycles, there can also be numerical "antibifurcations," an example of which is shown in Figure 7-8, consisting again of three straight-line segments, with their slopes reversed from those of the previous example (i.e., the middle segment has the least negative slope). The intersection is a stable focus, while at the extremeties the system reacts as if it contained an unstable focus. In between there is a closed contour which "repels" the recursions. This is precisely analagous to the situation of a continuous oscillator defined by nonlinear ordinary differential equations, in which the sys-

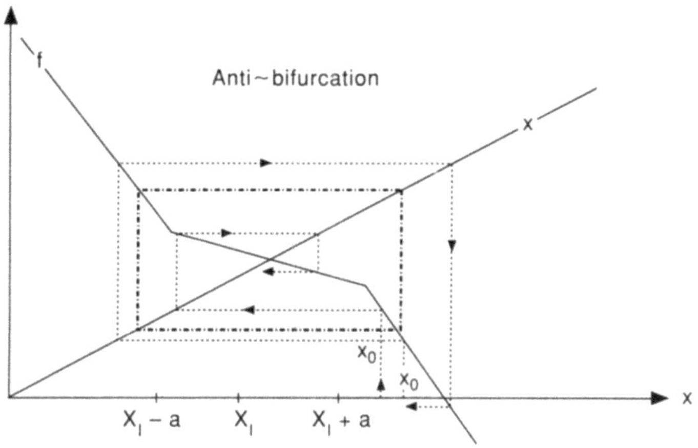

Figure 7-8. Anti-bifurcation.

tem is stable for small amplitudes and unstable for large amplitudes (see Problem 11).

7.3.4 Tangencies of Even and Odd Order

Tangencies of even and odd order are also possibilities in recursions, the former producing semi-stable (or semi-unstable) saddles and the latter producing either stable or unstable saddles, as shown in Figure 7-9.

7.3.5 Other Nonlinearities

The particular nonlinearity in Figure 7-10, which is asymmetric to the intersection, produces a center because the two slopes are reciprocals. In itself the recursion of Figure 7-10 is not so significant, but the slight modification shown in Figure 7-11 opens the door to far more interesting results.

With the aid of the foregoing functions, the composite three-segment recursion in Figure 7-12 is seen to produce a bifurcation. It is reasonable to speculate that variations of this composite function with more segments can produce other interesting results such as "trifurcations" (see Problems 12–15).

7.4 THE LOGISTICS MAP

7.4.1 The Arcane Behavior of the Logistics Map

We are now in a position to investigate the arcane Logistics Map, whose runic behavior is a delight (ref 4.). We note that the term map is equivalent to recur-

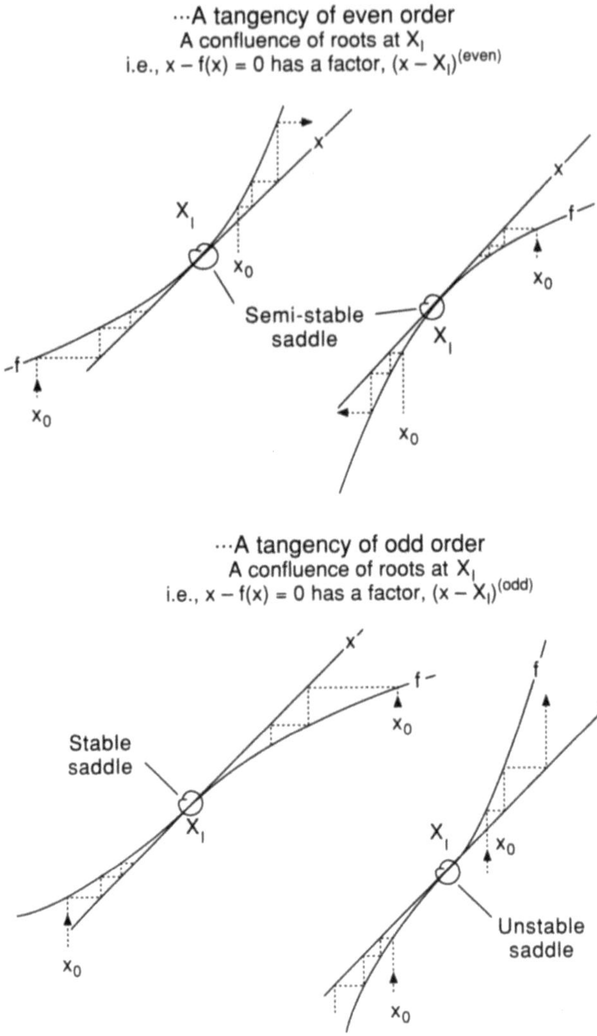

Figure 7-9. Various tangencies.

sion, iteration, or equation in this context.

$$x_{n+1} = C \cdot x_n \cdot (1 - x_n), \qquad \text{for } 0 < x_0 < 1$$

Part of its magic is created by the offhand restriction to initial values lying between 0 and 1, which plays an important part in preconditioning the mind of the player, who might otherwise be prompted to look for some rationality in an

$$f(x) = \begin{cases} f_L = m_L \cdot x + b_L, & x \le X_I \\ f_R = m_R \cdot x + b_R, & x \ge X_I \end{cases} \quad \cdots \text{ 2 straight lines}$$

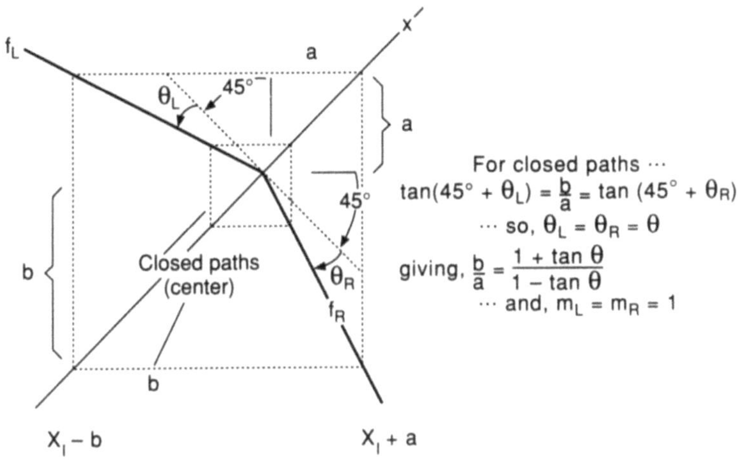

Figure 7-10. An "asymmetric" nonlinearity yields a center.

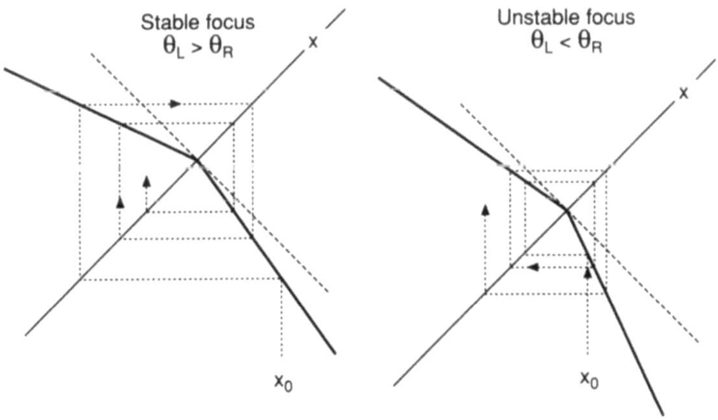

Figure 7-11. A "neighboring" function, yields Foci.

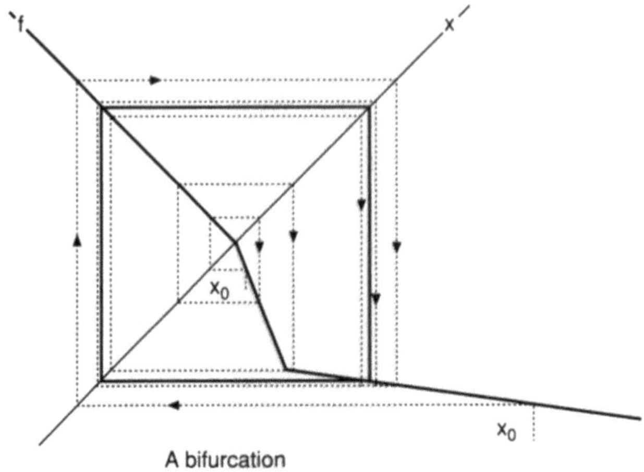

Figure 7-12. A bifurcation.

otherwise incredible arena. The magic comes with selecting various values of the coefficient, C, which yield seemingly inconsistent and beguiling results.

7.4.2 The Behavior Explained

Just as the recursion of the Mechanic's rule was displayed, so the Logistics Map can be plotted, the function, f, in this case is

$$f = C \cdot x \cdot (1 - x)$$

that is, a parabola opening downward which must intersect the straight line, x, at two points, provided $C \neq 0$. One of those points is always the origin, while the other lies to its "right" if $C > 1$ in the interval $0 < x < 1$. Should C take on values less than 1, this second intersection always lies to the "left," in the interval $x < 0$. These two facts, together with the conclusions drawn earlier regarding local stability, are enough to explain most of the mystery enshrouding the Logistics Map, as illustrated in Figures 7-13 through 7-22.

In the case of $0 < C < 1$, where the recursion seeks $x = 0$, Figure 7-13 illustrates the fact that the origin is a stable node, so any starting value $0 < X_0 < 1$ most certainly seeks it. The second intersection is at $X_i = -1$, and being an unstable node, it "repells" values in its neighborhood.

Values of $1 < C < 3$ all "seek" the single steady-state value $1 - (1/C)$, again provided the starting value lies within $0 < X_0 < 1$.

For

$$1 < C < 3 \cdot \cdot \cdot x_n \bigm| n \rightarrow \infty, \qquad \text{"seeks"}\ x_{ss} = 1 - 1/C$$

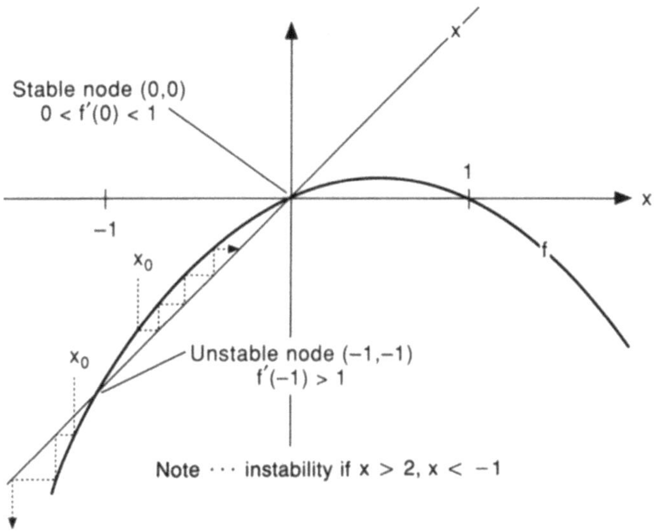

Figure 7-13. Logistics map, $C = 1/2$.

Let $C = 2$, which gives

$$x_0 = \begin{matrix} 1/4 \\ (.25) \end{matrix} \longrightarrow \begin{matrix} 3/8 \\ (.375) \end{matrix} \longrightarrow \begin{matrix} 15/32 \\ (.469) \end{matrix} \longrightarrow \begin{matrix} 255/512 \\ (.498) \end{matrix}$$

The explanation is apparent in Figure 7-14, where the intersection at the origin is now an unstable node, while the other, to its "right," is a stable node. The case shown, $C = 2$, places the parabola with its "top" precisely at the intersection, so the recursions always approach it from the "left," a situation reminiscent of the Mechanic's rule (see Problem 18).

Bifurcation is observed to take place for $3 < C < 3.44$ and for $C = 10/3$ (Figure 7-15).

Considering Figure 7-16, the origin is again an unstable node, while at the other intersection the slope of f is more negative than -1, making it an unstable focus. It is interesting that the formerly stable root ($1 < C < 3$), lying at the intersection has "disappeared," to be replaced by the two steady-state values which constitute the bifurcation. This results from the instability of the focus, the fact that x_i is confined to values in the range $0 < x < 1$, and the instability of the node at the origin. More will be said about the "disappearance" in connection with the existence of bifurcations.

A double bifurcation occurs in the range $3.45 < C < 3.5$, consisting of bifurcations of each of the previous bifurcations and suggesting the idea of "period-doubling" (Figure 7-17).

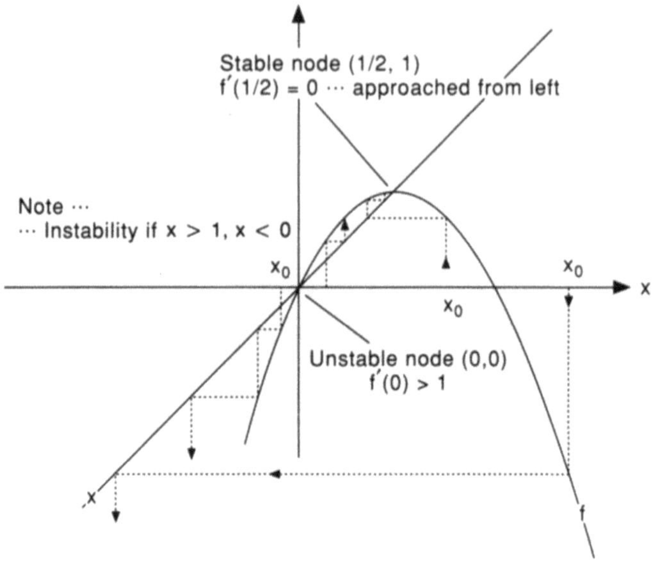

Figure 7-14. Logistics map, $C = 2$.

As illustrated in Figure 7-18 for $C = 3.5$, the steady-state limit cycle intersects itself, an oddity, because in continuous systems governed by ordinary differential equations there is never a way for this to happen. The Lorenz equations come to mind as a possible candidate, but they are three first-order equations, and, when viewed in three dimensions (ref. 5), their trajectories do not intersect.

Again, in Figure 7-18, the system does not relax into the unstable equilibrium point at the intersection because it is an unstable focus, unless, of course, the recursion is initiated at exactly the intersection value, in this case, $x_0 = 2.5/3.5$.

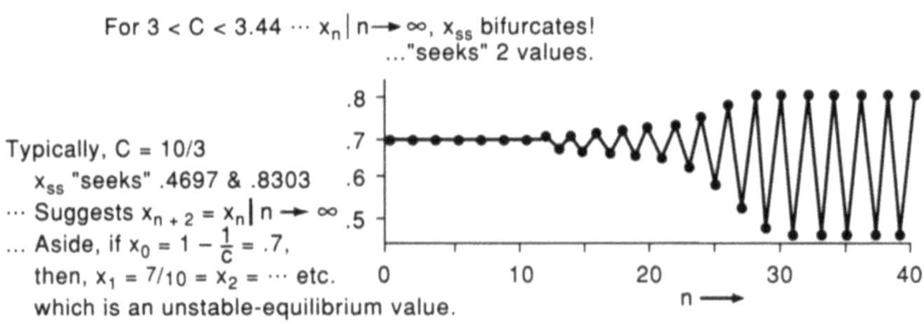

Figure 7-15. Bifurcation if $3 < C < 3.44$.

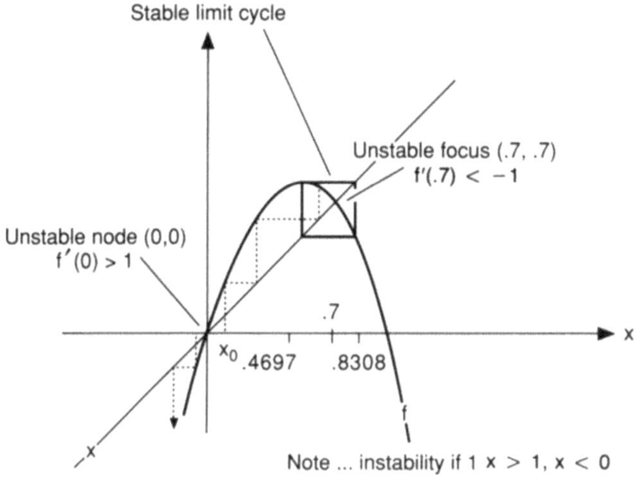

Figure 7-16. Bifurcation at $C = 10/3$.

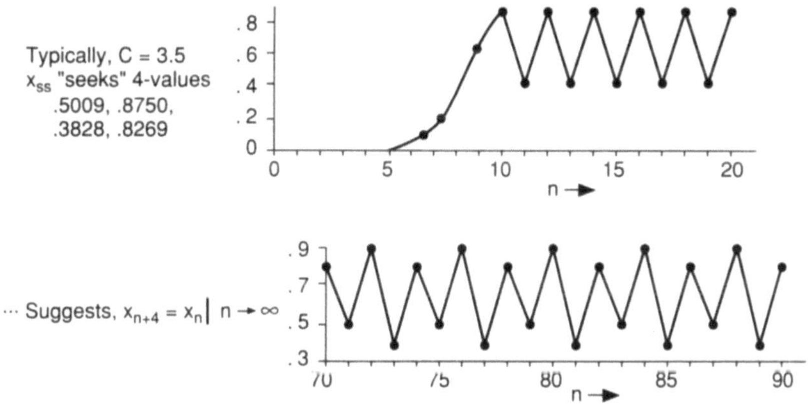

··· Another aside, if $x_0 = 1 - \frac{1}{C}\Big|_{c = 3.5} = \frac{2.5}{3.5}$, then $x_1 = \frac{2.5}{3.5} = x_2 = \cdots$ etc.

i.e., more evidence of an unstable-equilibrium value.

Universally ··· $x(0) = 1 - \frac{1}{c} \Rightarrow x_1 = x_2 = \cdots$ etc. $= x(0)$ iff $c \neq 0$

i.e., there is an equilibrium value for all c$^{\underline{s}}$, which is unstable if $3 < c < 1$ and stable otherwise.

Figure 7-17. Quatrofurcation if $3.45 < C < 3.54$.

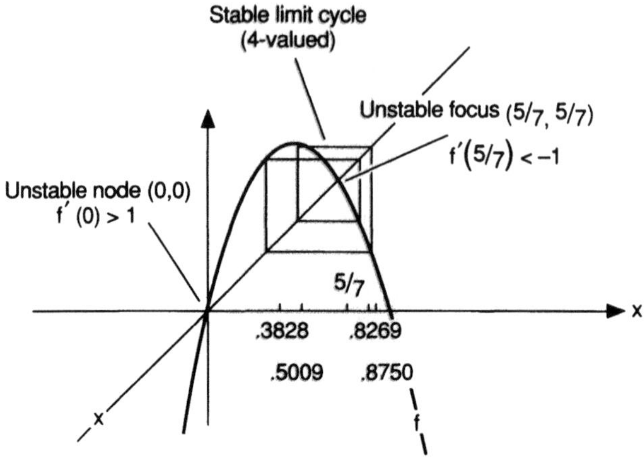

Figure 7-18. C = 3.5 double bifurcation (Quatrofurcation).

At the value, $C = 4.0$, the "nose" of the parabola has a value, $f = 1.0$, which guarrantees that a value of x greater than 1.0 can never be produced, and preventing negative values of x to ever occur, as shown in Figures 7-19 and 7-20. For no particularly obvious reason, the system doesn't seem to display a steady cyclic recursion, except for the interesting possibility that if it is initiated at $x = .5$, creating $f = 1.0$, then $x = 1.0$, and $f = 0.0$, whereupon $x = 0.0$ thereafter (see Problem 19). Except for this specific case, the idea that there is no repetition in the sequence suggests that its "period" has become infinite, which, if true, further suggests the means of generating repeatable pseudo-random (i.e., chaotic) sequences, each unique to its own starting value. The "randomness" of the numbers could be quantified (see Problem 20), but the data in Figure 7-19 appears to be equally distributed.

Should the value of C exceed 4.0, there are values of x near $\frac{1}{2}$ which recurse values of $f > 1.0$, and on subsequent recursions x exceeds 1.0, causing negative values thereafter, which grow because of the unstable node at the origin. This is shown in Figures 7-21 and 7-22 for a value of $C = 4.001$, which

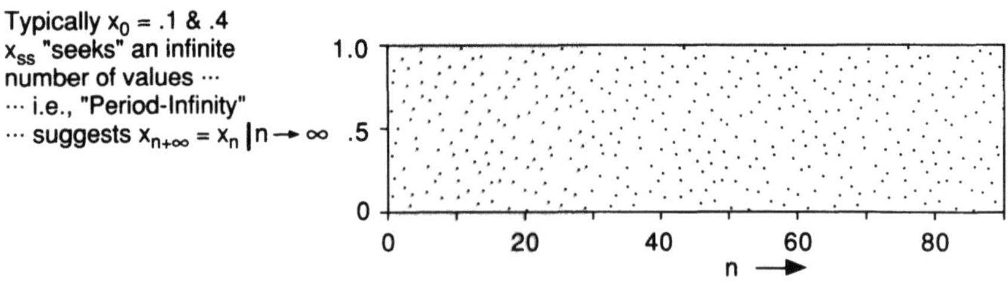

Figure 7-19. C = 4 No order, but repeatable, i.e., chaotic.

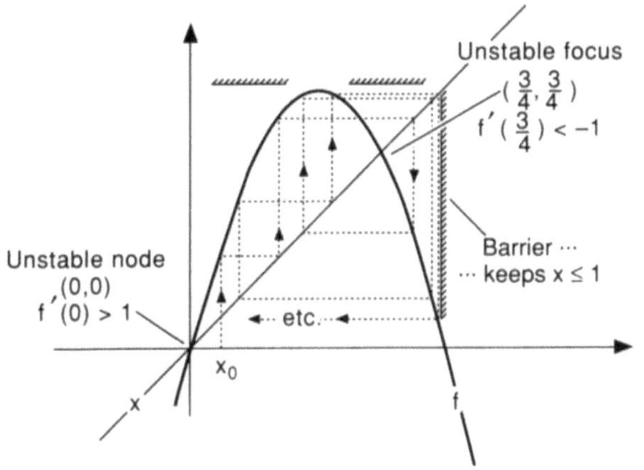

Figure 7-20. $C = 4$ Essential chaos, period infinity?

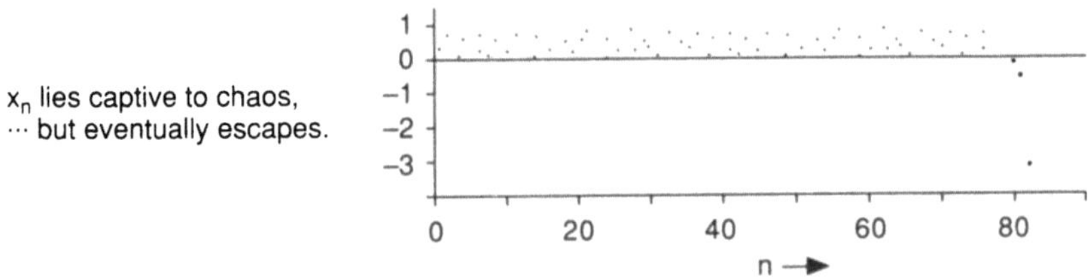

Figure 7-21. $C = 4.0001$ chaotic until it goes unstable.

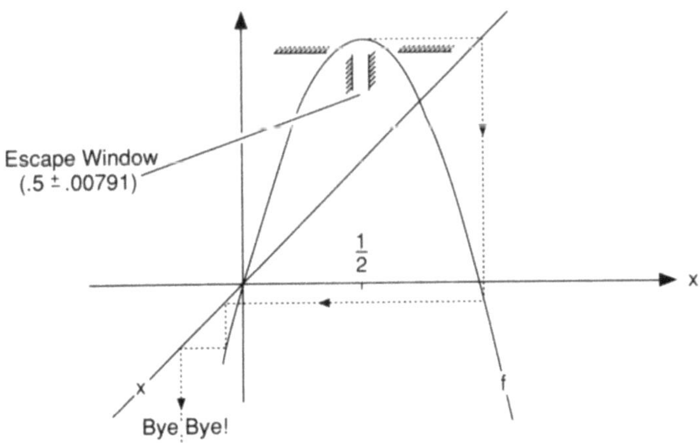

Figure 7-22. $C = 4.001$ Near chaos . . . insipient instability.

enters the region near $x = \frac{1}{2}$ on the 75th recursion, yielding a value exceeding 1.0 on the 76th recursion and increasingly negative values thereafter. Up to the 75th recursion, the system seems to be in "essential chaos."

7.4.3 The Anatomy of Bifurcations

Extending the graphical method which successfully portrays recursions in the Mechanic's rule, the generation of bifurcations is easily explained. To exist, the bifurcation is the stable repetition of every other value, suggesting that we examine the recursion formula relating the $x(n + 2)$ to $x(n)$, as

$$x_{n+2} = C \cdot x_{n+1} \cdot (1 - x_{n+1}) \mid x_{n+1} = C \cdot x_n \cdot (1 - x_n)$$

giving

$$x_{n+2} = C^2 \cdot x_n \cdot \{1 + a_1 x + a_2 x^2 + a_3 x^3\}$$

where

$$a_1 \equiv -(1 + C)$$

$$a_2 \equiv 2 \cdot C$$

$$a_3 \equiv -C$$

The right-hand side is a quartic, conveniently defined as $f_4(x)$, where the index, $4 = 2^2$, which is 2 raised to the power denoting the number of steps taken in the recursion. A plot of $f_4(x)$ is shown in Figure 7-23, along with the basic recursion formula, which can be denoted as $f_2(x)$ because it represents a one-step advance. The value, $C = 10/3$ is illustrated in Figure 7-23 and it will be

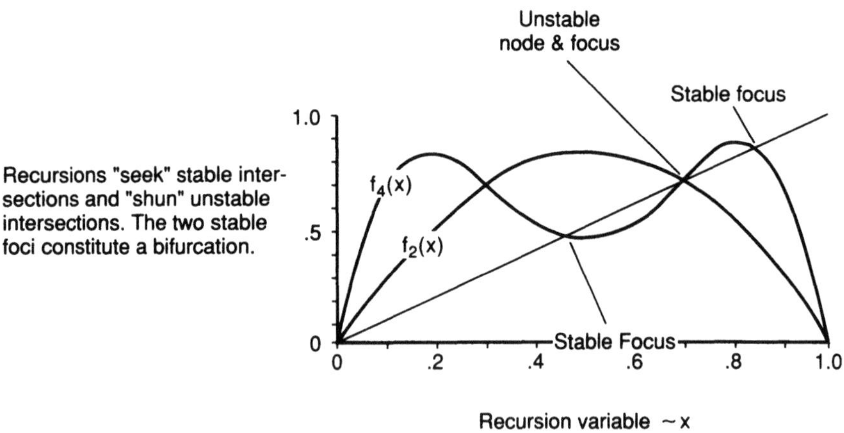

Figure 7-23. *f*-polynomials for C = 10/3.

noted that $f_2(x)$ intersects the "x-curve" with a slope more negative than -1, making it an unstable focus, repelling the recursion. This intersection occurs at the value, $x = 1 - 1/C = 0.7$, and Figure 7-23 also reveals that $f_4(x)$ intersects the "x-curve" at this point as well, but with a slope more positive than $+1$, making it an unstable node which repels every other point in the two-step recursion. However, there are two other intersections of $f_4(x)$ with the "x-curve," both of which have negative slopes lying in the region, $0 > f_4'(x) > -1$. These are stable foci, and both attract the two-step recursion, each being stable as they alternate, which explains the stable nature of the bifurcation.

We note that the intersections of Figure 7-23 are also defined by the relationships

$$x = x$$

$$x_{n+2} = f = C^2 x_n(1 + a_1 x_n + a_2 x_n^2 + a_3 x_n^3)$$

which when solved simultaneously for $x = x_{n+2} = x_n$ yields

$$1 + a_1 x + a_2 x^2 + a_3 x^3 = 0$$

which has the standard analytical cubic roots via Cardano's equations.

It is worth noting (ref. 5) that as C increases through 3, where the first bifurcation occurs, a point is reached, near $C = 3.45$, where each bifurcation branch also bifurcates, creating a "quatrofurcation."

Thus, the system's steady state proceeds as follows:

$0 < C < 1$	$x_{ss} = 0$
$1 < C < 3$	$x_{ss} = 1 - 1/C$
$3 < C < 3.44$	x_{ss} = 2 values—Bifurcation
$C = 3.45$	x_{ss} – 4 values—Quatrofurcation

This apparent "period-doubling" leads to the (mistaken) conjecture that all "P-furcations" will be powers of 2 (i.e., $P = 1, 2, 4, 8, \cdots$, etc.). Indeed, in ref. 5, periods of all the integers have been found. In particular, the "trifurcation," occurring around $C = 3.83$, was at one time thought not to exist. The proceedure used herein to find the "Period-2," bifurcation, discloses "Period-3" by a natural extension. Indeed, the approach yields any integer-period, and provides a basis for understanding the mysteries of the Logistics Map.

7.4.4 Period 3 Is a Snap!

The condition for Period-3, the trifurcation, to exist is that every third value repeats in the steady state. The recursion formula which generates this is constructured by relating $x(n + 3)$ to $x(n)$, as

$$x_{n+3} = C \cdot x_{n+2}(1 - x_{n+2}) \mid x_{n+2}$$
$$= C \cdot x_{n+1}(1 - x_{n+1}) \mid x_{n+1} = C \cdot x_n(1 - x_n)$$

giving,

$$x_{n+3} = C^3 \cdot x_n\left(1 + \sum_{i=1}^{z} b_j x_n^j\right)$$

of degree 8, where,

$$b \equiv (1 + C + C^2)$$
$$b_2 \equiv 2C \cdot (1 + C + C^2)$$
$$b_3 \equiv -C(1 + C + b \cdot C^2 + C^3)$$
$$b_4 \equiv 2 \cdot C^3(3 + 2 \cdot C)$$
$$b_5 \equiv -2C^3(1 + 2 \cdot C)$$
$$b_6 \equiv 4C^4$$
$$b_7 \equiv -C^4$$

According to the definition, the right-side is the polynomial, $f_8(x)$, so candidates for the values in Period 3 are the intersections between the "x-curve" and $f_8(x)$, of which, in the interval $0 < x < 1$, there must be at least one intersection with $x = x$ if $C > 1$, and there may be as many as seven. That $f_8(x)$ and its derivative, f_8', are polynomials is considerable help in understanding its nature. As noted in Problem 21, both show left–right symmetry about the $x = \frac{1}{2}$ vertical line and, therefore, there will be an odd number of intersections with the "$x = x$ curve," starting and ending with "downward" crossings, and alternating from "downward" to "upward" crossings proceeding to the right.

At each "upward" crossing the slope of $f_8(x)$ is greater than $+1$, making it an unstable node which repels recursion and thereby eliminating it as a candidate for a trifurcation value. Each "downward" crossing has a slope less positive than $+1$; and, provided the slope is not more negative than -1, it is either a stable node or a stable focus, either of which will "capture" the recursion. Thus, "downward" intersections with slopes more negative than -1 are also excluded as candidates for trifurcation values. The trick then, is to "intelligently" vary the value of C until precisely three "downward" intersections are found whose slopes lie in the interval, $-1 < f_8' < 1$. The process is aided by calculating the slope, f_8' (also a polynomial). The result (Figure 7-24), uncovers the value $C = 3.83$ as a trifurcation-producer (see also Problem 23).

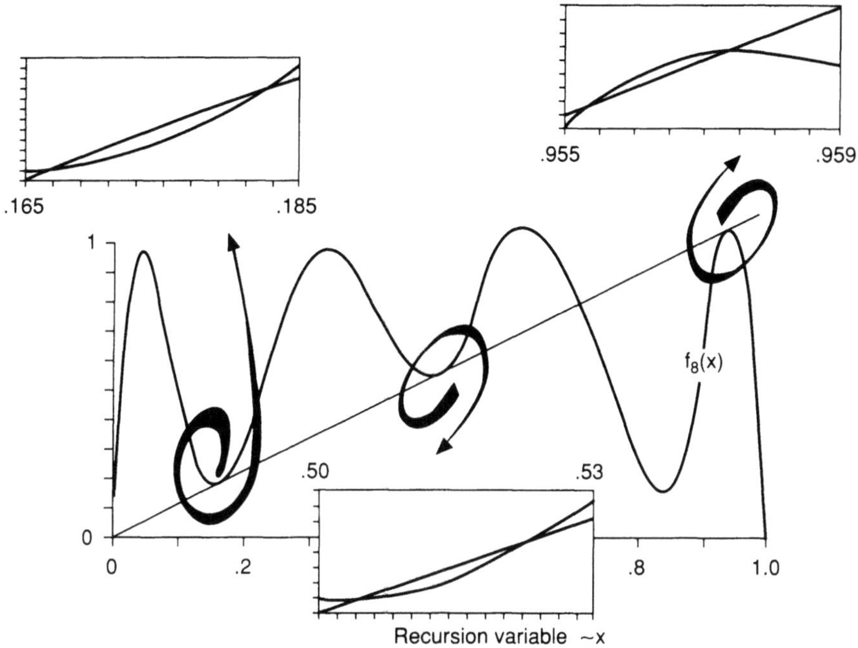

Figure 7-24. *f*-polynomial gives period 3 at 3 = 3.83.

It is of interest to note in Figure 7-24 that all three polynomials intersect the "x-curve" at the value $x = 1 - 1/C$, which is seen to be an unstable root for all, their slopes are either too high or too low. This suggests that $(x - 1 + 1/C)$ is a factor to all recursion formulae determined for integer-period-furcations (see Problem 25).

7.4.5 Higher Furcations

The observations just made for the trifurcation are true for any integer-furcation, and they form the basis for investigating recursions having larger periods. The recursion polynomial for any period, m, has degree $2^m = k$, is "even" about the vertical line, $x = \frac{1}{2}$; is zero at the "end-points" ($x = 0, 1$); and intersects "x-curve" an odd-number of times (which includes odd- and even-tangencies). All the "upward" crossings are unstable nodes which prevent steady recursion, and only the "downward" crossings are candidates for "furcation-values." The condition that the steady recursion relaxes toward them is that the slope of the recursion polynomial have a slope less negative than -1.

With these rules in mind, the first quatrofurcation is found by observing the intersections for the four-step recursion polynomial, $f_{16}(x)$, with the "x-curve," as the coefficient, C, is varied (see Problem 26, Figure 7-25).

IF(f16 > x, 1, −1)

x	f16(x)	HI/LO
0.433	0.433012	1
0.434	0.433999	−1
0.445	0.445007	1
0.446	0.445999	−1

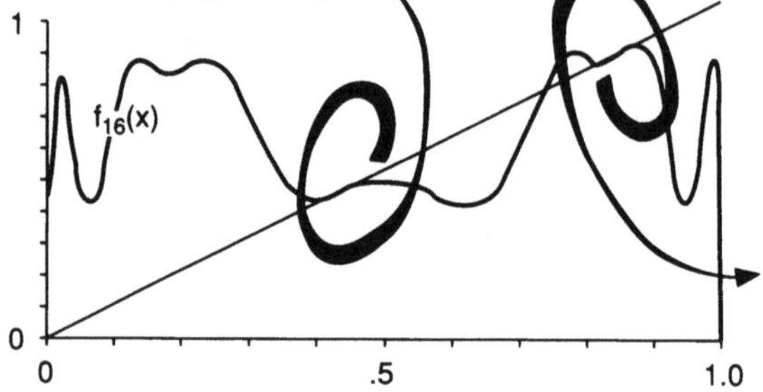

IF(f16 > x, 1, −1)

x	f16(x)	HI/LO
0.847	0.847006	1
0.848	0.847996	−1
0.852	0.852003	1
0.853	0.852992	−1

Figure 7-25. $C = 3.45$ quatrofurcates.

7.4.6 Is $C = 4$ Chaotic?

As shown earlier (Figure 7-23), the Logistics equation at $C = 4$ produces a seemingly random but quite repeatable sequence. Apparently, all the numbers lying between 0 and 1 are generated at one point or another: This repeatable-randomness is deserving of the name "Chaos," and it is easy to understand how it is generated. If one accepts the possiblity that at $C = 4$, all of the f-polynomials of degrees 2, 4, 8, 16, 32, 64, 128, 256, · · · etc., to infinity have the striking characteristic in the region, $0 < x < 1$, that after starting from zero, they grow to apogee at unity, subside to a valley at zero, and repeat this alternating pattern until finally exiting through zero at $x = 1$. Given this behavior, reminiscent of Tchebycheff polynomials, as the degree of the f-polynomial approaches infinity, the number of intersections it makes with the "x-curve" approaches infinity. And, as these intersections are all constrained to lie inside $0 < x < 1$, we have a pragmatic proof of our assertion. Graphic evidence is provided in Figures 7-26 and 7-27, the former illustrating a composite of the lower-degree f-polynomials and the latter illustrating a higher degree polynomial.

Should the coefficient, C, in the Logistics equation exceed 4, the sequence is "nearly" chaotic, as evidenced earlier in Figure 7-18, but it eventually escapes (to infinity) through a window created by one or more of the f-polynomials. By a window we mean an interval in x, say, $a < x < b$, such that the

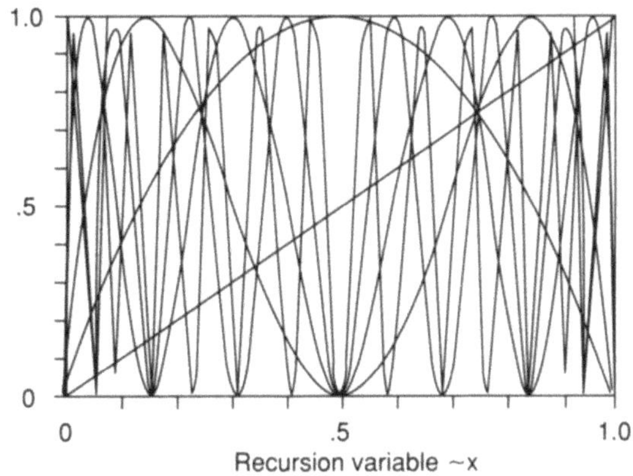

Figure 7-26. Chaotic composite of some *f*-polynomials at C = 4.

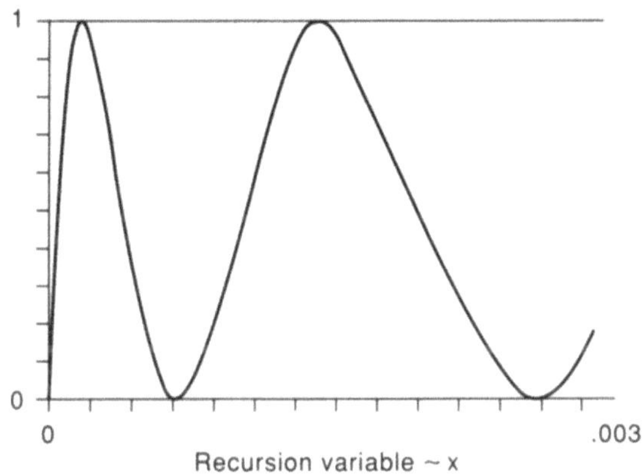

Figure 7-27. Close-up of $f_{128}(x)$ at c = 4.

magnitude of the polynomial $f_k(x)$ of order k is not within the range [0, 1]. (Figure 7-28 shows several such windows.)

7.5 RECURSING DIFFERENTIAL EQUATIONS

7.5.1 The Aircraft Takeoff Acceleration

In Chapter 1, the solution to a first-order linear ordinary differential equation with constant real coefficients (LODE/CRC) was found by recursing the equation from an initial guess. That it gave the correct answer was tacitly assumed, but the conditions of Existence and Uniqueness implicit in that solution are based on the uniform convergence of the recursion process. Indeed, this con-

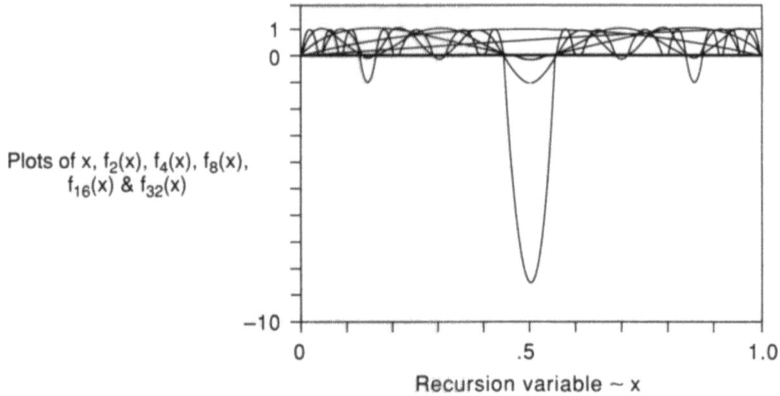

Figure 7-28. "Escape" from chaos through "windows" in the *f*-polynomials at $c = 4.05$.

vergence is another form of stability; and, because the equation is linear, the convergence can be guaranteed—a fact which is often taken lightly, because without it the method has no particular value. The method can be applied to nonlinear ordinary differential equations, just as Jacobi's iteration is applied to nonlinear algebraic equations, as illustrated by a form of the Ricatti equation which represents a simple model for the velocity of an aircraft during takeoff (see Problems 30–32).

During its takeoff, a jet-powered airplane, considered to have constant mass, is accelerated by the engine's thrust, usually constant, which overcomes the drag, taken to vary with the square of speed (as shown in Figure 7-29).

Assuming the aircraft starts from a dead stop, $V(0) = 0$, its acceleration is nearly constant because the drag is small; thus, a first guess at the solution, $V(t)$, is found by neglecting the V-squared term in the equation, giving V_1-dot

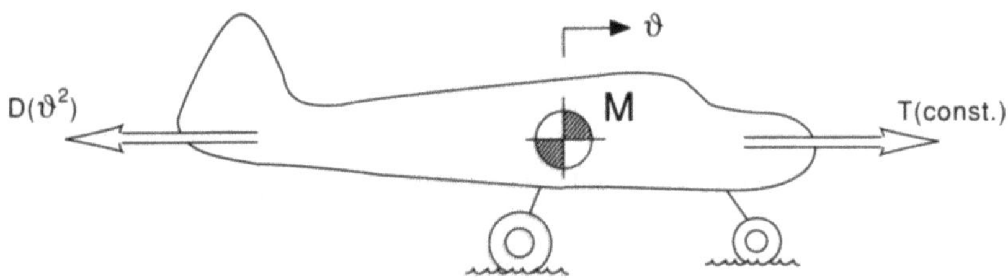

Defining ··· $\dot{\vartheta}_{max} \equiv \frac{T}{M}$ & $\vartheta_{ss}^2 \equiv \frac{T}{d}$ ··· $\frac{\dot{\vartheta}}{\dot{\vartheta}_{max}} = 1 - \left(\frac{\vartheta}{\vartheta_{ss}}\right)^2$

Normalizing ··· $\frac{\dot{\vartheta}}{\dot{\vartheta}_{max}} \equiv \dot{V}$ & $\frac{\vartheta}{\vartheta_{ss}} \equiv V$ ··· gives ··· $\dot{V} = 1 - V^2$

Figure 7-29. Aircraft takeoff.

= 1, or $V_1(t) = t$. With this as a first guess, the equation is used to recurse a "better" solution, $V_2(t)$, which in turn is recursed, giving $V_3(t)$, then $V_4(t)$, and so on (see Problems 29–32):

$$V_0(t) = t$$

$$V_1(t) = t - t^3/3$$

$$V_2(t) = t - t^3/3 + (2/15) \cdot t^5 - (1/63) \cdot t^7$$

$$V_3(t) = t - t^3/3 + (2/15) \cdot t^5 - (17/315) \cdot t^7 + \frac{114}{7505} t^9 - \frac{401}{155{,}925} t^{11} + \frac{4}{12{,}285} t^{13} - \frac{1}{59{,}535} \cdot t^{15}$$

$$V_4(t) = t - t^3/3 + (2/15) \cdot t^5 - (17/315) \cdot t^7 + \frac{62}{2835} t^9 - \frac{5{,}370{,}426}{705{,}130{,}275} t^{11} + \cdots \text{ to } t^{31} \text{ term}$$

Note that each succeeding recursion retains some, if not all, of the terms in its predecessor. In fact, each recursion seems to reaffirm a term from its predecessor and add a term which is then reaffirmed to be the next recursion. In this sense, the recursions appear to be generating a sequence; and the fact that the sequence is reproduced and extended is evidence that it is converging toward a solution. Whether this is true or or not can be verified by substitution (see Problem 29). The question would be satisfactorily answered were the exact solution known; and, indeed, it can be found by separating the variables and integrating, to give,

$$V(t) = \tanh(t) = \frac{1 - e^{-2t}}{1 + e^{-2t}}$$

$$= t - t^3/3 + (2/15) \cdot t^5 - (17/315) \cdot t^7 + \frac{62}{2835} t^9 - \frac{1382}{155{,}925} t^{11} + \cdots$$

which affirms the convergence previously conjectured.

7.5.2 Commentary

The method suggested here is due to Picard (refs. 6 and 7), which he used to establish the conditions for existence and uniqueness of solutions to linear ordinary differential equations. Those conditions are predicated on the proof that the "method of successive approximations" converges uniformly to a unique solution. The extension of the proof to nonlinear equations is generally unsuccessful; as a result, the use of Picard's recursion method may produce erroneous results—notably if the slope of the solution, x-dot(t) should become multivalued (see Problem 34), as this renders the "velocity" ambiguous, preventing its integration. Another difficulty arises if the slope, x-dot(t), becomes infinite, again inhibiting a meaningful integration, which is disallowed as a condition for both existence and uniqueness of solutions to ordinary differential equations.

Recursions need not be limited to numbers (as in Jacobi's iterations or the Mechanic's rule), nor to functions (as with Picard). An interesting situation is explored in Chapter 9, where Lin's method of extracting a quadratic factor from a polynomial is presented. It consists of dividing the polynomial by a trial quadratic, formed from coefficients in the polynomial. The division is repeated to show a first-remainder. The next trial divisor consists of the quadratic remainder found in the next-to-last step. This recursion is repeated and its convergence is measured by observing successive first-degree remainders. If they vanish, the recursion is stable. However the process not only converges in some cases, diverges in others, but bifurcates, trifurcates, and etc.. Aside from the sheer fun, these observations pose a significant problem in stability.

7.6 PROBLEMS

1. Formulate the Jacobi recursion for the case of three linear algebraic equations in three unknowns. Convert the equations into difference equations, using $z = \text{EXP}[sT]$, and draw the signal flow graph. Using Mason's rule, find the system's characteristic polynomial, the roots to which are its eigenvalues. This is a reduced cubic, which can be examined for roots which lie outside the unit curcle in the z-plane. Find the necessary and sufficient condition for stability of the recursion, and portray it as a three-dimensional volume with axes appropriately defined in terms of the coefficients in the equations.
2. Diagonal dominance is a sufficient but not necessary condition for stability. Show this to be the case for three equations in three unknowns, and compare your answer to the necessary and sufficient stability condition of Problem 1. (This is not easy, as it involves a cubic.)
3. Find a set of two algebraic equations, which are not diagonally dominant but display a stable Jacobi iteration.
4. In the case of two algebraic equations, there are two possibilities for a Jacobi iteration (i.e., solving the first for x_1 or solving it for x_2). (In general, there are $n!$ possibilities in n-equations.) For the equations,

$$x_1 - 2x_2 = -1$$
$$2x_1 - x_2 = 1$$

(a) Solve the first for x_1 and recurse.
(b) Solve the second for x_1 and recurse.
(c) Determine the condition of necessary and sufficient stability in each case, as well as the question of diagonal dominance.

5. The Gauss–Seidel method of iterating solutions from linear algebraic equations (ref. 3) consists of guessing only $n - 1$ initial values of the n variables and recursing only the equation of the "missing" variable. This value is used in the next equation, which is recursed just once, to give a new value for its variable, whereupon the two "new" values are used in the single recursion of a third equation. In the case of two equations in two unknowns, make a graph of the equations to be interated and compare the iteration proceedure with that of the Mechanic's rule in Figure 7-3.
6. The Gauss–Seidel iteration (ref. 3) discussed in Problem 5 is said to converge more rapidly to the solution than the Jacobi iteration, which suggests that it is more stable. Following the approach to stability taken in this chapter, determine the eigenvalues of the Gauss–Seidel and compare them to those found for the Jacobi iteration. What can you conclude regarding the relationship between rate of convergence stability of iterations in linear systems?
7. Show that the recursion formula developed in the text for the cube root is stable, in the sense that it converges to the correct solution.
8. Find a recursion formula for the nth root and show that it is stable, in the sense that it converges to the correct solulution.
9. Does the general formula found in Problem 8 for the nth root also apply for rational roots? For irrational roots?
10. Can the Mechanic's rule bifurcate? Discuss.
11. Can you rationalize or explain the presence of an unstable limit cycle in the form of Van der Pol's equation:

$$\ddot{x} - 2\epsilon_0(1 - \beta x^2)\dot{x} + x = 0$$

12. Find $f(x)$ which will "trifurcate."
13. Find $f(x)$ which will "quatrofurcate."
14. Find $f(x)$ which will "N-furcate."
15. Is it possible for N (in Problem 14) to approach infinity? This is what happens in the Logistics Map when the coefficient, C, approaches the value of 4, at which the recursion period becomes infinite and complete chaos is said to have been achieved.
16. In the Mechanic's rule, the starting value, $x(0) = 1 - (1/C)$, has been shown to reproduce itself, suggesting that it is a "root" associated with recursion. Examine the stability of this "root" by perturbing it slightly for the cases of $C = 0.5$, 2, 10/3, and 3.5.
17. In the case where $C = 10/3$, verify that a starting value of 0.1 will find its way to the unstable equilibrium value of 0.7. Are there any other starting values which find their way to 0.7?

18. In the case of $C = 2$, notice that the "top" of the parabola intersects the "x"-curve at zero slope. Prove that this is a stable point, using the arguments followed in the proof that the Mechanic's rule is stable. Note that in both cases the solution is approached from only one side.
19. Let $C = 4.0$, the case of "essential chaos," and recurse backward from the equilibrium value, $x = 0.0$ to $f = 0.0$ to $x = 1.0$ to $f = 1.0$, etc. What is the problem with this backward recursion?
20. In the case, $C = 4.0$, of "essential chaos," generate several sequences, each from a different starting value, and for each determine the probability distribution of the data, to affirm or deny the assertion that the numbers are pseudo-random.
21. At $C = 4.0$, plot the polynomials for 1, 2, 3, 4, $\cdots$, etc. steps, denoted by $f_2(x), f_4(x), f_8(x), f_{16}(x), \cdots$, etc., and note that they are all bounded in the range, $0 < x < 1$, by $0 < f_k(x) < 1$. Note also that they are all symmetric about the value $x = \frac{1}{2}$, and that they become extremely big outside. Is there any relation between these polynomials and the Tschebycheff polynomials?
22. Another way to look at bifurcations is to recognize that,

$$x(n + m) = x(n) \qquad \text{as } n \text{ goes to infinity for } m = 2, 3, \cdots$$

which gives the conditions for an "m-furcation." This amounts to the equality,

$$x = f_k(x), \qquad \text{where } k = 2^m$$

from which the polynomial, $R_m[x] = f_k(x) - x$ is formed. Values of x, denoted as $x(m)$, satisfying $R_m[x]$ are candidates for "steady-state" values, provided they are real. Form $R_1[x]$ and $R_2[x]$.
   (a) Verify that $R_1[x]$ *divides* $R_2[x]$.
   (b) The result of (a) is a quadratic, yielding two solutions. Verify that these are the bifurcation values for $C = 10/3$.
   (c) If $1 > C > 3$, say, $C = 2$, what are the bifurcation values?
   (d) $R_2[x]$ is a cubic. Examine its discriminant for the nature of its roots and verify that there can be no bifurcations outside the region, $-1 < C < 3$.
   (e) Examine $R_2[x]$ for $C = 3$. What are its roots?
   (f) Factor $R_2[x]$ for $C = 10/3$ and investigate the stability of its three roots.
   (g) What can you conclude about the recursions at $C < -1$?
23. Find the seventh-degree polynomial $R_3[x]$ as defined in Problem 22. Of its seven roots, three are trifurcation-values. Show that the first-degree polynomial, $R_1[x]$, *divides* $R_3[x]$ but that $R_2[x]$ does not.
24. Find the range of values for C near to 3.83 for which trifurcations occur,

and classify the intersections of $f_8(x)$ with the "x-curve" at the trifurcation values.

25. Does $f_2(x)$ *divide* $f_4(x)$? Does it *divide* $f_8(x)$? Does it *divide* $f_{16}(x)$? Does it *divide* all the f's?
26. Determine the furcation of period 5 in the Logistics equation by varying the coefficient, C, in the recursion polynomial, $f_{32}(x)$, and observing stable intersections between it and the "x-curve."
27. Recurse the following equation, for various values of C, $x(n + 1) = C \sin (x(n))$, from $x(0) = 1$. Among other things, this gives the following:

    C—steady-state value(s)
    1—0.28
    4.5—apparently chaotic
    4.78—bifurcates between 4.77 and −4.77
    5—bifurcates between 4.88 and −4.93
    5.2—bifurcates between 5.19 and −4.65

28. Recurse the following equation, for various values of C, $x(n + 1) = 2.5 \sin^2 (x(n))$, from $x(0) = 1.0$ This gives, among other things, the following:

    C—steady-state value(s)
    1.5—1.49
    2.0—1.71 and 1.96, i.e., a bifurcation
    2.5—0.90, 1.54, and 2.50, i.e., a trifurcation.
    3.0—0.0, i.e., it goes to the origin

29. Verify that the successively recursed solutions to the aircraft takeoff problem, in section 7.5.1 $V_0(t)$, $V_1(t)$, $V_2(t)$, $V_3(t)$, and $V_4(t)$ are increasingly better solutions by substitution.
30. Modify the aircraft takeoff problem to account for an unlevel runway, like that found in Seattle. Recurse the equation to show how a constant slope affects the takeoff time.
31. Modify the aircraft takeoff problem to account for a headwind (or tailwind). Recurse the equation to show how a constant headwind affects the takeoff time.
32. Use the results of the two preceeding problems to decide whether it is better to takeoff uphill and upwind or downhill and downwind, by evaluating:
    (a) The effect of 1° of slope (both uphill and downhill).
    (b) The effect of various windspeeds (both headwinds and tailwinds).
    (c) Find the "break-even" relation between slope and wind from the results found in (a) and (b).
    (d) At the values of found in (c), formulate the problem with *both* wind and slope, recurse the equation, and see if the "whole is the sum of its parts."

33. The Van der Pol equation contains a nonlinear "damping" term which either adds or subtracts energy as the value of x is either large or small, the result being a limit-stable limit cycle,

$$\ddot{x} - 2\epsilon_0(1 - \beta x^2)\dot{x} + x = 0$$

Should ϵ be very small, the equation is just that of a linear undamped oscillator,

$$\ddot{x} + x = 0$$

giving

$$x(t) = x_1 \sin t | x_1(0) = 0$$

where X_1 is established by either another initial condition or some other condition. One possibility is that, in lieu of the hypothesized cyclic nature of the steady-state solution, the energy dissipated must just balance the energy added, suggesting that the damping term averages out to zero in each cycle. This is established by forcing its integral over one cycle to vanish, where the approximate solution, $x_1(t)$, is assumed to be a valid description of the motion,

$$\int_0^{2\pi} (1 - \beta x^2)\dot{x} \cdot dt = 0$$

giving

$$x_1 = 2/\sqrt{\beta}$$

and the "first-harmonic" solution to the steady state of Van der Pol's equation is

$$x_1(t) = \frac{2}{\sqrt{\beta}} \cdot \sin t$$

The equation can then be recursed from its second derivative to produce a (hopefully) better solution:

$$\ddot{x} = 2(1 - \beta x_1^2)\dot{x}_1 - x_1$$

(a) Show that this recursion produces a solution which retains the first-harmonic term and adds a third-harmonic term proportional to ϵ.
(b) Find a second recursion based on the solution to (a), and show that it adds a fifth-harmonic term proportional to ϵ^2. (This is not for the faint at heart!)
(c) Can you justify the absence of the even-harmonics in the recursions?
(d) When ϵ takes on a large value, say, 10–100, the steady wave-shape looks very much like a square-wave. Does the absence of even harmonics support this observation?

34. In applying Picard's method of successive approximations, a difficulty is encountered when the slope of the trajectory is multivalued. This is the case in the nonlinear equation studied in Chapter 1,

$$|\dot{x}| = -x$$

Try to use Picard's recursion method to find the solution and discover the ambiguity.

7.7 REFERENCES

1. Shields, Paul C., *Elementary Linear Algebra*, Worth, New York (1980).
2. Gelfand, S. I., Gerver, M. L., Kirillov, A. A., Konstantinov, N. N., and Kushnirenko, A. G., *Learn Limits Through Problems*, Gordon & Breach, New York (1969).
3. Smith, W. Allen, *Elementary Numerical Analysis*, Reston Englewood Cliffs, New Jersey (1986).
4. Gleick, James, *Chaos, Making a New Science*, Viking, New York (1987).
5. Campbell, David K., Nonlinear science, from paradigms to practicalities, *Los Alamos Science* **15** (special issue) pp. 218–262 (1987).
6. Spiegel, Murray, *Applied Differential Equations*, Prentice-Hall, Englewood Cliffs, New Jersey (1981).
7. Wylie, C. Ray, *Differential Equations*, McGraw-Hill, New York (1979).
8. Gille, J.-C., Pelegrin, M. J., and DeCaulne, P., *Feedback Control Systems*, McGraw-Hill, New York (1959).

Chapter 8
Digital Simulation

8.0 BACKGROUND

Simulation, once the exclusive domain of analog computation, substitutes a model—usually differential equations—for a real dynamic device whose motion is too large or remote for convenient study. Originally, the operational amplifier was the only device capable of "real-time" calculation because its integrators could be adjusted to an R-C time constant of 1 second. Despite its noise, saturation, and inability to adequately multiply or generate functions, there was no other means in the 1940s and 1950s fast enough. Eventually, however, digital technology overcame its slowness, permitting discrete integrators to run in real time, and so digital simulation has largely replaced analog methods. An example (Figure 8-1) is an aircraft's flight-control system, where the human pilot deflects the surfaces causing motion simulated by a computer. Visual and motion cues are presented to the pilot closing the loop.

Today, the concept of simulation has been broadened to include any system of equations, solved in "real-time" when necessary, and adjoined by hardware or people as appropriate. The condition that a digital simulation run at "real-time" requires that the execution time of each recursion be less than the elapse of problem time it represents. Most mechanical, thermal, hydraulic, and pneumatic devices fall into this category. Real time calculations are not always necessary, particularly when the system analysis is not instantaneously required. We do not address these real time considerations here, rather we present techniques of analyzing nonlinear systems numerically.

8.1 RECURSION FORMULAE: FUNDAMENTAL TO DIGITAL SIMULATION

By its very nature, digital simulation involves discrete data, which is combined using the operations of common algebra (i.e., addition, subtraction, multiplication, and division). Combinations form algorithms which approximate irrational algebraic functions (such as logarithms, exponentials, and circular and

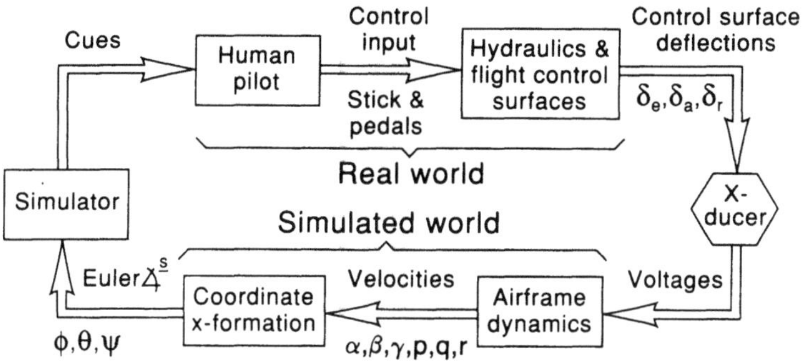

Figure 8-1. Flight simulation.

hyperbolic functions), often requiring recursions, which prompts two important questions. First, is the recursion stable (i.e., does it approach a specific value)? And second, does it converge to the desired value?

Addressing stability, consider a set of linear algebraic equations,

$$[A]_{n\times n}\cdot x]_{n\times 1}=b]_{n\times 1}$$

An explicit solution can be found by either Cramer's rule or the Gauss Elimination to yield symbolically,

$$x]_{n\times 1}=[A]^{-1}_{n\times n}\cdot b]_{n\times 1}$$

provided the matrix is not singular. Jacobi's iteration (ref. 1) is an implicit recursion generating successively closer approximations, in which case it is said to be stable. However, it may generate successively worse approximations, diverging from the solution, in which case it is unstable. Jacobi's iteration is illustrated by a system of two algebraic equations,

$$a_{11}\cdot x_1+a_{12}\cdot x_2=b_1$$

$$a_{21}\cdot x_1+a_{22}\cdot x_2=b_2$$

Each of these is "solved" for a different unknown,

$$x_1=\frac{1}{a_{11}}\cdot(b_1-a_{12}\cdot x_2)$$

$$x_2=\frac{1}{a_{22}}\cdot(b_2-a_{21}\cdot x_1)$$

Note there is a second possibility (i.e., solving the first for x_2 and the second for x_1). In more general cases, there will be $n!$ such possibilities. Regardless of which is chosen, a recursion algorithm is generated by guessing values of x_1

and x_2 to be entered on the right side. The equations then produce new estimates which are entered into the right side to produce yet another new estimate. It is an implicit recursion which follows the algorithm,

$$x_1^{(n+1)} = \frac{1}{a_{11}} \cdot (b_1 - a_{12} \cdot x_2^{(n)})$$

$$x_2^{(n+1)} = \frac{1}{a_{22}} \cdot (b_2 - a_{21} \cdot x_1^{(n)})$$

where $x_1^{(0)}$, $x_2^{(0)}$ are guessed.

A specific example of Jacobi's iteration demonstrates both the stable and unstable cases. The equations to be solved are:

$$2x_1 - x_2 = 1$$

$$x_1 - 2x_2 = -1$$

which are straight lines intersecting at the point (1, 1) as shown in Figure 8-2. Solving the first for x_1 yields a stable iteration as indicated in Figure 8-3 for two different initial guesses. Solving the first for x_2 yields an unstable iteration as indicated in Figure 8-4 for two different initial guesses.

A criterion of stability for Jacobi's iteration is that [A] be diagonally dominant; its diagonal elements having larger magnitudes than the sum of the magnitudes of all other row elements. This is a sufficient condition, meaning that [A] need not always be diagonally dominant for stability. And, of course, it holds only for linear equations, leaving the question of stability of nonlinear equations unanswered.

The necessary and sufficient condition for a stable Jacobi iteration can be found by considering the equations to represent a feedback system. Let us in-

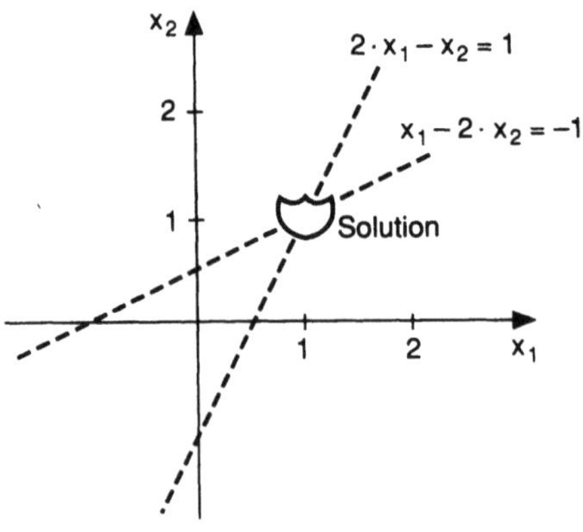

Figure 8-2. Solution to two algebraic equations.

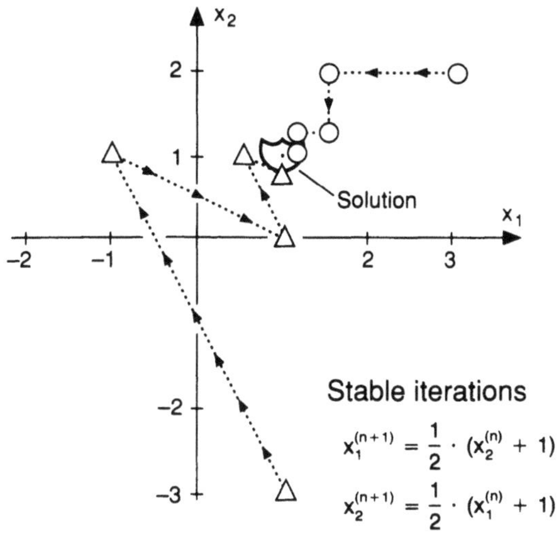

Figure 8-3. Stable iterations of two algebraic equations.

terpret the variables x_1 and x_2 as functions of some independent variable, say, t. Then we can pursue stability investigations in terms of the behavior relative to t. In this way, we look for methods of analyzing stability using our existing bag of tricks, namely Laplace transforms.

We recall that if the characteristic equation has right half-plane poles, we have an unstable system. By rewriting the algebraic equations as a function of

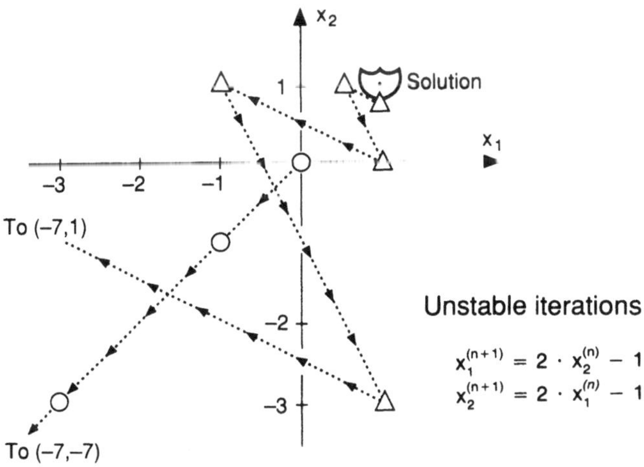

Figure 8-4. Unstable iterations of two algebraic equations.

t, we get

$$a_{11}x_1(t) + a_{12}x_2(t) = b_1$$

$$a_{21}x_2(t) + a_{22}x_2(t) = b_2$$

These equations, in recursion form are

$$x_1(n + 1) = [b_1 - a_{12}x_2(n)]/a_{11}$$

$$x_2(n + 1) = [b_2 - a_{22}x_1(n)]/a_{22}$$

where we recognize that for an advance increment T we move from condition "n" to condition "$n + 1$." Thus, when we investigate Laplace transforms, we find that

$$X_1(s) = \mathcal{L}\{x_1(t)\}$$

and

$$\mathcal{L}\{x_1(t + T)\} = X_1(s)e^{sT}$$

Thus, applying these rules we have

$$X_1(s)e^{sT} = [b_1/s - a_{12}X_2(s)]/a_{11}$$

$$X_2(s)e^{sT} = [b_2/s - a_{21}X_1(s)]/a_{22}$$

which yields, while dropping the notation $X_1(s)$ and using X_1,

$$X_1[e^{sT} - (a_{12}a_{21}/a_{11}a_{22})e^{-sT}] = \frac{b_1}{a_{11}s} - b_2\frac{a_{12}}{a_{11}^2 s} \cdot e^{-sT}$$

provided that $x_1 \neq 0$. Now, in the dynamic system [i.e., $x = x(t)$], we have rules for interpreting system stability without forcing functions. Thus, it is permissible to investigate the basic system with $b_1 = b_2 = 0$ to see if the system is stable. Under this condition we get

$$e^{sT} - (a_{12}a_{21}/a_{11}a_{22})e^{-sT} = 0$$

However, $s = \sigma + j\omega$ and stability requires $\sigma < 0$. Substituting and setting reals = reals and imaginaries = imaginaries, we get

$$e^{2\sigma T} \cos 2\omega T = a_{12}a_{21}/a_{11}a_{22}$$

$$e^{2\sigma T} \sin 2\omega T = 0$$

from which the second yields $\omega = n\pi/2T$, creating values of $\cos 2\omega T = \pm 1$, so that

$$e^{2\sigma T} = \pm(a_{12}a_{21}/a_{11}a_{22})$$

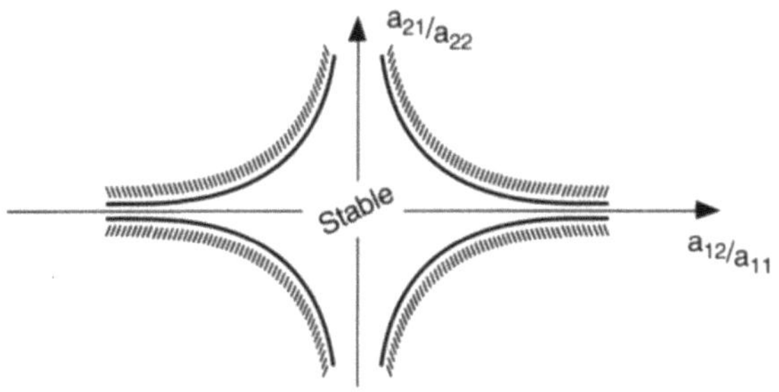

Figure 8-5. Boundaries for stable iterations.

and at the stability boundary, $\sigma = 0$, $e^{2\sigma T} = 1$, yielding the equation

$$1 \pm a_{12}a_{21}/a_{11}a_{22} = 0$$

which is the necessary and sufficient condition separating stable and unstable regions (Figure 8-5).

8.1.1 An Outline of the Path to Stable Iterations

Before addressing the specific issue of stability, we must introduce some mathematical concepts which will enable us to reach some conclusions about stability. In particular, we will investigate the sampling process (which converts an analog or continuous signal into discrete data allowing digital computer techniques to be employed), conversion of an impulse function into a finite pulse having specific widths and amplitudes (resulting in a box car generator), the need for Fourier analysis, a hold circuit, defining sample times, and Z transforms. This all leads to a transfer function analysis in the Z domain (as compared with the Laplace s domain) and, hence, stability and convergence considerations.

The use of a sampler implies that we periodically extract a piece of data from the stream of data, allow this one piece of data to act on the system dynamics, and then assess the result. Symbolically, this is shown as in Figure 8-6, where it is usual to have the impulse train with equidistant sample times (i.e., every T seconds a new sample is taken). We define the impulse as the Dirac delta function,

$$u_0(t), \quad \text{or} \quad \delta(t)$$

As the ability to work with a zero-width pulse is physically impossible, we accept reality and introduce a device which "holds" the data level we have

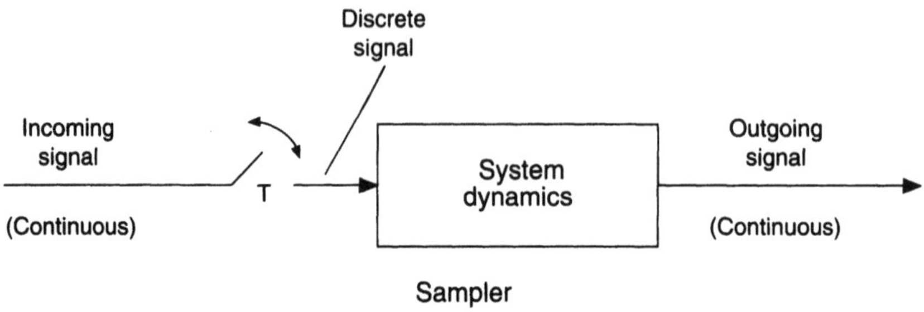

Figure 8-6. Sampler creates modulated impulses as sampled data operation.

sampled mathematically as an impulse for a finite time T, until we take the next sample (Figure 8-7).

The output "boxcar" signal is then evaluated by means of the Laplace transform of its Fourier series, yielding a means of applying Laplace transforms to the input/output of the system dynamics. However, because the math gets a bit difficult, we introduce Z transforms to simplify our mathematical life.

So now let's back up, define what we mean by each of these terms, and then retrace our steps allowing us to develop the theory.

The Dirac delta function (or impulse function), $u_0(t)$, is a mathematical oddity which has certain properties, and we will only suggest a few here. We define a unit pulse $P(t)$, such that

$$P(t) = [u_{-1}(t) - u_{-1}(t - h)]1/h$$

as shown in Figure 8-8.

As we shrink h to zero, the width decreases while the height increases, maintaining a unit area. In the limit, we obtain

$$u_0(t) = \lim_{h \to 0} P(t) = \text{Dirac delta function}$$

We note that

$$\int_{-\infty}^{\infty} u_0(t)\, dt = 1$$

and

$$\int_{t_1}^{t_2} x(t)\, u_0(t - \tau)\, dt = x(\tau), \qquad t_1 < \tau < t_2$$

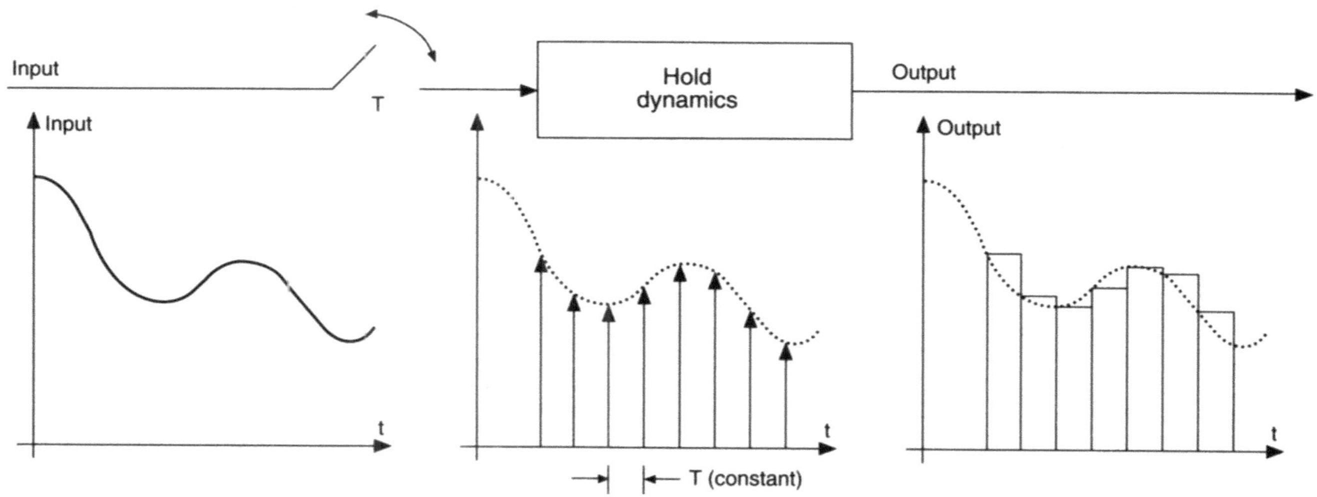

Figure 8-7. Hold reconstructs piecewise—continuous signal.

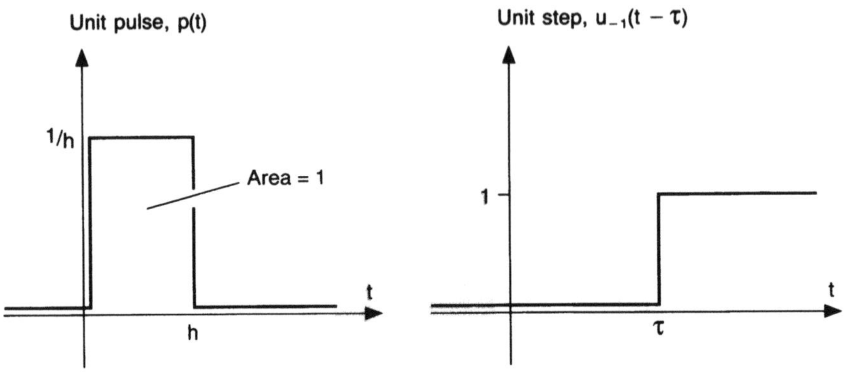

Figure 8-8. The unit pulse and unit step defined.

Now suppose we need to digitize a signal and have abstracted only one sample point. Then we would have insufficient information since the sample point corresponds to only one Dirac delta function times the "height" of the actual signal. Compare this with one frame of a motion picture. To get a better "picture" we need a train of such impulses/samples as is depicted in Figure 8-9 and is designated $\delta_T(t)$.

The total picture is better approximated by using an impulse train to take the samples, and the analog signal $x(t)$ is approximated by $x^*(t)$, where

$$x^*(t) = \delta_T(t)\, x(t)$$

To apply Laplace transforms requires some familiarity with Fourier series of these impulse trains. Previously, we used Fourier series to study the Describing Function as applied to continuous signals. Now we apply the same concepts to discrete pulse trains. We recognize that the signal being analyzed is assumed to be periodic, even if it isn't; however, as the pulse train is periodic of period T, this assumption isn't necessary, and it is possible to apply the series directly. Instead of sine and cosine expansions, it is preferable to use the exponential form, and so we say,

$$\delta_T(t) = \sum_{n=-\infty}^{\infty} C_n e^{jn\omega_s t}$$

where ω_s = sampling frequency $= 2\pi/T$

$$C_n = \frac{1}{T}\int_{-T/2}^{T/2} \delta_T(t)\, e^{-jn\omega_s t}\, dt = \frac{1}{T}$$

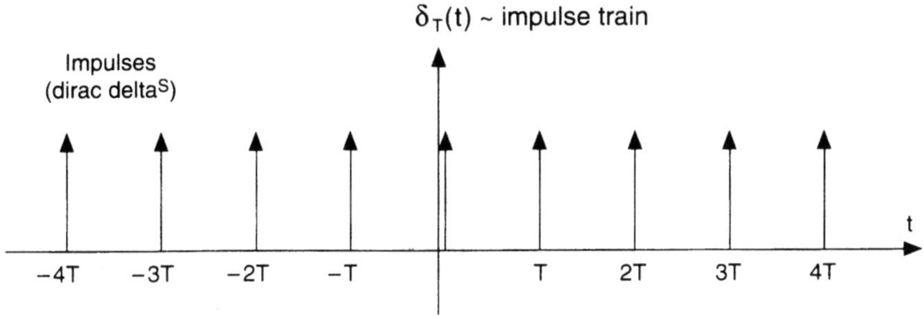

Figure 8-9. The impulse train.

Consequently,

$$x^*(t) = x(t)\frac{1}{T}\sum_{n=-\infty}^{\infty} e^{jn\omega_s t}$$

As this presents a neat way to express the sampled/digitized signal, we now are ready to analyze the stability characteristics. Hence, we apply the Laplace transform; and using the complex-shift property, we find

$$x^*(s) = \frac{1}{T}\sum_{-\infty}^{\infty} x(s + jn\omega_s)$$

We also note that since the shift theorem applies for any argument s, let the argument be $s + jm\omega_s$ and then

$$x^*(s + jm\omega_s) = \frac{1}{T}\sum_{-\infty}^{\infty} x[s + j(m + n)\omega_s]$$

and since the index n contains all integers, positive and negative, if m is also an integer, it is already included in the summation. Therefore, we can set $m = -n$ and obtain

$$x^*(s) = x^*(s + jn\omega_s)$$

As we are digitizing signals at more than one point in the system; in between such sample points, the system dynamics continue to act. Thus, we have situations which result in a sequence of information which have: (a) continuous signal, (b) sampling, (c) sampled signals (pulses) driving dynamical elements yielding a continuous signal, and (d) the "sampled" continuous signal

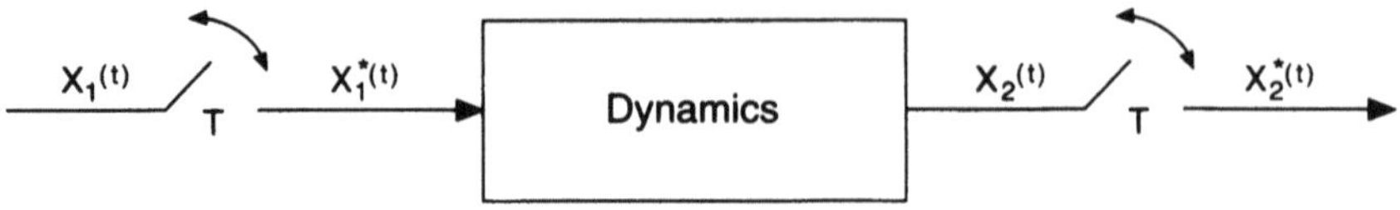

Figure 8-10. Surrounding continuous dynamics with samplers.

being sampled and thus digitized (see Figure 8-10). Hence, we need to evaluate this sequence of events mathematically.

The issue then, is moving from $x_1^*(t)$ to $x_2^*(t)$. We assume the dynamics to be $G(s)$, so that

$$x_2(s) = x_1^*(s)\ G(s)$$

and

$$x_2^*(s) = [x_2(s)]^* = [x_1^*(s)\ G(s)]^*$$

Since the sampling process implies we superimpose an impulse train on the signal, we get

$$x_2^*(s) = \frac{1}{T} \sum_{-\infty}^{\infty} [x_1^*(s + jn\omega_s)\ G(s + jn\omega_s)]$$

Furthermore, we recognize that the signal x_2 is not digitized after it is acted upon by the dynamics G, but is continuous. Thus, since

$$x_1^*(s) = x_1^*(s + jn\omega_s)$$

then

$$x_2^*(s) = \frac{1}{T} \sum_{-\infty}^{\infty} x_1^*(s)\ G(s + jn\omega_s) = \frac{1}{T} x_1^*(s) \sum_{-\infty}^{\infty} G(s + jn\omega_s)$$

$$x_2^*(s) = x_1^*(s)\ G^*(s)$$

As we will see as we progress, Z transforms are a convenient way to manipulate the sampled data (i.e., the * quantities).

8.2 THE SAMPLING PROCESS CREATES DISCRETE DATA

The sampling process is the means by which a continuous signal is transformed into discrete data. Once discrete, they are handled by finite difference equations and organized as recursion formulae. Driving in the rain is an example of sampling, where the wiper provides a series of instantaneous views of the road. Another is movie photography, which produces sequential images of a scene, where the repetition rate must be fast enough to record all action as otherwise intersample events would go unnoticed. From an analytic standpoint, this rate must be at least double the highest frequency contained by the continuous signal, and is known as its Nyquist rate, which guarantees that no information is lost by sampling.

The movie camera suggests that a model for sampling should be some form of pulse modulation, where the pulse height and width would be analogous to the aperture and time settings, respectively. The resulting sampled data take the form of finite duration pulses, whose width is small compared to the sample spacing, T, of successive leading edges. Analytically, this is hard to deal with unless the width is arbitrarily small. But to preserve a latent image on film as the exposure is shortened, the aperture must grow, and while Mother Nature prescribes a limit on aperture, analytically there is no constraint since our pulse approaches the unit impulse (Dirac delta), $\mu_0(t)$. With it the value of a function at any time, $t = \tau$, can be extracted by integration (Figure 8-11).

The integrand is a weighted impulse and, therefore, a time function, while $x(\tau)$ is purely a number associated with a particular time. The former is referred to as "sampled-data," while the latter is referred to as "discrete-" or "digital-data" in order to preserve the difference.

The movie camera advances the film 24 times each second so that every 41 2/3 milliseconds a new sample of the action is taken. The equivalent sampling of a function employs an impulse train, $\delta_T(t)$, to be modulated as shown in Figure 8-12. Since $\delta_T(t)$ is cyclic it can be expanded in a Fourier series to produce a basic relation for the sampled data, x^*, and its Laplace transform, $X^*(s)$.

8.3 AN APPROACH TO DIGITAL SIMULATION: INTRODUCE SAMPLERS

Given a system of cascaded filters, an attempt to create a digital simulation would seem to be the introduction of samplers between filters in order to extract sampled data for each function (as shown in Figure 8-13).

8.3.1 The Problem

Each filter is driven not by its continuous input, but rather by the sampled data of that input. That is, f_i sees a modulated impulse train and not a continuous signal. Thus, there is a contradiction because the samplers, while needed for sampled data (and thus digital data), by their very presence destroy the validity of the representation.

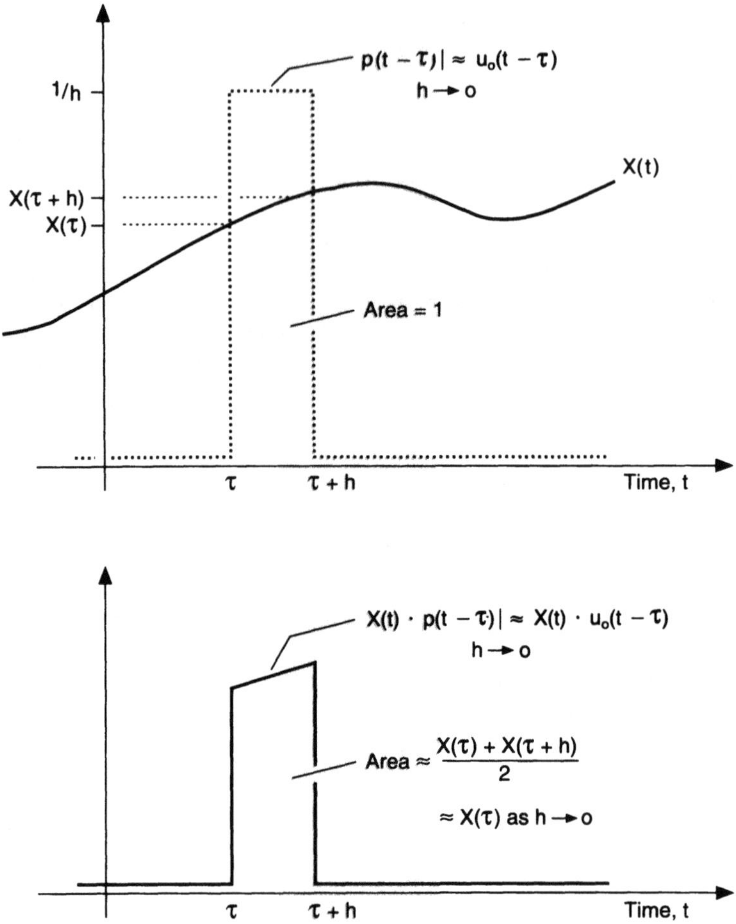

Figure 8-11. Extracting *x(t)* with an impulse.

8.3.2 The Resolution—A Boxcar

Since the samplers are needed for sampled data but the filters need continuous inputs, a hold device is added, converting the sampled data back to a piecewise continuous signal, approximating the unsampled signal (Figure 8-14).

The validity of a simulation based on sample/hold insertions rests with how closely the reconstructed signal approximates its continuous parent signal. To examine how good this approximation can be, a specific hold device is chosen. The simplest and most utilitarian is the Zero-Order Hold (ZOH), and its combination with the sampler is called a Boxcar generator (Figure 8-15).

The sampler ···

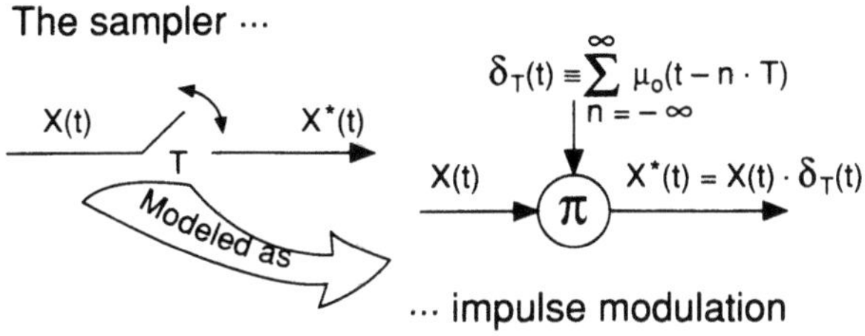

··· impulse modulation

Observe ··· δ_T is cyclic.

It's Fourier series ··· $\delta_T = \sum C_n \cdot e^{j \cdot n\,\omega_s\,t}$

$$\text{Where } C_n = \frac{1}{T} \int_{-T/2}^{T/2} \delta_T(t) \cdot e^{-j\,n\,\omega_s\,t}\,dt = \frac{1}{T}$$

$$\text{So} \cdots \delta_T = \frac{1}{T} \sum_{-\infty}^{\infty} e^{j\,n\,\omega_s\,t}$$

$$\text{The sampled data} \cdots x^*(t) = \frac{1}{T} \sum_{-\infty}^{\infty} x(t) \cdot e^{j\,n\,\omega_s\,t}$$

And ··· $X^*(s) = \frac{1}{T} \sum_{-\infty}^{\infty} X(s + jn\omega_s)$ by complex shift theorem.

Figure 8-12. The sampling process.

Given ··· cascaded analog filters

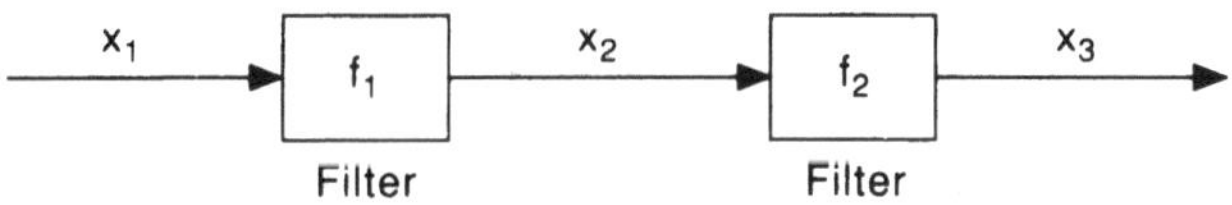

Separate ··· by samplers (synch'd)

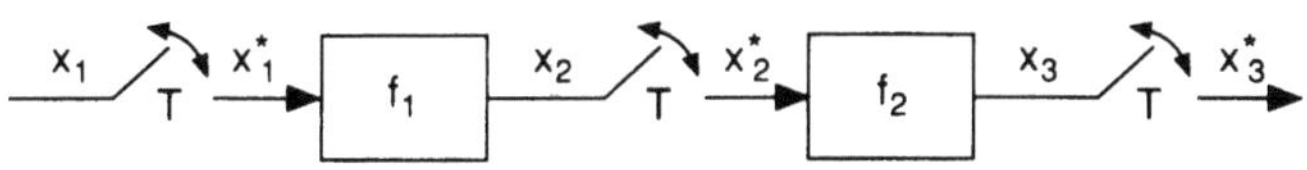

Where ... $X_2^*(s) = X_1^*(s) \cdot F_1(s)$

So ... $X_2^*(s) = [X_1^*(s) \cdot F_1(s)]^* = X_1^*(s) \cdot F_1^*(s)$

Likewise ... $X_3^*(s) = X_2^*(s) \cdot F_2^*(s)$

Figure 8-13. Cascade sampling.

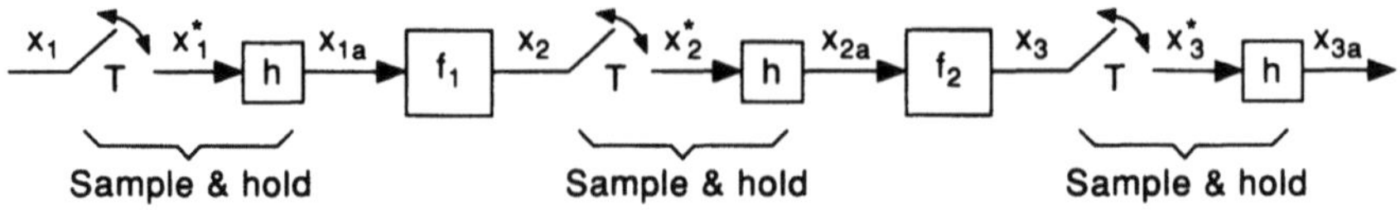

Question ~ Does $x_{1a} \approx x_1$? ···

··· if so, we have a basis for simulation.

Figure 8-14. Add a "hold" to each sampler.

The ZOH holds the value of the last sample until a new value arrives. How well the Boxcar output approximates its input may be determined by the relation between the continuous signals x_i and x_{i_a} in Figure 8-14. With reference to Figure 8-15: x_1 relates to x_1^* by

$$X_1^*(s) = \frac{1}{T} \cdot \sum_{n=-\infty}^{\infty} X_1(s + jn\omega_s)$$

and, x_1^* relates to x_{1_a} by,

$$X_{1_a}(s) = \frac{1 - e^{-sT}}{s} \cdot X_1^*(s)$$

so, x_1 relates to x_{1_a} by,

$$X_{1_a}(s) = \frac{1 - e^{-sT}}{s} \cdot \frac{1}{T} \sum_{n=-\infty}^{\infty} X_1(s + jn\omega_s)$$

(irrational!)

8.3.3 The Magic

(i) Mother Nature forces x_1 to be "low-pass."
(ii) We adopt "rapid sampling," T is very small.

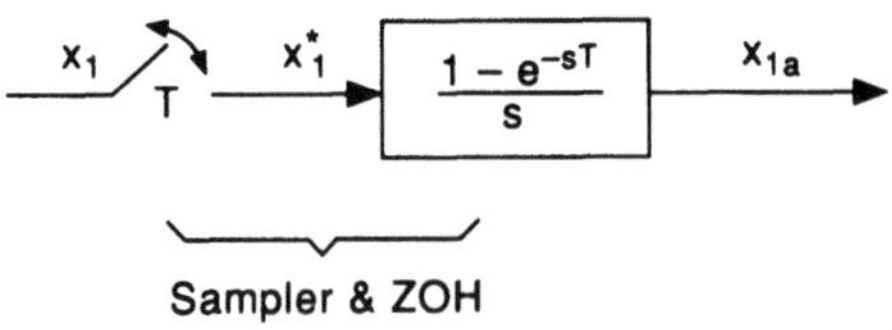

$h \equiv h_0(t) = \mu_{-1}(t) - \mu_{-1}(t - T)$

Figure 8-15. A boxcar generator—a sampler and zero-order hold.

Together

$$X_1^*(s) \approx \frac{X_1(s)}{T}$$

giving

$$X_{1_a}(s) \approx \frac{1 - e^{-sT}}{s} \cdot X_1(s)$$

$$\approx X_1(s) \cdot \left[1 - \frac{sT}{2!} + \cdots\right]$$

$$\approx X_1(s) \cdot e^{-sT/2} \approx X_1(s)$$

T is very small.

8.3.4 Conclusion

An approximate digital simulation of an analog system consists of: separating each analog filter by a Boxcar; finding the results of the *sample times only* (i.e., $t = k \cdot T$).

8.4 CONCEPT OF PULSE FILTERS IMPLICIT IN DIGITAL SIMULATIONS

With reference to Figure 8-16, the sampled-data outputs of successive samplers surround the analog filter and its associated hold. Thus, the outputs of successive samplers can be thought of as surrounding a device which is stimulated by pulses and in turn produces pulses (i.e., a pulse filter).

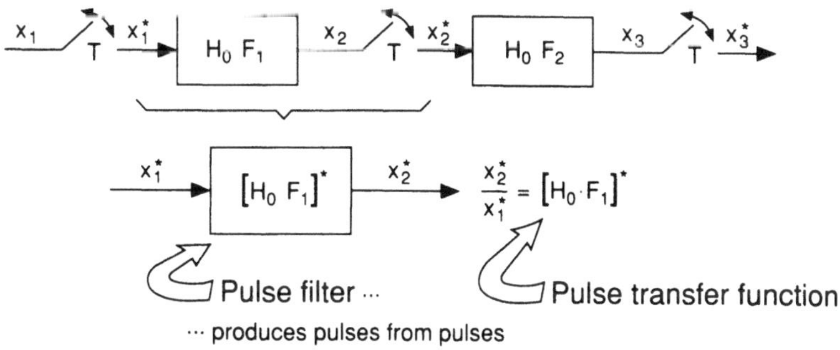

Figure 8-16. Concept of pulse filters . . . implicit in digital simulations.

Examining the transfer function of the pulse filter using a ZOH,

$$
\begin{aligned}
[H_0 \cdot F_1]^* &= \frac{1}{T} \cdot \sum_{n=-\infty}^{\infty} H_0(s + jn\omega_s) \cdot F_1(s + jn\omega_s) \\
&\approx \frac{1}{T} \cdot H_0(s) \cdot F_1(s), \qquad \text{iff } T \rightarrow 0 \text{ and ``lo-pass''} \\
&\approx \frac{1}{T} \cdot \frac{1 - e^{-sT}}{s} \cdot F_1(s) \\
&\approx e^{-sT/2} \cdot F_1(s) \approx F_1(s)
\end{aligned}
$$

8.5 INTRODUCING Z-TRANSFORMS

Almost any hold device contains delays which manifest themselves as exponentials, e^{-sT}. Mixing a rational, $F(s)$, with an irrational prevents the use of common algebra, rendering operational calculus seemingly impossible. If, however, we concern ourselves solely with the sampled data, ignoring the intersample dynamics, a neat solution is presented. For all functions, $x(t)$, having rational Laplace transforms, $X(s)$, the sampled function, $x^*(t)$ has as its Laplace transform $X^*(s)$ in which the argument, s, appears only in the exponential, e^{sT}. Rational Laplace transforms, $X(s)$, derive from Linear Ordinary Differential Equations with Constant Coefficients (LODE/CC), and therefore virtually all linear models of systems are included. The proof that $X^*(s) = X^*(e^{sT})$ follows.

8.5.1 Assertion

$X^*(s)$ is rational in e^{-sT} if $X(s)$ is rational in s.

8.5.2 A Preliminary

Since $X(s)$ is rational, then it expands in partial fractions and it contains only the terms

$$
\frac{1}{s + \alpha}, \quad \frac{1}{(s + \alpha)^2}, \quad \ldots, \quad \frac{1}{(s + \alpha)^n}; \qquad \alpha \in C
$$

8.5.3 Proof of Assertion

Each term is found by residues (ref. 2):

$$
X^*(s) = \sum_R \text{residues of } \Phi(\psi) \text{ in poles of } X(\psi)
$$

where

$$\Phi(\psi) \equiv \frac{X(\psi)}{1 - e^{(\psi - s)\cdot T}}$$

which is true for

$$X(s) = \frac{1}{s + \alpha} \Rightarrow \Phi(\psi) = \frac{1}{\psi + \alpha} \cdot \frac{1}{1 - e^{(\psi - s)\cdot T}}$$

$$X^*(s) = (\psi + \alpha) \cdot \Phi(\psi)|_{\psi = -\alpha} = \frac{1}{1 - e^{-\alpha T} \cdot e^{-sT}}$$

which is true for

$$X(s) = \frac{1}{(s + \alpha)^2} \Rightarrow \Phi(\psi) = \frac{1}{(\psi + \alpha)^2} \cdot \frac{1}{1 - e^{(\psi - s)\cdot T}}$$

$$X^*(s) = \frac{d}{d\psi} \{(\psi + \alpha^2 \cdot \Phi(\psi)\}|_{\psi = -\alpha} = \frac{d}{d\psi} \left\{\frac{1}{1 - e^{(\psi - s)\cdot T}}\right\}\Bigg|_{\psi = -\alpha}$$

which is true for

$$X(s) = \frac{1}{(s + \alpha)^n} \Rightarrow \Phi(\psi) = \frac{1}{(\psi + \alpha)^n} \cdot \frac{1}{1 - e^{(\psi - s)\cdot T}}$$

$$X^*(s) = \frac{d^{n-1}}{d\psi^{n-1}} \{(\psi + \alpha)^n \cdot \Phi(\psi)\}|_{\psi = -\alpha} = \frac{d^{n-1}}{d\psi^{n-1}} \left\{\frac{1}{1 - e^{(\psi - s)\cdot T}}\right\}\Bigg|_{\psi = -\alpha}$$

Q.E.D.

Thus, $X^*(s) = X^*(e^{sT})$ and a very natural substitution is $z = e^{sT}$, which gives the z-transform,

$$X(z) \equiv \mathcal{L}\{x^*(t)\}|_{e^{sT} = z} = X^*(e^{sT})|_{e^{sT} = z}$$

In this way, the z-transform is defined through the mapping from the s-plane to a new complex z-plane.

Every point in the s-plane maps to a single point in the z-plane, but the converse is not true as is evident in the inverse relation,

$$s = \frac{1}{T} \cdot \log(z) = \frac{1}{T}\{\ln |z| + j\, z_p + j \cdot 2 \cdot k \cdot \pi\}$$

Now, using an obvious relation denoting $x^*(t)$ as a series of weighted impulses, the z-transform is,

$$x^*(t) = x(0) \cdot \mu_0(t) + x(T) \cdot \mu_0(t - T) + \cdots$$

$$+ x(n \cdot T) \cdot \mu_0(t - n \cdot T)$$

$$= \sum_{k=0}^{n} x(k \cdot T) \cdot \mu_0(t - k \cdot T)$$

$$X^*(s) = \sum_{k=0}^{n} x(k \cdot T) \cdot e^{-ksT}$$

$$X(z) = \sum_{k=0}^{n} x(k \cdot T) \cdot z^{-k}$$

8.6 THE PULSE FILTER NOW BECOMES A Z-TRANSFORM

The substitution (mapping) of z for the complex exponential constitutes the z-transform in which the pulse filter is expressed as a rational function of the new complex variable z (Figure 8-17).

8.7 INTRODUCTORY EXAMPLE OF DIGITAL SIMULATION

A solution by digital simulation is sought for the linear equation with step input from rest,

$$\dot{x}_2 + x_2 = x_1, \qquad x_1 = \mu_{-1}(t), \qquad x_2(0) = 0$$

$$x_2 = 1 - e^{-t} \qquad \text{exact solution}$$

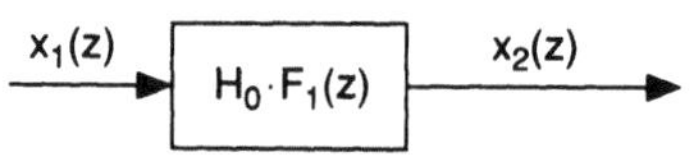

$$\mathcal{Z}\left\{[H_0(s) \cdot F_1(s)]^*\right\}$$

$$\text{Where, } H_0(s) \cdot F_1(s) = \frac{1 - e^{-sT}}{s} \cdot F_1(s)$$

$$\text{So, } H_0F_1(z) = \mathcal{Z}\left\{(1 - e^{-sT}) \cdot \left[\frac{F_1(s)}{s}\right]^*\right\}$$

$$= (1 - z^{-1}) \cdot \mathcal{Z}\left\{\left[\frac{F_1(s)}{s}\right]^*\right\}$$

Figure 8-17. The pulse filter now becomes a z-transform.

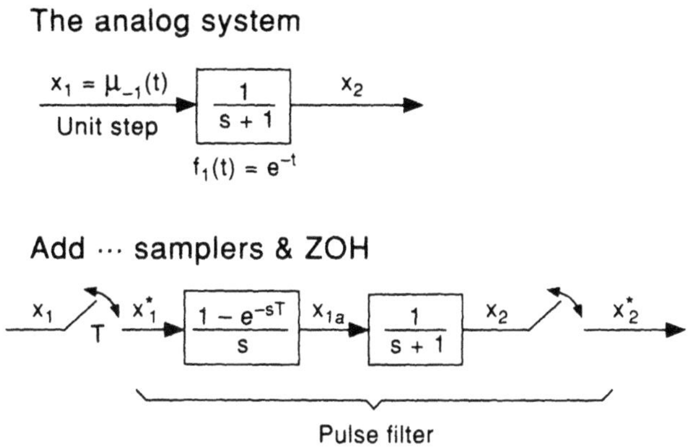

Figure 8-18. Simulation example.

The simulation develops by adding a Boxcar and sampling the output as shown in Figure 8-18. The pulse transfer function is,

$$\frac{X_2^*}{X_1^*}(s) = \left[\frac{1 - e^{-sT}}{s} \cdot \frac{1}{s + 1}\right]^*$$

And the pulse filter becomes, using partial fractions

$$\mathcal{Z}\left\{\left[\frac{1 - e^{-sT}}{s} \cdot \frac{1}{s + 1}\right]^*\right\} = \frac{1 - e^{-T}}{z - e^{-T}} = \frac{x_2}{x_1}(z)$$

Cross-multiplying gives the recursion formula, using z as an advance in time,

$$x_2^{(n+1)} = e^{-T} \cdot x_2^{(n)} + (1 - e^{-T}) \cdot x_1^{(n)}$$

Starting from "rest", $x_2(0) = 0$ and with a unit step input,

$$x_2^{(1)} = e^{-T} \cdot 0 + (1 - e^{-T}) \cdot 1 \qquad = 1 - e^{-T} = x_2(T)$$

$$x_2^{(2)} = e^{-T} \cdot (1 - e^{-T}) + (1 - e^{-T}) \cdot 1 \qquad = 1 - e^{-2T} = x_2(2 \cdot T)$$

$$x_2^{(3)} = e^{-T} \cdot (1 - e^{-2T}) + (1 - e^{-T}) \cdot 1 \qquad = 1 - e^{-3T} = x_2(3 \cdot T)$$

$$\vdots \qquad\qquad\qquad\qquad \vdots$$

$$x_2^{(k)} = e^{-T} \cdot (1 - e^{(n-1)T}) + (1 - e^{-T}) \cdot 1 = 1 - e^{-kT} = x_2(k \cdot T)$$

$$\vdots$$

Observe—This agrees exactly.
Conjecture—A lucky chance.

It is appropriate to examine the pulse filter for stability, which is possible as the system is linear. In the s-plane, stable poles lie inside the left half-plane, which maps to the interior of a unit circle in the z-plane. The system's pole is at e^{-T}, which lies inside the stable circle for $T > 0$; hence, the recursion is always stable.

8.8 DIGITALLY SIMULATING A FEEDBACK SYSTEM

The introductory example of Section 8.7 is based on an explicit relation between the system's input and output. That same system can be interpreted (Figure 8-19) as a feedback system, providing the basis for an implicit relation between input and output. The simulation proceeds as follows.

8.8.1 Simulating the Loop Dynamics

Insertion of samplers and hold generates the recursion formula (Figure 8-20).

8.8.2 Simulating the Adder

The adder contains no dynamics, so there is no need to reconstruct data. The fact sampling distributes over addition confirms that it is a linear process. It is surrounded by samplers (Figure 8-21).

8.8.3 Closing the Loop

The loop is closed as shown in Figure 8-22.

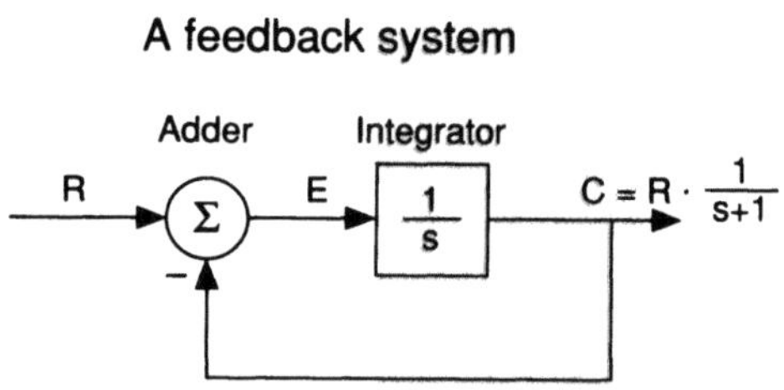

Its response to ···

··· an impulse, $\mu_0(t)$	$C(t) = e^{-t}$
··· a step, $\mu_{-1}(t)$	$C(t) = 1 - e^{-t}$
··· a ramp, $t \cdot \mu_{-1}(t)$	$C(t) = t - 1 + e^{-t}$
··· a quadratic, $t^2 \cdot \mu_{-1}(t)$	$C(t) = t^2 - t + 1 - e^{-t}$

Figure 8-19. Implicit simulation.

Add samplers & ZOH

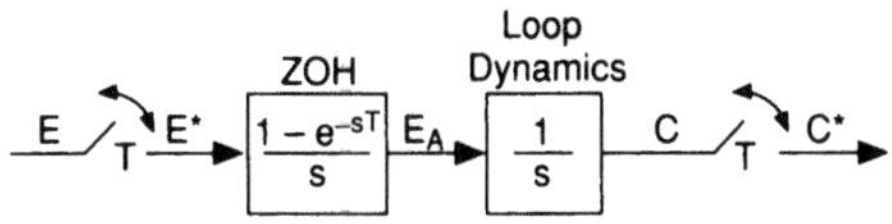

$$C(z) = \mathcal{Z}\{C\} = E(z) \cdot \mathcal{Z}\left\{\left[\frac{1-e^{-sT}}{s} \cdot \frac{1}{s}\right]^*\right\}$$

$$L(z) = \frac{C(z)}{E(z)} = \frac{T}{z-1}$$

Recursion ~ $C^{(n+1)} = C^{(n)} + T \cdot E^{(n)}$

Figure 8-20. Simulating loop dynamics.

··· Surround it with samplers

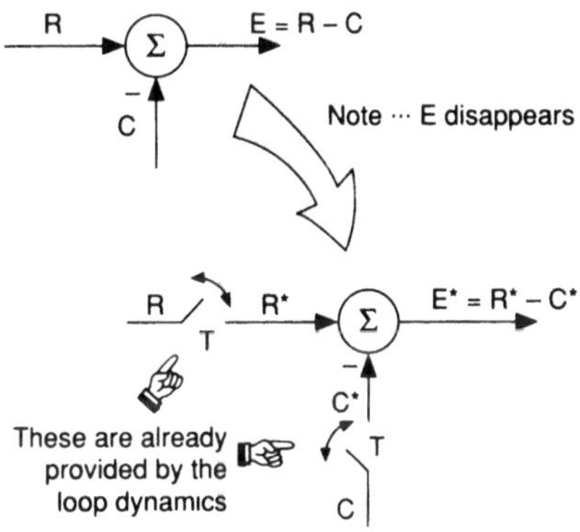

Figure 8-21. Simulating the adder.

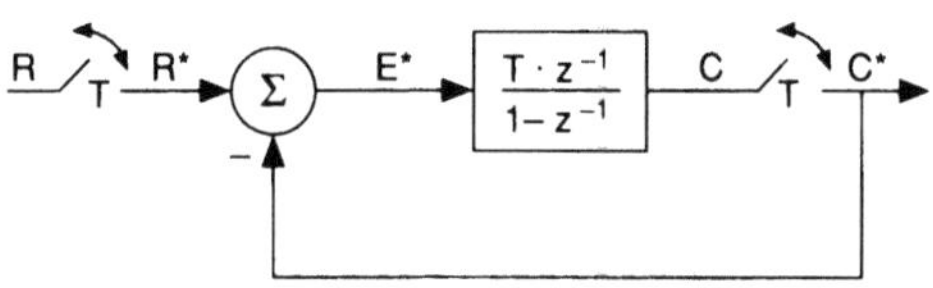

··· Recursion formulae

1. Loop dynamics $C^{(n+1)} = C^{(n)} + T \cdot E^{(n)}$
2. Initial condition $C^{(0)} = C(0)$
3. Adder $E^{(n)} = R^{(n)} - C^{(n)}$

Solved either ···

··· simultaneously

or ··· by eliminating $E^{(n)}$

$C^{(n+1)} = (1 - T) \cdot C^{(n)} + T \cdot R^{(n)}$

Figure 8-22. Closing the loop.

8.8.4 The System Pulse Filter

The system pulse filter is represented as

$$\frac{C}{R}(z) = \frac{T}{z + (T - 1)}$$

There is a pole at $z = 1 - T$, indicating stability for $0 < T < 2$, and an unstable recursion for $T > 2$. Figure 8-23 verifies this instability by the impulse response, and provides a comparison with the explicit simulation of Section 8.7 and the exact result (taken from ref. 7).

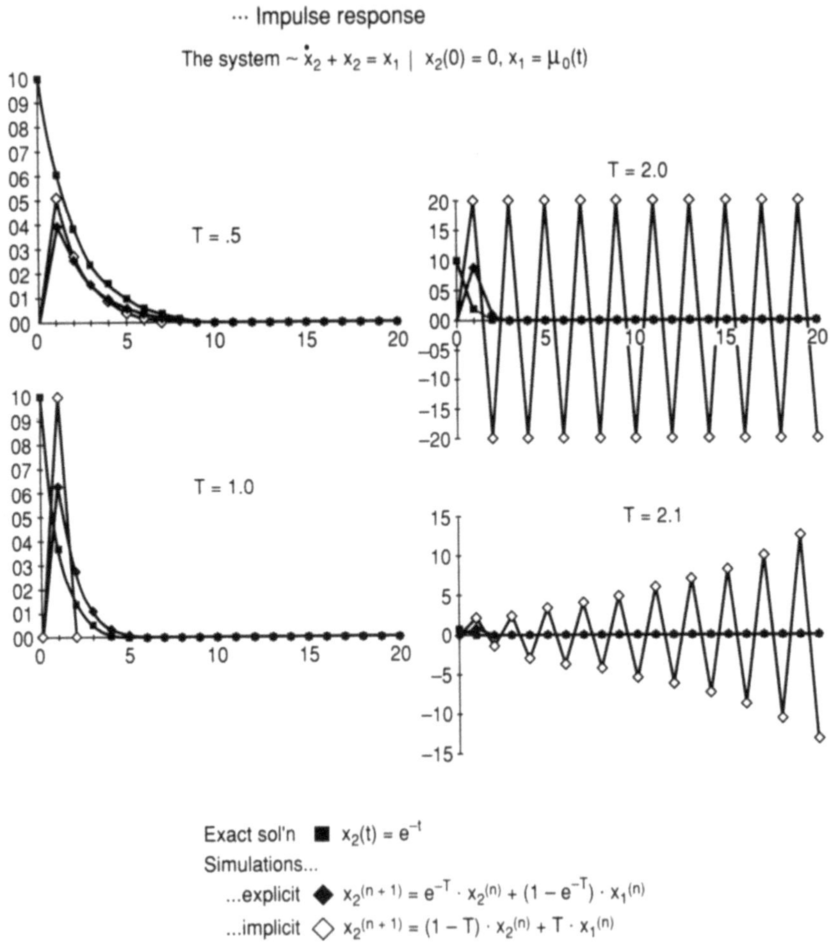

Figure 8-23. Comparison of simulations . . . impulse response.

8.9 SIMULATING HYSTERESIS DUE TO BACKLASH IN GEARS

The tooth-to-tooth freeplay in gears, combined with their mechanical compliance, as described in Chapter 3, Section 3.2.4.2, is a system characterized by hysteresis. Its block diagram and approximating equations are shown in Figure 8-24. The Lissajous pattern approached for infinite integrator gain is the hysteresis loop shown in Figure 8-25.

The digital simulation of hysteresis starts by inserting samplers and a hold as in Figure 8-26, where the recursion equations are written for the summer, the deadband, and the integrator, respectively. The validity of the recursions, which may be executed on a spreadsheet, is indicated on Figure 8-27.

Further evidence of the fidelity of the digital simulation is found by comparing its frequency response with the first-harmonic of the idealized trapezoidal hysteresis characteristic, as shown in Figure 8-28.

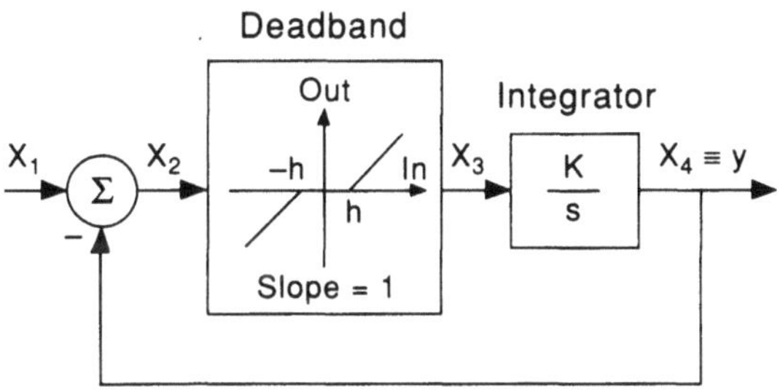

··· 3 eq'ns ~ $\dot{X}_4 = K \cdot X_3$

$$X_2 = X_1 - X_4$$

$$X_3 = \begin{cases} X_2 - h, & X_2 > h \\ 0, & \text{otherwise} \\ X_2 + h, & X_2 < -h \end{cases}$$

··· eliminate X_2 & X_3

$$\frac{\dot{X}_4}{K} = \begin{cases} X_2 - h, & X_2 > h \\ 0, & \text{otherwise} \\ X_2 + h, & X_2 < -h \end{cases} \Bigg| \; X_2 = X_1 - X_4$$

··· transform ~ $\mathcal{L}\{X_4\}$

$$X_4 = \begin{cases} \dfrac{K \cdot (X_2 - h)}{s + K}, & X_2 > h \\ \text{Last C} \sim \text{constant, otherwise} \\ \dfrac{K \cdot (X_2 + h)}{s + K}, & X_2 < -h \end{cases}$$

Figure 8-24. Hysteresis approximation.

… for $K \rightarrow$ BIG $\underset{\text{shakily}}{\Longrightarrow}$ $K >> s$

… giving $X_{4_{ss}} = \begin{cases} X_1 - h, & X_2 > h \\ \text{Last C}, & -h < X_2 < h \\ X_1 + h, & X_2 < -h \end{cases}$

… which plots as ~

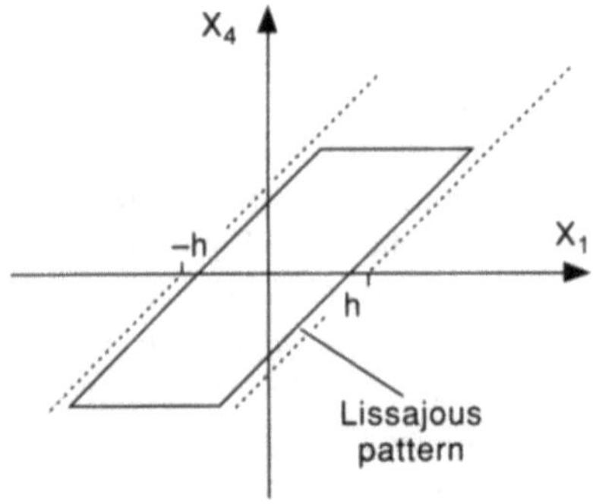

Figure 8-25. Hysteresis characteristic.

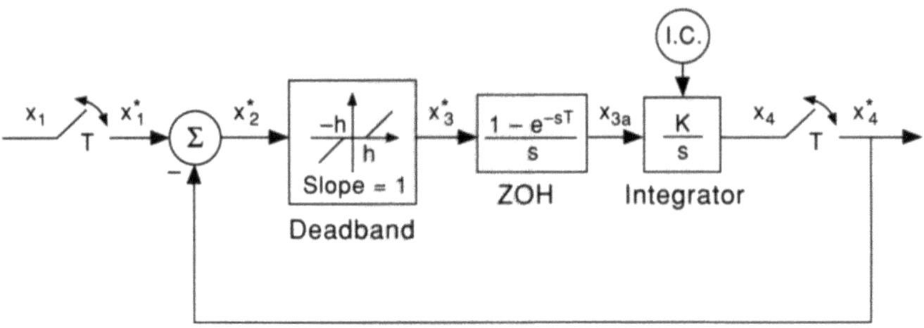

… The recursions are

~ Summer $x_2^{(n)} = x_1^{(n)} - x_4^{(n)},\ n > 0$

~ Deadband $x_3^{(n)} = \begin{cases} x_2^{(n)} - h, & x_2^{(n)} > h \\ 0, & -h \le x_2^{(n)} \le h \\ x_2^{(n)} + h, & x_2^{(n)} < -h \end{cases}$

~ Integrator … from x_4 0, its initial condition

$$\frac{x_4^*}{x_3^*} = \mathcal{Z}\left\{\left[\frac{1-e^{-sT}}{s} \cdot \frac{K}{s}\right]^*\right\} = K \cdot (1 - Z^{-1}) \cdot \mathcal{Z}\left\{\left[\frac{1}{s^2}\right]^*\right\} = \frac{K \cdot T}{Z - 1}$$

$$\text{So} \ldots\ x_4^{(n+1)} = x_4^{(n)} + K \cdot T \cdot x_3^{(n)}$$

Figure 8-26. Simulation of hysteresis . . . insert samplers and hold.

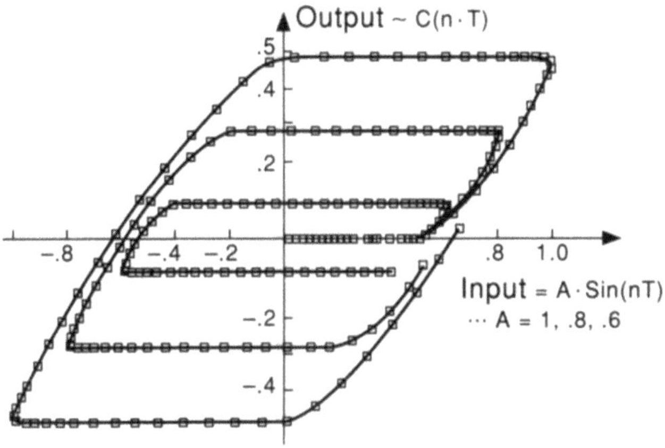

Figure 8-27. Lissajous pattern . . . supports the approximation.

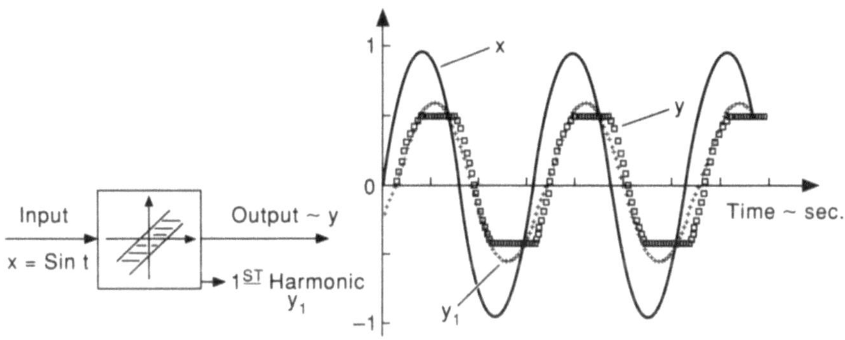

Figure 8-28. Compare 1st harmonic to frequency response.

8.10 SIMULATING A SYSTEM WITH HYSTERESIS

The system shown in Figure 8-29 represents an actuator with hysteresis due to play in its gears, driving a linearized integrating second-order plant. The describing function (Chapter 4) represents a nonlinearity with sinusoidal input, whose steady cyclic output is approximated by the first harmonic and is used to investigate cyclic behavior, typified by limit cycles. It is a complex function if the nonlinearity lacks odd-quadrant symmetry, causing a phase delay symptomatic of hysteresis in Figure 8-28, and is developed in Figure 8-30.

Intersections of the describing locus, $-1/N$, and the linear loop frequency response, $L(j\omega)$, predict limit cycles, whose existence can only be verified by

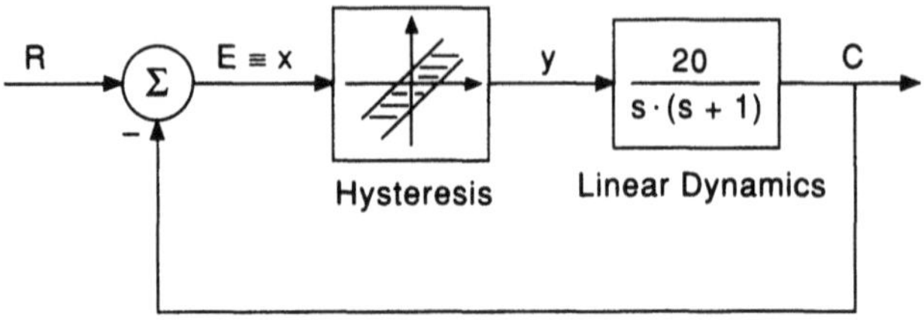

Figure 8-29. The system.

Its characteristics

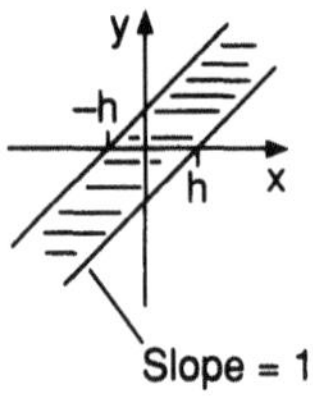

Input ⋯ x

$x = x_m \cdot \text{Sin}\omega t$

Output ⋯ $y \equiv e$

$$y = \begin{cases} x - h, & 0 \le \omega t \le \pi/2 \\ x_m - h, & \pi/2 < \omega t \le \pi - \propto \\ x + h, & \pi - \propto < \omega t \le \pi \end{cases}$$

where ⋯ $\propto \equiv \text{Sin}^{-1}\left\{\frac{x_m - 2h}{x_m}\right\}$

Its first harmonic ~ y_1 ⋯

⋯ $y_1 \equiv A_1 \cdot \text{Cos}\omega t + B_1 \cdot \text{Sin}\omega t$

Note ⋯ y has 1/2-wave symmetry.

i.e., integrate over 1/2-cycle.

$$A_1 = \frac{2}{\pi}\int_0^{\pi} y(t) \cdot \text{Cos}\omega t \cdot d\omega t = \frac{4h}{\pi} \cdot \left\{\frac{h}{x_m} - 1\right\}$$

$$B_1 = \frac{2}{\pi}\int_0^{\pi} y(t) \cdot \text{Sin}\omega t \cdot d\omega t =$$

$$= \frac{x_m}{\pi} \cdot \left\{\frac{\pi}{2} + \propto + \frac{2h}{x_m} \cdot \left(1 - \frac{2h}{x_m}\right) \cdot \sqrt{\frac{x_m}{h} - 1}\right\}$$

So ⋯ $N_1(x_m) \equiv \dfrac{B_1 + j \cdot A_1}{x_m}$ describing function.

Figure 8-30. Describing function for hysteresis.

test or simulation. Figure 8-31 is a stylized plot indicating two intersections: one a stable limit cycle, and the other an unstable limit cycle.

The system simulation is a matter of writing the recursions for the hysteresis (which was done in Section 8.9) and the linear loop dynamics (consisting of a dc gain, a real pole, and an integrator) as shown in Figure 8-32. The complete simulation is a set of seven recursion equations in eight unknowns, one being the input, *R*, which is specified separately. The seven equations to

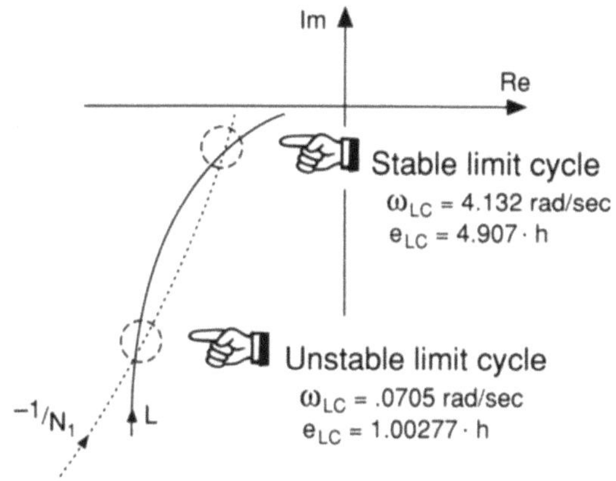

Figure 8-31. Limit cycles predicted.

The linear dynamics are $\frac{C}{x_4}(s) = \frac{20}{s\cdot(s+1)}$

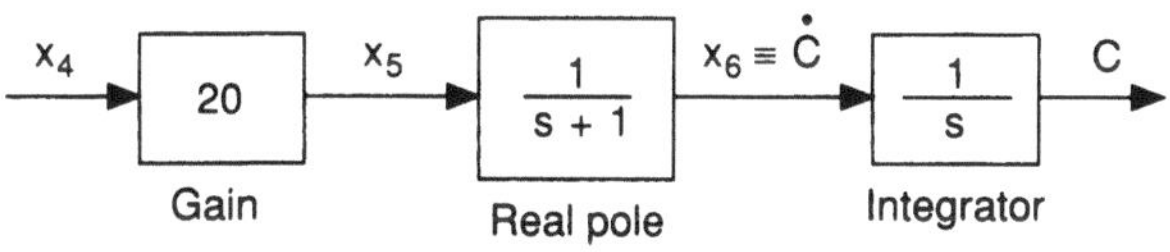

The simulation recursions are ···

··· the gain ~ $x_5 = 20 \cdot x_4$ i.e., $x_5^{(n)} = 20 \cdot x_4^{(n)}$

··· real pole ~ $\frac{x_6}{x_5}(s) = \frac{1}{s+1}$ so ··· insert boxcars.

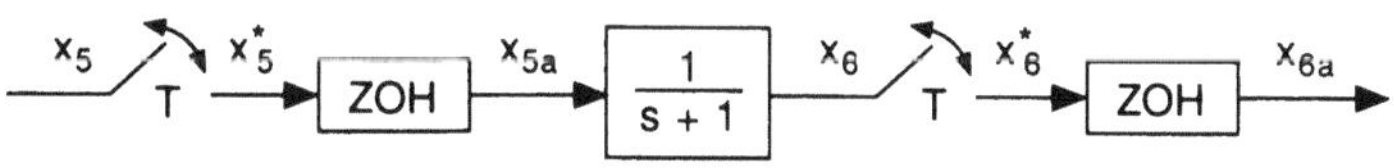

$$\frac{x_6^*}{x_5^*} = \frac{x_6}{x_5}(z) = \mathcal{Z}\left\{\left[\frac{1-e^{-sT}}{s}\cdot\frac{1}{s+1}\right]^*\right\} = \frac{1-e^{-T}}{z-e^{-T}}$$

i.e., $x_6^{(n)} = e^{-T}\cdot x_6^{(n-1)} + (1-e^{-T})\cdot x_5^{(n-1)}$

··· integrator ~ same as x_4/x_3 ··· i.e., $C^{(n)} = C^{(n-1)} + T\cdot x_6^{(n-1)}$

Figure 8-32. Recursions for simulating the linear dynamics.

be recursed are:

$$x_1^{(n)} = R^{(n)} - C^{(n)} \qquad \text{where } x_1 \equiv e, \text{ system error}$$

$$x_2^{(n)} = x_1^{(n)} - x_4^{(n)}$$

$$x_3^{(n)} = \begin{cases} x_2^{(n)} - h, & x_2^{(n)} > h \\ 0, & \text{otherwise} \\ x_2^{(n)} + h, & x_2^{(n)} < -h \end{cases}$$

$$x_4^{(n)} = x_4^{(n-1)} + K \cdot T \cdot x_3^{(n-1)}$$

$$x_5^{(n)} = 20 \cdot x_5^{(n)}$$

$$x_6^{(n)} = e^{-T} \cdot x_6^{(n-1)} + (1 - e^{-T}) \cdot x_5^n$$

$$C^{(n)} = C^{(n-1)} + T \cdot x_6^{(n-1)}$$

with $R^{(n)}$ specified a priori and $IC^s \cdots C^{(0)} \equiv C(0)$, $x_6^{(0)} \equiv \dot{C}(0)$, $x_4^{(0)} \equiv x_4(0)$. Finally, x_1 and x_5 can be eliminated, leaving five equations to recurse (assuming neither x_1 nor x_5 are sought):

$$x_2^{(n)} = \cancelto{0}{R^{(n)}} - C^{(n)} - x_4^{(n)}$$

$$x_3^{(n)} = \begin{cases} x_2^{(n)} - h, & x_2^{(n)} > h \\ 0, & \text{otherwise} \\ x_2^{(n)} + h, & x_2^{(n)} < -h \end{cases}$$

$$x_4^{(n)} = x_4^{(n-1)} + K \cdot T \cdot x_3^{(n-1)}$$

$$x_6^{(n)} = e^{-T} \cdot x_6^{(n-1)} + (1 - e^{-T}) \cdot 20 \cdot x_s^{(n)}$$

$$C^{(n)} = C^{(n-1)} + T \cdot x_6^{(n-1)}$$

The simulation can be used to investigate the existence of the stable limit cycle predicted by the describing function (Figure 8-31). In doing so, it is started from an initial condition on the output integrator, $C(0)$, with no input (i.e., $R = 0$). By trial and error, the solution is found to converge from $C(0) = 4$ and diverge from $C(0) = 3$ (Figure 8-33), indicating a stable limit cycle exists somewhere in between. Further investigation establishes the limit cycle at $C(0) = 3.62$ by the closure of the phase plane trajectory. The predicted limit cycle in Figure 8-31 is compared to the simulation result in Figure 8-34.

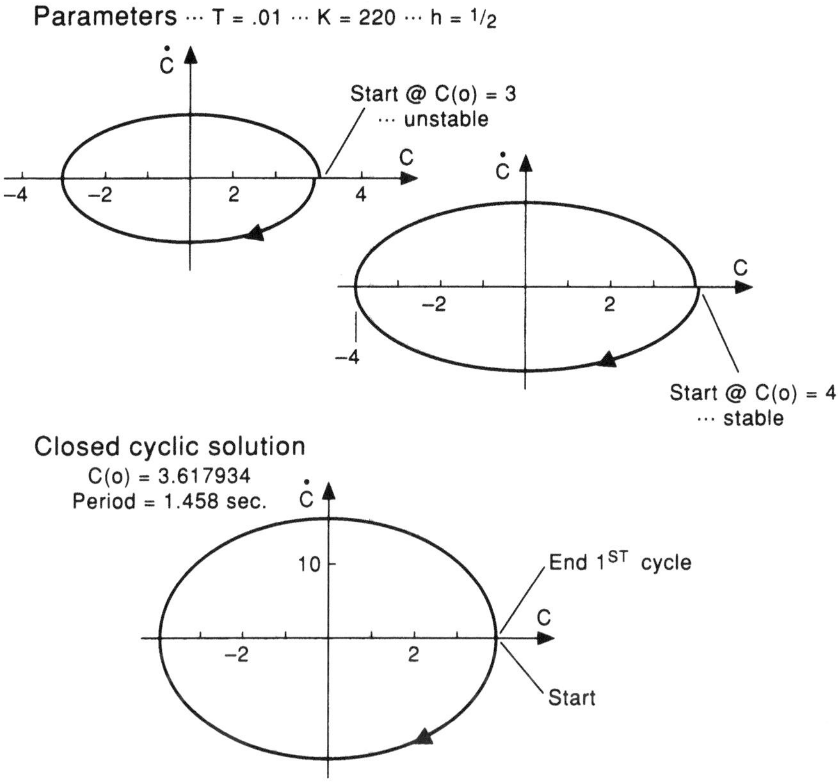

Figure 8-33. Simulation uncovers stable limit cycle.

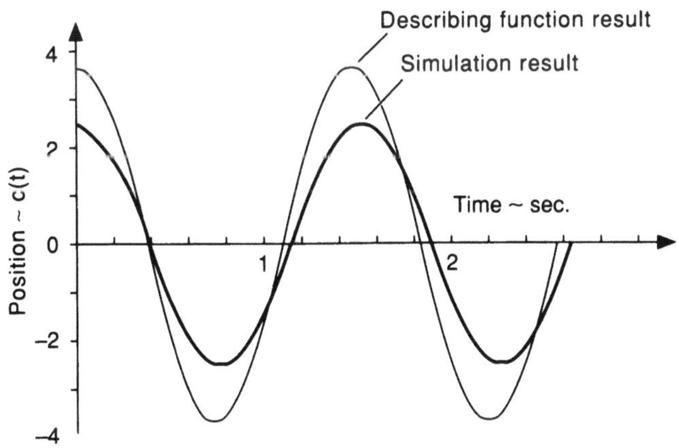

Figure 8-34. Comparing results . . . simulation and describing function.

8.11 PROBLEMS

1. Repeat the introductory example of digital simulation (Section 8.7), without the zero-order hold as in Figure 8-35, in order to show the necessity for a hold-device.

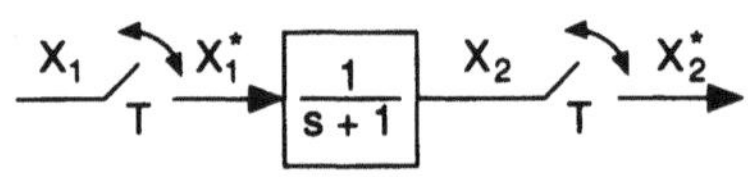

$$\cdots \text{ where } \frac{X_2^*}{X_1^*}(s) = \mathcal{Z}\left\{\left[\frac{1}{s+1}\right]^*\right\} = \frac{z}{z - e^{-T}}$$

2. Repeat the introductory example of digital simulation for:
   (a) A ramp input, $x_1(t) = tu_{-1}(t)$.
   (b) A squared input, $x_1(t) = t^2u_{-1}(t)$.
   Compare with the exact solution to show how well the recursions work.
3. Repeat the simulation of Section 8.10 using a double pole as the linear dynamics,

$$L(s) = \frac{20}{(s+1)^2}$$

Investigate both the stable and unstable limit cycles by starting the simulation from appropriately selected initial conditions at its output, $C(0)$.

8.12 REFERENCES

1. Gerald, Curtis F., *Applied Numerical Analysis*, Addison-Wesley, Reading, Massachusetts (1978).
2. Tou, Julius T., *Digital and Sampled-Data Control Systems*, McGraw-Hill, New York (1959).
3. Larkin, Peter, Unpublished paper, 35.451 Discrete time systems, The Hartford Graduate Center (Spring 1986).

Chapter 9
Spreadsheet Simulation–A Tutorial

9.1 INTRODUCTION

Many spreadsheets are available, among which are EXCEL, LOTUS 1-2-3, ENABLE, QUATRO, and SYMPHONY (refs. 1–7). All offer mathematical tools suitable for digital simulation, and yet there is a dearth of literature addressing the use of desk- and laptop personal computers and spreadsheets for scientific and engineering applications—particularly the modeling of dynamic systems governed by Ordinary Differential Equations. This may be due in part to the spectacular acceptance of LOTUS by the Management community, which has established it as preeminent among all the spreadsheets, as it commands over half the market. Accordingly, we choose LOTUS 1-2-3 as the basis for the tutorial that follows.

The reluctance (prejudice?) of the Engineering community to use a "management tool," despite its many virtues, should be assuaged, which is the motivation for this tutorial. And, the fact that the many spreadsheet packages so readily available at both the office and at home makes it a shame to neglect their use in investigating dynamic processes, especially those of a nonlinear nature which do not easily, if at all, admit to analytic solutions.

As prerequisites for what follows, the student should have an elementary knowledge of integration and differentation, ordinary differential equations, and numerical methods. In addition, a reference manual for LOTUS 1-2-3 (or the spreadsheet of choice), should be available to interpret and verify keystrokes.

We will examine *only* two LOTUS commands—COPY and GRAPH. The elementary commands for writing statements and formulae, and moving from cell to cell within the spreadsheet are assumed. And, this tutorial can be interpreted for other spreadsheets, such as ENABEL, EXCEL, QUATRO, and SYMPHONY, having a set of equivalent commands.

9.2 USE OF THE "COPY" COMMAND

We will first list the set of integers, (0, 1, 2, . . . , 10), labeling them "n". To do this, the following steps are necessary.

NOTE. It is instructive and helpful to scan the menu before and after each keystroke, for the prompts and queries. After very little practice, this information is retained and the user rapidly develops proficiency with the spreadsheet.

Step 1. Enter "n into cell A11, resulting in the desired heading, n, which appears right-justified in the cell at Row 11 and Column A. The keystroke string is— "n↵

Step 2. Enter +0 into A12, resulting in the number "zero" being displayed in the cell at Row 12 and Column A. The keystroke string is— +0↵

Step 3. Enter +A12 + 1 into A13, resulting in 1. The keystroke string is— +A12+1↵

Step 4. Copy the result of Step 3. into cell A14 by typing the string— /C↵↓↵ —which results in the entry "2" displayed in cell A14. In COPYING, notice how the formula has been incremented as displayed above the spreadsheet.

Step 5. At cell A14, type the string—/C↵.↓↓↵—which COPIES the formula in A14 to A15 and A16 successively. Examine the formulae in A15 and A16, which yield the values 3 and 4, respectively.

Step 6. At cell A16 enter the string—/C↵.↓↓↓↓↓↓↵ —which generates 5 through 10 in cells A17 through A22.

REMARK. You now have a column of integers headed by "n". In the process you have found that the COPY command can (as we have just used it) increment formulae "downward" row-by-row, which is very convenient if we want to successively recurse an algorithm, such as the present one which creates the set of increasing integers . . .

> . . . "add +1 to the current cell value and place the result into the cell immediately below."

FURTHER. Digital simulations of dynamic systems are basically applications of algorithms such as this, which approximate such processes as integration and filtering by recursion formulae. Thus, the COPY command might be considered the soul of a digital simulation.

ROW	COLUMN A	
11	n	
12	0	This is the column of integers we sought.
13	1	
14	2	
15	3	

COLUMN A ROW		
16	4	
17	5	
18	6	
19	7	This is the column of integers we sought.
20	8	
21	9	
22	10	

9.2.1 More on the COPY Command

We next create an adjacent column consisting of multiples of the first column, the multiple being a number (+ or −), which we would like to be able to arbitrarily select, called ''k''. Let's call this multiple by the letter, ''k'', and create a column headed by ''kn'' to denote the product, k times n, with entries giving the values. This is done according to the following steps.

Step 1. Enter ''k= into A9, giving a means of identifying or labeling the value of ''k'' which has been selected as the multiplier.

Step 2. Enter +.5 into B9, thereby arbitrarily selecting the multiplier of 1/2.

Step 3. Enter ''kn into B11, giving heading, kn.

Step 4. Enter +A12*B$9 into B12, giving 0, which is the desired multiple of 0 in A12. Check the formula in cell B12.

Step 5. At B12 enter /C↵.↓↓↓↓↓↓↓↓↓↓↵ which COPIES the formula in B12 into cells B13 thru B22 in such a way that the rows of ''A'' (in the formula) are successively incremented, but keeping the rows of ''B'' ANCHORED. The anchoring feature is achieved by the insertion of the ''$-sign AFTER the ''B.'' Check the formulae in cells B12 thru B22 to verify. The two columns are shown below.

ROW	COLUMNS → A	B	
9	k =	0.5	
10			
11	n	kn	
12	0	0	
13	1	0.5	
14	2	1	These are the two columns containing data. The second is k-times the first, and k can be selected in cell B9 to be any value.
15	3	1.5	
16	4	2	
17	5	2.5	
18	6	3	
19	7	3.5	
20	8	4	
21	9	4.5	
22	10	5	

REMARK. You now have two columns of numbers which should plot as a straight line. In a more general case you may not know what the relation would be, and, typically, an engineer's insatiable curiosity drives him to plot the information, a rather laborious task, and a topic that is seldom offered in undergraduate school. For a more interesting plot, go to the first EXERCISE (at the end). It is a neat attribute of LOTUS 1-2-3 that its GRAPH command easily facilitates plotting variables against one another, as is discussed in the next section.

9.2.2 An Investigation

Suppose you wanted to copy the column headed by kn from B to C. Try the string,

/C↓↓↓↓↓↓↓↓↓↓↓↵ →↵

and notice the result—all zeros! Now examine each formula in the C-column. You will note that they are not what you expected, although their evaluation is correct. The problem is that in COPYing to the right, the B-row was not anchored, so the formula, when copied, looks for the value k in cell C13 (rather than B13), which of course has no entry, so 1-2-3 calls it the number 0 (zero) by default. To correct the difficulty, it is necessary to "anchor" the value of k in cell B13 not only going "down" (which was done by inserting $ after B), but also going to the "right," which is done by inserting $ ahead of B. Thus, all the formulae in the B-column must be revised, the first entry in cell B15 being +A15*B13. Verify that this works after modifying the B-column and COPYing into column C.

You will gain proficiency with 1-2-3 (or any other spread sheet) by adding further to this problem. Typically, you can develop another column which lists the squares of the integers, by entering, say, into cell D11 the formula,

+A11^2

and then COPYing it into cells D12 through D22. You might then develop a column of the square roots of the integers in column E with the formula,

+@SQRT(A11)

9.3 USING THE GRAPH COMMAND

Our aim is to plot the values in the two preceding columns so see what relation, if any, exists between them. To do so, we employ the GRAPH command, as indicated in the following steps.

Step 1. Enter /G and *do not* hit RETURN. A "menu" appears as well as the "prompt,"

Set graph type.

Step 2. Enter T but NOT ↵. Another menu appears and another prompt,

Line graph

This is a cosmetic choice, but we want to plot one set of numbers (representing *Y*) against another set (representing *X*) so, the choice is "*XY*".

Step 3. Enter *X* but NOT ↵ . You have now selected a *Y* vs. *X* plot (*Y* being the vertical axis and *X* being the horizontal axis), and you will next assign which cells are to be plotted as *X* and *Y*, respectively.

Step 4. Enter *X* but NOT ↵ . You are now going to assign the cells which represent the *X*-values. You are prompted by,

Enter X-axis range:

with the current cell as the candidate. There are two ways of addressing the *X*-values (in their proper order): by a direct entry as described in step 5a, or by means of the cursor, as described in Step 5b.

either Step 5a: Enter the string,

A12 . . A22↵

You have now selected these cells to constitute the *X*-values (in that order).

or Step 5b: Move the cursor to the "top" of the column containing the desired *X*-values. "Anchor" the cursor with a period [.], and then move the cursor down the column, noticing that those cells are "highlighted." When all the desired cells are highlighted, enter ↵ , and you have completed entering all the *X*-values to be plotted. You are now presented with the same menu as in Step 1.

Step 6. You are now about to select the cells to be plotted for the *Y*-range, however, they are not designated as "*Y*". Rather, there are *six* (6) possible sets of values, each of which can be plotted against *X*. These are designated as A, B, C, D, E, and F, allowing as many as 6 sets of cell-data to be plotted against *X*. Enter A and repeat either Step 5a or Step 5b, using the desired range of cells for the *Y*-values. You have now completed the data-entry for your graph.

Step 7. Enter V and *voila*!!! a graph appears, not the nicest, but to an engineer the cosmetics are unimportant (or should be). A reproduction is shown in Figure 9-1. To return to the spreadsheet, enter ↵ .

TO RECAPITULATE. The entire construction of the graph consists of the following entry string,

/GTXXA12 . . A22 AB12 . . B22 V

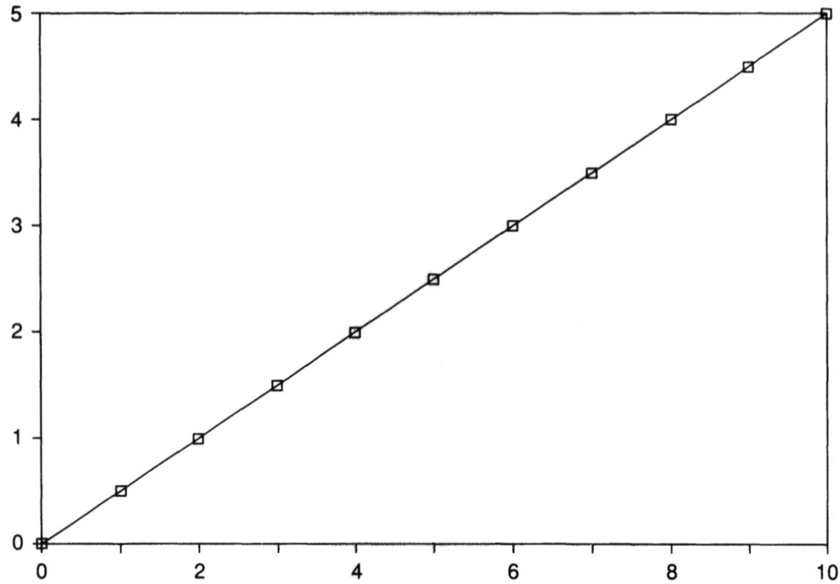

Figure 9-1. "Graph" of integers (column A) against their multiples (column b).

It is worth going through this string and observing the various defaults and prompts which are displayed. The ease of entering the string is based on 1-2-3's ability to interpret the first letter of a menu entry as the entry itself, eliminating the need to move the cursor to the desired entry and enter it.

9.3.1 More on "GRAPH"—Its Cosmetics

The graph can be spruced up by adding two titles, legends, and what-have-you, most of which the enterprising student can easily find in the menu under OPTIONS. We leave it as an exercise to verify the additions made to our original graph as shown in Figure 9-2. In the process, the student will discover how helpful the menu is and how easy it is to investigate the many alternatives available.

Finally, the graph may be SAVED for future use to be printed on any compatible printer using PRINTGRAPH, a package of LOTUS graphics, usually available separate from 1-2-3, containing many options such as size, rotation, and fonts. It may also be NAMED, and indeed there are a host of cosmetics which can seduce, and possibly addict, the most fastidious engineer. This is not an idle statement because there are so many intriguing possibilities that the user is easily sidetracked by the piper.

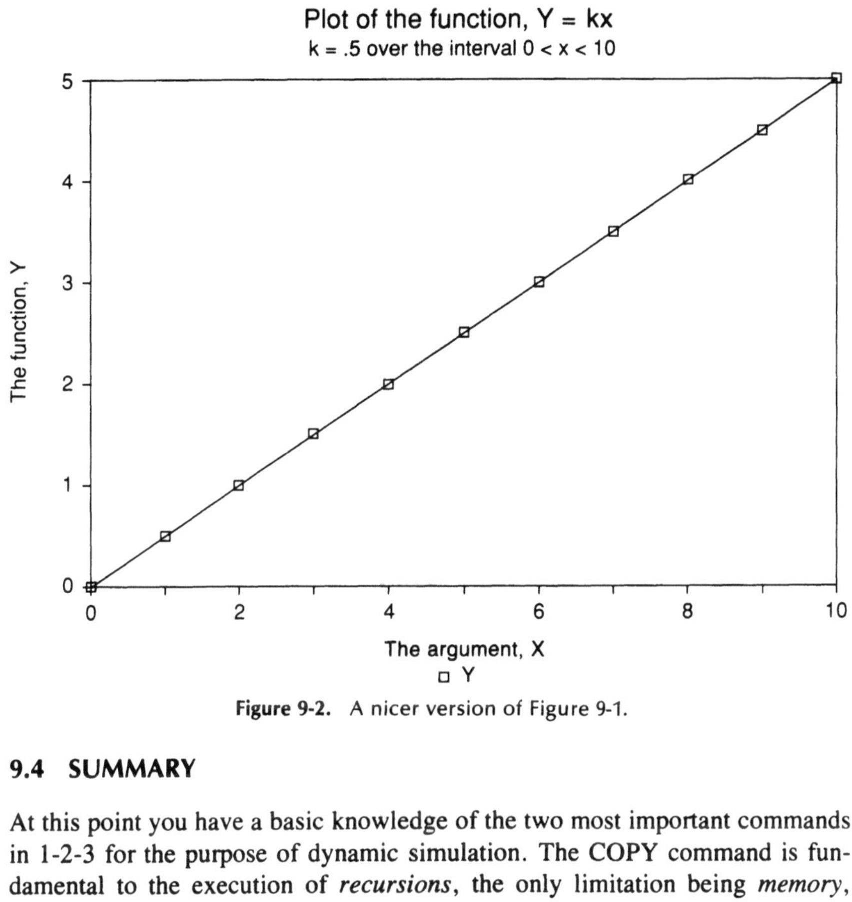

Figure 9-2. A nicer version of Figure 9-1.

9.4 SUMMARY

At this point you have a basic knowledge of the two most important commands in 1-2-3 for the purpose of dynamic simulation. The COPY command is fundamental to the execution of *recursions*, the only limitation being *memory*, which determines how many times an algorithm can be iterated. The GRAPH command gives a quick way of looking at the result of the recursion.

A further advantage of using spreadsheets is the fact that they are "menu-driven," which provides the user with a display after each keystroke to indicate the status and the alternatives. To the faint of heart, there is always the Escape key, which will "undo" anything done by the last keystroke. As you develop a proficiency with the menu, you will find that it comes rather naturally, requiring only the keystroke of leading letter of a selection, which is especially convenient for those with some typing skill.

Not only is the menu easy to follow, but, should you return to the use of a spreadsheet after a prolonged absence, it is a great help in regaining your proficiency.

With a rudimentary capability in the use of the COPY and GRAPH commands, you are in a position to proceed with the exercises that follow. They are intended to further develop your skill in the use of LOTUS 1-2-3 in particular, and spreadsheets in general, to analyze problems of increasing complexity, culminating in the simulation of a nonlinear feedback system. Associated with each exercise are some homework problems, which you are encouraged to solve as they provide additional practice in developing your skill.

9.5 SOME HEURISTIC EXERCISES—BRIEF DESCRIPTIONS

We provide here a brief description of each of the exercises so the student can execute them if he or she feels adequately proficient. Detailed descriptions of suggested solutions are given in the next section.

Exercise 1. Plot sine and cosine waves, each with its own adjustable frequency. Then they can be plotted against one another at the same frequency to produce a circle. They can also be plotted against one another at different frequencies to produce Lissajous figures.

Exercise 2. Recurse the Logistics Equation and plot the result for several values of its coefficient, C.

C = 2 gives convergence
C = 10/3 gives a bifurcation
C = 3.83 gives a trifurcation
C = 4 gives essential chaos

Exercise 3. Recurse Lin's method of factoring polynomials. This is an algorithm involving long division of the polynomial by a trial quadratic. Its quadratic remainder is then used as the next divisor and, thereafter, the algorithm (hopefully) converges. Examples of convergence, divergence, bifurcation, and essential chaos are demonstrated.

Exercise 4. Discuss Integration and Differentiation and show some simple digital approximations which yield recursion formulae. Apply these recursions to a sine-wave and compare the results with known solutions.

Exercise 5. Solve a Linear Ordinary Differential Equation with constant coefficients (whose solution is known by analysis), using a recursion algorithm to approximate integration. This approximate solution can be compared to the exact solution for various values of the "sample time." Typically, we choose the first-order equation

$$\dot{X} + X = 0, \qquad \text{with initial condition, } X(0) = X_o$$

The value of this equation is that it is representative of the dynamics often encountered in real systems which are simple enough to be linearly modeled, such as the response of a thermometer when dipped into a cold bath, an R-L or R-C circuit excited by a battery, a dc motor's speed response to armature voltage excitation, or the linearized model of a steel ball "falling" through a viscous medium, such as oil.

Exercise 6. Solve a *nonlinear* ordinary differential equation by an approximate discrete integration algorithm. Just like the previous exercise, there is an analytic solution with which to compare the result, but it is not so easy to find, nor is it possible in all cases. A respectable approximation can be obtained by "fairing" a curve by eye when the slopes of the solution are identified as in the method of isoclines.

The equation we choose is,

$$\dot{X}^2 = X, \qquad \text{with arbitrary initial condition.}$$

Exercise 7. Investigate the transient behavior of a feedback control system consisting of a real pole and an integration driven by a saturation.

9.6 SUGGESTED SOLUTIONS TO THE EXERCISES

An Admonition. The student is heartily encouraged to observe the prompts and cues displayed on the screen after each keystroke in order to gain a proficiency with LOTUS 1-2-3. The letters in each string are the first letter of each member of the displayed menu, and with practice this interpretation becomes second nature.

9.6.1 Exercise 1

Plot sine and cosine waves, each with its own adjustable frequency. Then they can be plotted against one another at the same frequency to produce a circle. They can also be plotted against one another at different frequencies to produce Lissajous figures.

SUGGESTED SOLUTION. Our goal is to make plots of the sine- and cosine-functions, with the capability of choosing the frequency of each independently. These will be plotted against the argument, time. As a separate issue, we will also make plots of the two functions against one another (which are commonly referred to as Lissajous Figures).

To start with, the functions are specified formally as:

The sine-function, sin (ωt)
The cosine-function, cos (ωt)

Both of these are defined in terms of a series, which could be evaluated were it not for the fact that these series are *infinite* in length, making them impossible to be evaluated exactly (for which they are called *irrational* functions).

Aside. You may find it interesting to look up the definition of irrational functions.

This fact presents a formidable barrier—just how can we evaluate these (or any other) irrational functions? The answer is, we can't, it must be approximated by a finite number of terms. But, even this is, at most, laborious. Luckily LOTUS has liquidated our difficulties by giving us, in one command, what is advertised as an acceptable approximation to not only the sine and cosine, but the logarithm, the exponential, and many other irrational functions. They are grouped under the general heading of "@ FUNCTIONS," which includes:

MATHEMATICAL FUNCTIONS	DATE FUNCTIONS
LOGICAL FUNCTIONS	DATABASE STATISTICAL FUNCTIONS
SPECIAL FUNCTIONS	
FINANCIAL FUNCTIONS	STATISTICAL FUNCTIONS

The description of these functions is available in any 1-2-3 *Reference Manual*, and under the "HELP" command.

We are now in a position to proceed with the generation of the sine- and cosine-waves. First, we must list the argument, "time," for which the functions are to be evaluated. Then each function is to have its own frequency, which is a multiple of "time," to give the "angle" of the sine and cosine. Lastly, the @SIN( ) and @COS( ) are used to evaluate the functions.

To list "time," the following steps are suggested, and the student can modify them as proficiency accumulates.

Step 1. Create a heading to read "time" in cell A11. This is done with the string "time where opening the quote "right-adjusts".

Step 2. Create a column of values of time in the cells A12, A13, . . . etc., in which each exceeds its predecessor by a specified increment. So, in A8 write,

"Dtime =

Then in A9 write the increment of time you desire, noting that it is easily changed. We will use +.02. Next, enter +0 in A11. In A12 enter the string,

+A11+A9

Finally, COPY this formula down as far as you like (or until you run out of memory), using the string

/C↵ .↓↓↓↓↓↓ n-times ↓↓↓↓↵

You now have the following:

Column Row	A	
8	Dtime	
9	0.02	
10		
11	time	These are the times at
12	0	which the sine and cosine
13	0.02	are to be evaluated
14	0.04	
15	0.06	
16	0.08	

To list the sine- and cosine-functions, we make use of both time and frequency to indicate their arguments, and then use the @-functions to generate their approximations, as indicated in Step 3, which follows.

Step 3. To identify the two frequencies, enter the headings in B8 and C8,

''omega-1 ''omega-2

In B9 and C9 their values are entered, noting that these are ''radian-frequencies'' [i.e., they are the ''cyclic-frequency'' in Hertz (or CPS) multiplied by 2*PI = 6.28nnn]. The notation, ''nnn'', implies the use of additional digits for greater accuracy. Now, PI, just like the sine and cosine, is irrational, but LOTUS 1-2-3 provides us with an acceptable approximation with a function denoted as @PI (or we can enter our own approximation for it such as 22/7 or 3.14159). In any event, there is enough confusion between the two frequencies to give it special attention. We presume the ''cyclic-frequency'' is more common, so we create two more headings in B5 and C5, ''f1-CPS and ''f2-CPS, below which our values are entered, B7 and C7. Then the ''radian-frequencies'' are calculated in B9 and C9 as,

+2*@PI*B7 and +2*@PI*C7

We choose values of 1 in each case and the table is,

Column Row	A	B	C	
5		f1-CPS	f2-CPS	
6		1	1	
7				
8	Dtime	omega-1	omega-2	
9	0.02	6.283185	6.283185	Two frequencies
10				are now available for
11	time			use with sine and/or
12	0			cosine.
13	0.02			
14	0.04			
15	0.06			
16	0.08			

Step 4. The sine- and cosine-functions are now added. First, their headings are placed in D11 and E11 by entering

''Sin(w1*t) and ''Cos(w2*t)

Then in D12 and E12, their formulae are entered as,

+@SIN(C12*d9) and +@COS(C12*e9)

Finally, these formulae are each COPIED into the column headed by cell C12, giving,

Column / Row	A	B	C	
5		f1-CPS	f2-CPS	
6		1	1	
7				
8	Dtime	omega-1	omega-2	
9	0.02	6.283185	6.283185	These are the data we seek.
10				
11	time	SIN(w1*T)	COS(w2*t)	
12	0	0	1	
13	0.02	0.125333	0.992114	
14	0.04	0.248689	0.968583	
15	0.06	0.368124	0.929776	
16	0.08	0.481753	0.876306	

Try the following string

/RFF3↵

observing the cues and prompts after each keystroke. By doing so, you can see what possibilities are available for modifying the data display.

Now that all the necessary data are available, the desired graphs of the two functions are constructed using the GRAPH command. Provided you have previously COPIED enough data (through line 68), the following string will produce a graph, as shown in Figure 9-3. It is instructive to observe the result of each keystroke while entering the string,

/GTXXA12 . . A68↵ AB12 . . B68↵ BC12 . . C68↵ V

It is sometimes desirable to plot one function against the other, the argument used as a running parameter. Accordingly, we plot the the cosine vertically and the sine horizontally with the following string, again emphasizing that the student will benefit by observing the screen and interpreting the result of each keystroke:

/GTXXB12 . . B68↵ AC12 . . C68↵ V

resulting in the graph shown in Figure 9-4.

Some interesting plots can be made by using different frequencies for the sine and cosine, respectively. These are called Lissajous plots, one of which is shown in Figure 9-5, where the cosine's frequency is triple that of the sine. We leave it to the student to generate the string which makes up this plot and to investigate other combinations of frequencies. This can be done by inserting arbitrarily selected values of frequency into the cells B6 and C6. As a point of interest, as long as these values are rational numbers, the Lissajous figure will

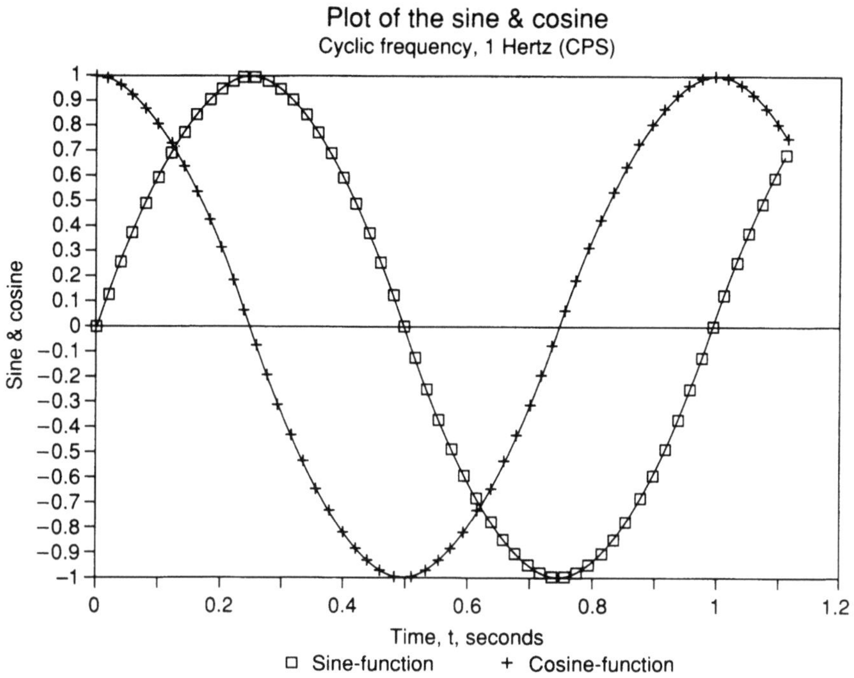

Figure 9-3. "Graph" of cosine and sine.

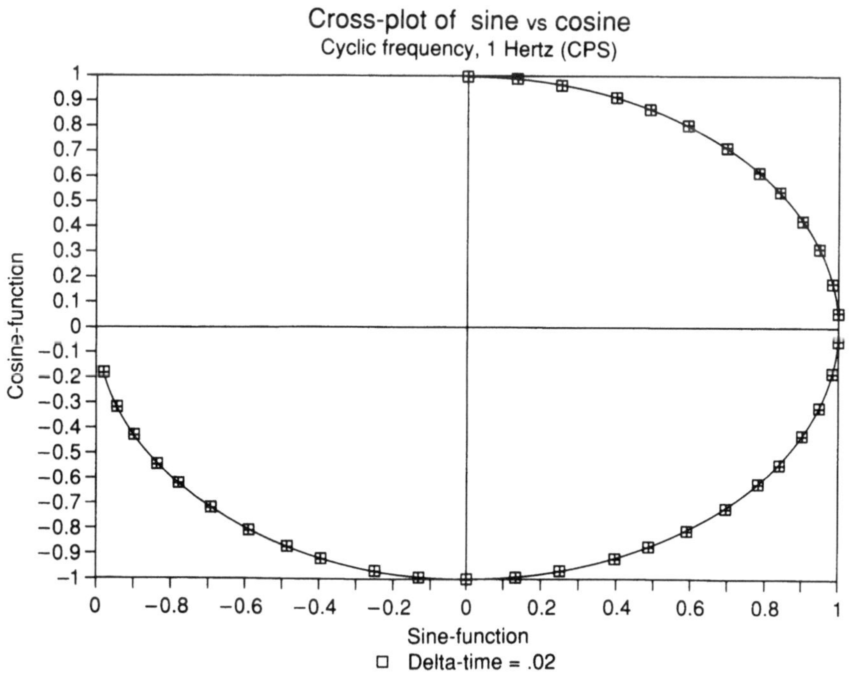

Figure 9-4. "Graph" of cosine versus sine.

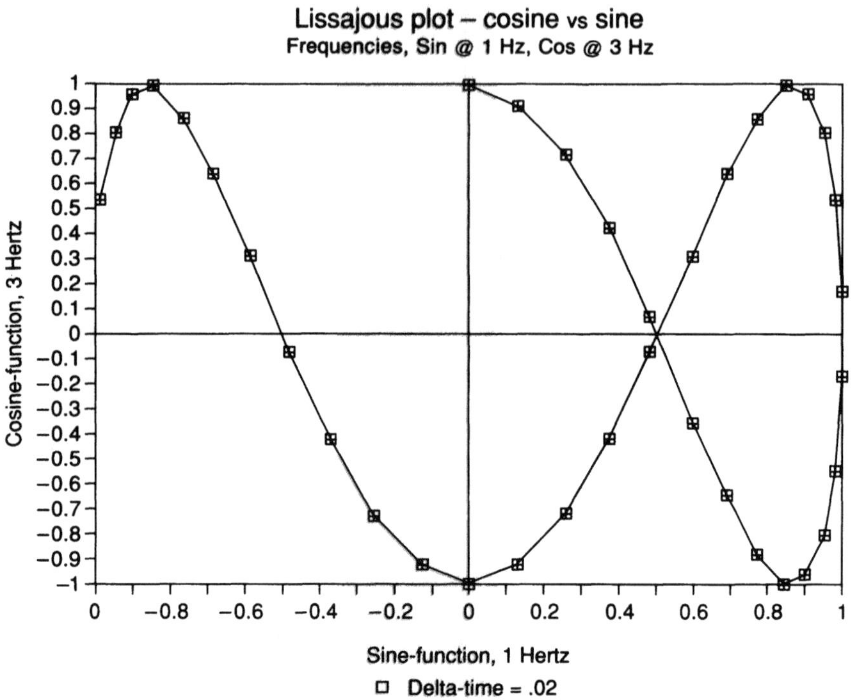

Figure 9-5. "Graph" of Cos (3ft) versus sin(ft).

eventually repeat itself, although it may take a devious route in doing so. However, should either or both the frequencies be irrational, the Lissajous figure will never repeat. You can explore this issue very easily with the 1-2-3 simulation.

9.6.2 Exercise 2

Recurse the Logistics Equation and plot the result for several values of its coefficient, C.

C = 2 gives convergence
C = 10/3 gives a bifurcation
C = 3.82 gives a trifurcation
C = 4 gives essential chaos

Suggested Solution. The Logistics Equation (Map) is a recursion formula employed empirically in population studies, which in recent years has come under the scrutiny of Chaos enthusiasts because of its arcane behavior.

The recursion is,

$$X(n+1) = C*X(n)*[1-X(n)], \qquad \text{with } 0 < X(0) < 1$$

and by properly choosing C, rather abnormal results are found. The mystery is veiled by the insistence on the initial condition, inhibiting an overall examination which is otherwise enlightening. In order to examine the behavior, we first make C available as an identified number, enter ''C = in cell A1, and its value in cell B1. Then a column headed by ''n is initiated in cell A3, and a running index of integers is entered below it, starting with 0 (zero) in cell A4. Next, the initial value is identified by placing ''X(n) into cell B3 and its value [in this case X(0) = .1] is entered into cell B4. The recursion formula is entered in B5 with the string,

+B4*B1*(1−B4)

and this is copied as many times as you like with the string,

/C↵ .↓↓↓↓↓↓↓↓↓ · · · ↓↓↓↓↓↓↓↵

giving the following spread sheet for C = 2.

C =	2	
n	X(n)	
0	0.1	
1	0.18	These are the data
2	0.2952	for recursions of
3	0.416113	the logistics map.
4	0.485926	
5	0.499603	Notice that they
6	0.499999	converge to a
7	0.5	single value, in
8	0.5	this case, .5.
9	0.5	
10	0.5	

A graph of the recursion can now be made using the string,

/GTXXA4 . . A14↵ AB4 . . B14↵ V

and the graph, including titles and legend, is shown in Figure 9-6.

Homework Problem. In the GRAPH shown in Figure 9-6, the scales for the X- and Y-axes were automatically set to correspond to the maximum and minimum values found in each column. For reasons which will become clear later, it is desirable to scale the Y-axis between a minimum of 0 and a maximum of 1. Simply by following the MENU, introduce this change. If need be, consult the 1-2-3 manual.

Another Homework Problem. As a point of interest, you may wonder if the logistics map converges to the same value for different initial values, X(0).

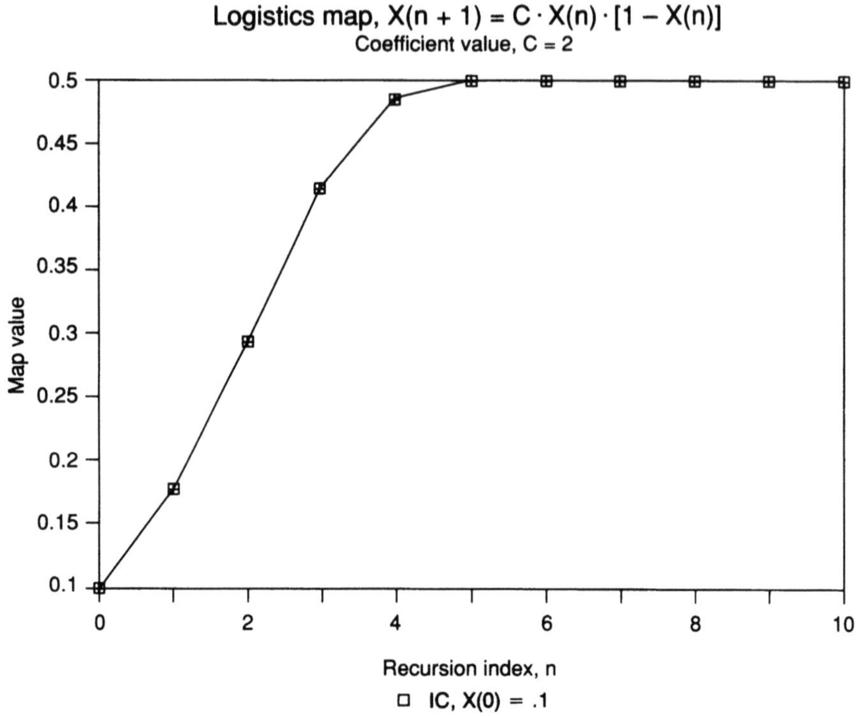

Figure 9-6. "Graph" of logistics map.

Try this by altering the value in cell C4, and perhaps even be brave enough to investigate values *outside* the specified limit.

One of the interesting results obtained from the Logistics equation is its ability to seek not just one, but two numbers in its steady state. This is found for a value of the coefficient, C = 10/3. To illustrate it, we add a value +10/3 into cell C1, enter the heading, "X(0) into cell C3, and its initial value (now we choose .2; see the Homework Problem) into C4. The recursion values are then determined in column C by entering the recursion formula,

+C4*C1*(1−C4)

into cell C5 and then COPYING it with,

/C↵.↓↓↓↓ · · · ↓↓↓↓↵

Giving the following spreadsheet:

C =	2	3.333333	
n	X(0)	X(0)	
0	0.1	0.2	These are the data
1	0.18	0.533333	for recursions of
2	0.2952	0.829629	the logistics map
3	0.416113	0.471147	at 2 values of its
4	0.485926	0.830558	coefficient.
5	0.499603	0.469103	
6	0.499999	0.830151	Notice that one
7	0.5	0.470000	converges to a single
8	0.5	0.830333	value, while the other
9	0.5	0.469599	converges to two values.
10	0.5	0.830252	

The GRAPH command plots the data with the string,

/GTXXA4 .. A14 AB4 .. B14 BC4 .. C14 V

giving the plot shown in Figure 9-7.

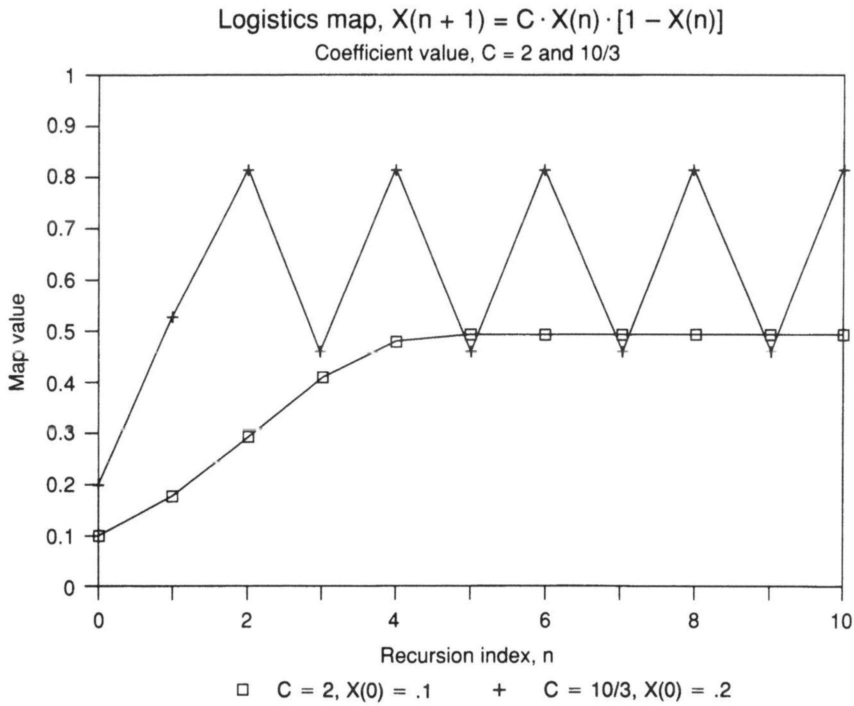

Figure 9-7. "Graph" of two logistics maps.

Homework Problem. For those who are intrigued by the Logistics equation, try the initial value of X(0) = .1 (in column C at cell C4) and note the unexpected result. You may verify this behavior by working the Logistics equation "backward" to verify that the initial condition can influence the nature of the iterations.

Homework Problem. Can you find another way to generate column C by copying the entirety of column B to the "right"? (*HINT:* The answer lies in how the value of the coefficient B1 is anchored when it is first copied "down" column B.)

Not only can the Logistics map seek one or two steady-state values, but it can also seek THREE! with the proper choice of its coefficient.

Homework Problem. Add another column of data to the spreadsheet in which the coefficient is taken as 3.83 and plot the data. The graph of the data will appear as in Figure 9-8, where the plot consists of only recursions 98 through 108.

You might expect, based on the preceding results, that the Logistics equation can reach a steady state in which every 4th, 5th, 6th, etc., values repeat. Not

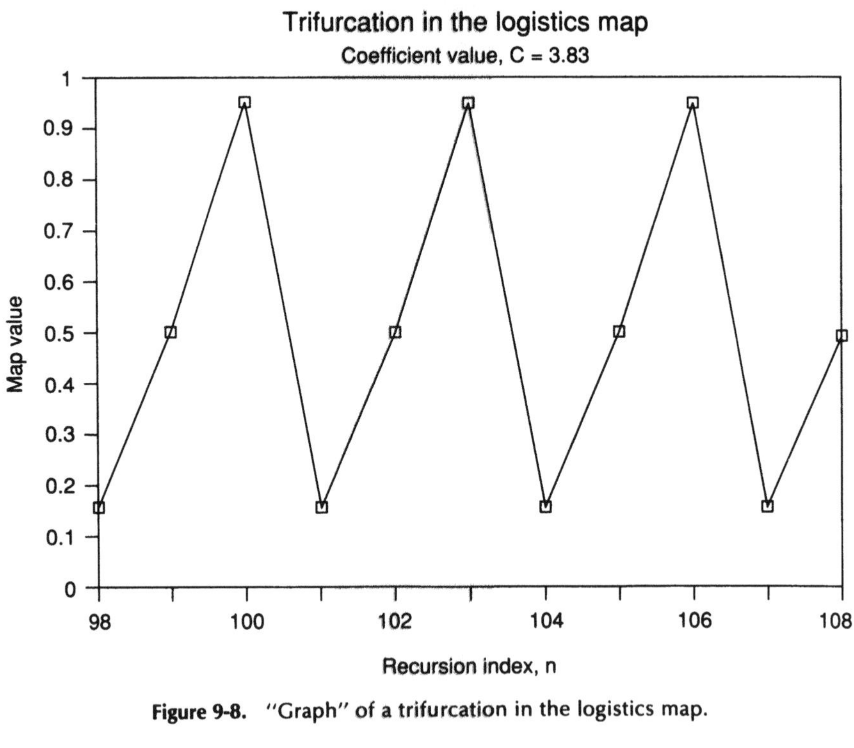

Figure 9-8. "Graph" of a trifurcation in the logistics map.

only is this true, but its "PERIOD" of "furcation" can be INFINITE, implying that it has lost any proclivity for periodicity. The numbers seemingly have no relation to one another, so we are tempted to call them "random." However, they *always* repeat for the same initial condition, X(0), so we can't truly use the word random. Instead, the word "chaos" seems to fit. As the value of the coefficient, C, is allowed to approach 4 (from below), the period is observed to lengthen, and at precisely 4, it apparently becomes infinite.

Homework Problem. Add a column of data representing the Logistics map for C = 4, in column E and plot the values for any initial value, X(0). Your plot should appear as in Figure 9-9.

Homework Problem. The data in Figure 9-9 appear to be random, for any choice of initial value, X(0), but since we cannot manage an infinite number of values, there is no absolute proof of this conjecture. You may, however, gather evidence by determining the probability that a number falls inside a specified interval. As an exercise in the use of 1-2-3, divide the interval from 0.0 to 1.0 into 10 subintervals and determine the number of values falling into each of the latter. Dividing this number by the total gives a measure of the prob-

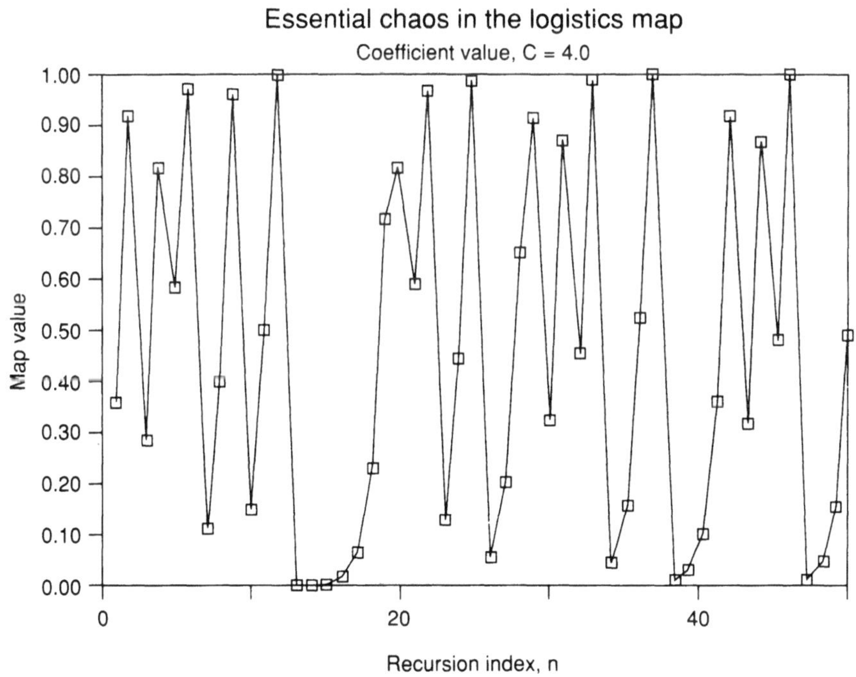

Figure 9-9. "Graph" of essential chaos in the logistics map.

ability. Plot this to see if the numbers have the same probability of falling into each of the subintervals.

The propensity of the Logistics map to produce furcations can be investigated by recognizing:

First, in the steady state, each value repeats, or every second value repeats, or every third value repeats, etc.

Second, successive repetitions imply, X(n + 1) = X(n) as n increases. This is the same as forcing, X = C*X*(1 − X), which can be evaluated by plotting *both* sides of the equation against X, and looking for the intersection.

Third, alternate repetitions imply X(n + 2) = X(n) as n increases. Now, the recursion requires both

$$X(n + 2) = C*X(n + 1)*[1 - X(n + 1)]$$

and

$$X(n + 1) = C*X(n)*[1 - X(n)]$$

Substituting the second into the first gives, X = f(X), where f(X) is a quadratic. If a steady state exists, it will manifest itself as an intersection between X and f(X), both plotted against X.

To illustrate another use of the COPY command, let us investigate the case of the bifurcation that has been shown to occur at C = 10/3. The spreadsheet is set up to evaluate in respective columns, X, (10/3)*X*[1 − X], and finally, f(X), which is a quartic found numerically by COPYING all the formulae in the 2nd column into the 3rd, giving the spreadsheet below. (The student should verify.)

3.333333 = C

X	CX(1 − X)	f(X)
0	0	0
0.1	0.3	0.7
0.2	0.533333	0.829629
0.3	0.7	0.7
0.4	0.8	0.533333
0.5	0.833333	0.462962
0.6	0.8	0.533333
0.7	0.7	0.7
0.8	0.533333	0.829629
0.9	0.3	0.7
1	0	0

These data enable the search for a bifurcation in the logistics map.

The data are plotted in Figure 9-10, where it will be seen that the quartic, f(X), intersects X not just twice, but THRICE! Notice that two of the intersec-

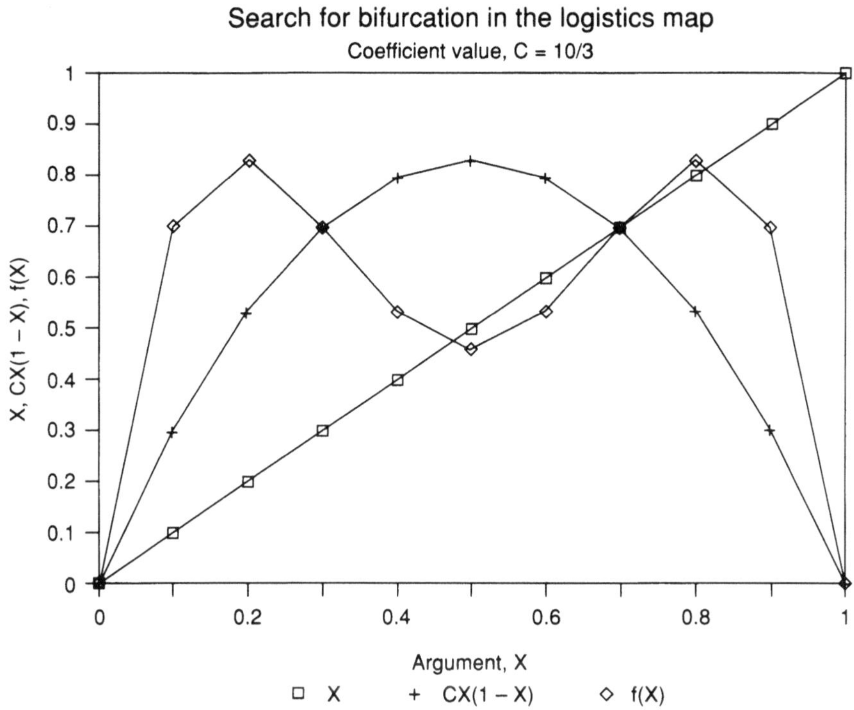

Figure 9-10. Analysis of a bifurcation in the logistics map.

tions correspond to the bifurcation values. The mystery is, what significance, if any, does the third intersection have? The answer has to do with the stability of these intersections, which is a topic we avoid here as it doesn't directly make use of a spreadsheet analysis; however, it is discussed in Chapter 7.

Homework Problem. In the case of a trifurcation, the Logistics map seeks, in its steady state, a sequence of three values, so we conclude that it must satisfy the condition, X(n + 3) = X(n), as n increases, for the coefficient, C = 3.83. Using the recursion formula, this results in the relation, X = f8(X), where the right side is an 8th-order polynomial. Intersections of it with X are candidates for the trifurcation values. Use the COPY command to evaluate f8(X) in the previous spreadsheet, plot the results, and compare to the recursion values obtained earlier for C = 3.83.

9.6.3 Exercise 3

Recurse Lin's method of factoring polynomials (ref. 8). This is an algorithm using long division of the polynomial by a trial quadratic. Its quadratic remainder is then used as the next divisor, and, thereafter the algorithm (hopefully) converges. Examples of convergence, divergence, bifurcation, and essential chaos are demonstrated.

Suggested Solution. The Newton–Raphson method (ref. 9) of factoring a polynomial derives from a truncation of its Taylor expansion. In its common form, the expansion is truncated after the term containing its first derivative, and, because the roots are sought, this is equated to zero,

$$f(X) = f(X_1) + [X - X_1]f'(X_1) = 0$$

This is solved for X, and a recursion formula is created by denoting X as the newly recursed value,

$$X(n + 1) = X(n) - [f(Xn)/f'(Xn)]$$

Clearly, there is no reason to restrict this method to polynomials, but polynomials or not, the recursion can only extract one root at a time. Notice that the right side of the recursion is real, so the root being extracted is also real. This uncovers the basic pitfall in Newton–Raphson: it fails to address complex roots. This hinderance is avoided if the Taylor series is truncated after the term containing the second derivative. Following the same proceedure, a rather more complicated recursion formula is obtained, which is the solution to a quadratic, allowing the generation of complex roots. This quadratic is,

$$f''X^2 + 2X(f' - X_1f'') + (f''X_1^2 - 2f'X_1 + 2f) = 0$$

and its complexity is the reason it is seldom used.

Lin recognized this shortcoming and set about finding his own trial quadratic; however, instead of trying to extract the actual root, he sought to extract the entire quadratic factor by a recursive means (which would hopefully converge). He argued that if a polynomial contained two roots which were smaller than all others (and the smaller the better), then, in the neighborhood of those two roots, the polynomial would be close to vanishing and would be dominated by its three terms of lowest degree, which become a logical approximation for these roots. Therefore, he selected these, which form a quadratic, and proceeded with long division until a first-degree remainder was found. For an nth-degree polynomial, this would require $n - 1$ divisions, and he also specified the division be performed with descending powers dividing descending powers. Should the remainder be zero, he would have found a quadratic factor of the polynomial, but normally such success is not expected. Lin's next problem was to somehow specify another new quadratic divisor which would hopefully be a better approximation than the original guess. He selected the next-to-last remainder, which is a quadratic, on the basis of the following reasoning. Were the first-degree remainder exactly zero, the quadratic from which it was formed would be a factor of the polynomial; and, if this iterative process were to converge, this condition would eventually be approached. Although he had no assurance that his iteration would converge, the argument made sense.

What is appealing about Lin's method is the ease with which the first trial quadratic is found. Unfortunately, as noted, there is no guarantee of convergence, but if the algorithm diverges, a suggested modification is to replace the argument, X, with its reciprocal, Y = 1/X, and proceed. Should the original divergence be caused by the absence of a "small enough" pair of roots, this transformation effectively converts "large" roots into small ones, and possibly allows convergence. However, the question of convergence is rather difficult to address, and the examples which follow are designed to foster interest in pursuing its answer.

We will consider a fourth-degree polynomial, of the form,

$$f(X) = a_4X^4 + a_3X^3 + a_2X^2 + a_1X + a_0$$

from which the first trial divisor is the quadratic making up the last three terms,

$$q(X) = a_2X^2 + a_1X + a_0$$

Then, after dividing twice, a quadratic remainder is obtained,

$$r_1(X) = b_2X^2 + b_1X + b_0$$

and this is used as the next divisor, after which the same algorithmic process is repeated over and over. Each cycle of the iteration produces a quadratic remainder which is used as the next divisor. As a means of measuring the rate of convergence, the first-degree remainder can be computed. Should both its coefficients approach zero, convergence is clearly taking place; and should they get progressively bigger, the process is diverging. Since this is a nonlinear operation, it is difficult to predict ahead of time whether the iterations will converge or diverge; and, indeed, because of its nonlinear nature, there is the possibility that something intermediate will occur. As will be seen, it may bifurcate, producing alternately the same quadratic remainder when it reaches its steady state. This is not unlike the bifurcations which occur in the Logistics map. And, as will also be seen, the iterations can repeat in higher periods, and even achieve a state of essential chaos, where the repetition period is infinite, analogous to the results obtained from the Logistics map corresponding to the coefficient value of 4.

The long division is performed according to the rule shown in Figure 9-11.

The long division is carried out as indicated in Figure 9-12. The leading coefficients of both the original quartic polynomial and Lin's quadratic trial divisor have been normalized to simplify the display. The example is taken from ref. 10, where a modification to Lin's method is presented.

LIN'S METHOD

FACTORING A POLYNOMIAL

Given $f(x) = a_n x^n + a_{n-1}x^{n-1} + \cdots + a_3x^3 + \underbrace{a_2x^2 + a_1x + a_0}_{q(x) \cdots \text{ First Trial Divisor.}}$

For f(x) of 4th-degree $\cdots$ $n = 4$

Division $\cdots$ $f(x)/q(x) = \underbrace{\alpha_2 x^2 + \alpha_1 x + \alpha_0}_{\text{Dividend}} + \dfrac{R(x)}{q(x)}$ ← Remainder @ 1st-degree.

$$\begin{array}{r|lllll}
 & a_4/a_2 & \dfrac{a_3 - \dfrac{a_1a_4}{a_2}}{a_2} & b_2/a_2 & & \\
\hline
a_2 \quad a_1 \quad a_0 & a_4 & a_3 & a_2 & a_1 & a_0 \\
 & a_4 & a_1a_4/a_2 & a_0a_4/a_2 & & \\
\hline
 & & a_3 - \dfrac{a_1a_4}{a_2} & a_2 - \dfrac{a_0a_4}{a_2} & a_1 & \\
 & & a_3 - \dfrac{a_1a_4}{a_2} & \dfrac{a_1}{a_2}\left(a_2 - \dfrac{a_1a_4}{a_2}\right) & \dfrac{a_0}{a_2}\left(a_2 - \dfrac{a_1a_4}{a_2}\right) & \\
\hline
\end{array}$$

These form the next → b_2 $\quad b_1 \quad b_0 = a_0$

Trial Divisor $\cdots$ $\quad \dfrac{b_2a_1}{a_2} \quad \dfrac{b_2a_0}{a_2}$

$b_2\cdot x^2 + b_1\cdot x + b_0$ $\quad b_2$

$\underbrace{c_1 \quad c_0}_{\text{Remainder}}$

$R(x) = c_1\cdot x + c_0$

Figure 9-11. The long division process used in Lin's method.

Homework Problem. You can generate the formula inserted into each cell in Figure 9-12 using the PRINT command, and the string,

/PPRMij .. Nkl

which establishes the RANGE of cells to PRINT, bordered by Mij Nkl (say M18), to the upper-left cell and Nkl (say N93), to the lower-right. Next, the following string,

/PPOOCQAG

commands the printer to print the formula assigned to each cell rather that its numerical value.

Thus far, a single cycle of Lin's recursion has been accomplished. To get the next cycle it would be possible to key-in all the same information. However,

```
LIN's METHOD OF FACTORING POLYNIOMIALS BY RECURSIONS ...
     ... from ref 10, "Modern Flight Dynamics", where Lin's
          iterations converge for G1(x) given below ...

GIVEN:    ... A QUARTIC POLYNOMIAL.

    G1(x) = x^4 + 1.85*x^3 + 10.429*x^2 + 11.756*x + .243

   ... which has the following roots, via Tartaglia's equations,
    x1 = -.02106, x2 = -1.19750, x3 & x4 = -.31572 +\- j*3.08782

LONG DIVISION PERFORMED BELOW ... DESCENDING POWERS / DESCENDING POWERS
  i.e., Quadratic Divides Quartic, with coefficients A0 ... A4,

              1 = A4
           1.85 = A3
         10.429 = A2       ... QUARTIC FROM REFERENCE 8.
         11.756 = A1
          0.243 = A0

COMPUTATIONS:  QUARTIC IS DIVIDED TWICE, 1st-DEGREE REMAINDER OBSERVED.

FIRST ITERATION ...                  QUADRATIC QUOTIENT
                                     1 0.722758 9.590976
   TRIAL QUADRATIC DIVISOR        **********   QUARTIC DIVIDEND   *********
   A2/A2    A1/A2    A0/A2     A4/A4    A3/A4     A2/A4     A1/A4     A0/A4
       1 1.127241 0.023300         1     1.85    10.429    11.756     0.243
                                   1 1.127241 0.023300
                                     0.722758 10.40569    11.756
                                     0.722758 0.814723 0.016840
   QUADRATIC REMAINDER, IS NEXT DIVISOR ->    9.590976 11.73915     0.243
                                              9.590976 10.81134 0.223473
                           1st-DEGREE REMAINDER ->     0.927814 0.019526
```

Figure 9-12. Spreadsheet for one cycle of Lin's method.

the "COPY" command obviates this labor, provided provision is made to use the quadratic remainder and the next divisor in the next cycle. We leave it as an exercise for the student to generate the second cycle, and thereafter repeat cycle-after-cycle, using the COPY command to show that the recursion converges. You may check the solution by using the Tartaglia equations to find the exact answer, or simply substitute (using 1-2-3).

Just the results can be displayed to eliminate the lengthy computation as shown in Figure 9-13, where the divisor, quotient, and remainder quadratics, and the first-degree remainder are listed.

Homework Problem. Reference 10 gives a quartic purporting to diverge under Lin's iteration. In actual fact, enough iterations reveal a bifurcation, seeking not one, but alternately two quadratics.

$$f(X) = X^4 + 1.85X^3 + 1.811X^2 + 2.285X - .243$$

which has the real quadratic factors,

$$f(X) = (X^2 - 1.54918X - .16132)(X^2 + 3.0082X + 3.74599)$$

and roots, $X = .09794,\ -1.64712$, and $-1.5041 \pm 1.21806j$

Check your answer against the tabulation shown in Figure 9-14.

```
FIRST ITERATION ...                           QUADRATIC QUOTIENT
                                              1 0.722758 9.590976
  TRIAL QUADRATIC DIVISOR         **********  QUARTIC DIVIDEND   *********
  A2/A2    A1/A2    A0/A2         A4/A4   A3/A4    A2/A4     A1/A4    A0/A4
      1 1.127241 0.023300           1      1.85   10.429    11.756    0.243
                                    1 1.127241 0.023300
                                      0.722758 10.40569   11.756
                                      0.722758 0.814723 0.016840
  QUADRATIC REMAINDER, IS NEXT DIVISOR ->      9.590976 11.73915    0.243
                                               9.590976 10.81134 0.223473
                        1st-DEGREE REMAINDER ->         0.927814 0.019526

SECOND ITERATION ...                          QUADRATIC QUOTIENT
                                              1 0.626020 9.637427
  TRIAL QUADRATIC DIVISOR         **********  QUARTIC DIVIDEND   *********
  B2/B2    B1/B2    B0/B2         A4/A4   A3/A4    A2/A4     A1/A4    A0/A4
      1 1.223979 0.025336           1      1.85   10.429    11.756    0.243
                                    1 1.223979 0.025336
                                      0.626020 10.40366   11.756
                                      0.626020 0.766236 0.015861
  QUADRATIC REMAINDER, IS NEXT DIVISOR ->      9.637427 11.74013    0.243
                                               9.637427 11.79601 0.244176
                        1st-DEGREE REMAINDER ->         -0.05587 -0.00117

THIRD ITERATION ...                           QUADRATIC QUOTIENT
                                              1 0.631818 9.634116
  TRIAL QUADRATIC DIVISOR         **********  QUARTIC DIVIDEND   *********
  B2/B2    B1/B2    B0/B2         A4/A4   A3/A4    A2/A4     A1/A4    A0/A4
      1 1.218181 0.025214           1      1.85   10.429    11.756    0.243
                                    1 1.218181 0.025214
                                      0.631818 10.40378   11.756
                                      0.631818 0.769669 0.015930
  QUADRATIC REMAINDER, IS NEXT DIVISOR ->      9.634116 11.74006    0.243
                                               9.634116 11.73610 0.242916
                        1st-DEGREE REMAINDER ->         0.003963 0.000083

FOURTH ITERATION ...                          QUADRATIC QUOTIENT
                                              1 0.631406 9.634349
  TRIAL QUADRATIC DIVISOR         **********  QUARTIC DIVIDEND   *********
  B2/B2    B1/B2    B0/B2         A4/A4   A3/A4    A2/A4     A1/A4    A0/A4
      1 1.218593 0.025222           1      1.85   10.429    11.756    0.243
                                    1 1.218593 0.025222
                                      0.631406 10.40377   11.756
                                      0.631406 0.769428 0.015925
  QUADRATIC REMAINDER, IS NEXT DIVISOR ->      9.634349 11.74007    0.243
                                               9.634349 11.74035 0.243005
                        1st-DEGREE REMAINDER ->         -0.00027 -0.00000

FIFTH ITERATION ...                           QUADRATIC QUOTIENT
                                              1 0.631435 9.634332
  TRIAL QUADRATIC DIVISOR         **********  QUARTIC DIVIDEND   *********
  B2/B2    B1/B2    B0/B2         A4/A4   A3/A4    A2/A4     A1/A4    A0/A4
      1 1.218564 0.025222           1      1.85   10.429    11.756    0.243
                                    1 1.218564 0.025222
                                      0.631435 10.40377   11.756
                                      0.631435 0.769444 0.015926
  QUADRATIC REMAINDER, IS NEXT DIVISOR ->      9.634332 11.74007    0.243
                                               9.634332 11.74005 0.242999
                        1st-DEGREE REMAINDER ->         0.000019 0.000000

SIXTH ITERATION ...                           QUADRATIC QUOTIENT
                                              1 0.631433 9.634333
  TRIAL QUADRATIC DIVISOR         **********  QUARTIC DIVIDEND   *********
  B2/B2    B1/B2    B0/B2         A4/A4   A3/A4    A2/A4     A1/A4    A0/A4
      1 1.218566 0.025222           1      1.85   10.429    11.756    0.243
                                    1 1.218566 0.025222
                                      0.631433 10.40377   11.756
                                      0.631433 0.769443 0.015926
                                               9.634333 11.74007    0.243
                                               9.634333 11.74007 0.243000
NOTE ... 1st-DEGREE REMAINDER HAS VANISHED!  ->         -0.00000 -0.00000
                   ... LIN's METHOD HAS CONVERGED.
```

Figure 9-13. Summary of iterations in Lin's method.

```
*******     LIN'S METHOD ... RECURSIVELY FACTORING A QUARTIC     *******

REF 8, HILDEBRAND, "INTRO TO NUMERICAL ANALYSIS" McGRAW-HILL, 1956

LONG DIVISION, RESULTS  BELOW ... DESCENDING POWERS / DESCENDING POWERS

  i.e., QUADRATIC is divided TWICE into the quartic giving
        a QUADRATIC remainder, which is used as the NEXT DIVISOR.

       1 = A4     QUARTIC = A4x^4 + A3x^3 + A2x^2 + A1x + A0 = P(x)
    1.85 = A3     & TRIAL QUADRATIC = A2x^2 + A1x + A0 = Q(x)
   1.811 = A2     DIVISION GIVES, P(x)/Q(x) = B2x^2 + B1x + ...
   2.285 = A1     ... + [QUADRATIC REMAINDER]/Q(x)
  -0.243 = A0     NEXT, DIVIDE BY QUADRATIC REMAINDER ... REPEAT!

SUMMARY OF RECURSIONS IS AS FOLLOWS ...
```

~~ DIVISOR ~~~~			~~~~ QUOTIENT ~~~~			~ REMAINDER ~	
x^2	x	constant	x^2	x	constant	x	constant
1	1.261733	-0.13418	1	0.588266	1.202944	0.846137	-0.08158
1	1.965122	-0.20200	1	-0.11512	2.239233	-2.13862	0.209334
1	1.010052	-0.10851	1	0.839947	1.071128	1.294254	-0.12676
1	2.218362	-0.22686	1	-0.36836	2.855024	-4.13204	0.404701
1	0.771072	-0.08511	1	1.078927	1.064181	1.556269	-0.15242
1	2.233482	-0.22834	1	-0.38348	2.895845	-4.27038	0.418250
1	0.758822	-0.08391	1	1.091177	1.066903	1.566973	-0.15347
1	2.227535	-0.22776	1	-0.37753	2.879735	-4.21569	0.412894
1	0.763616	-0.08438	1	1.086383	1.065802	1.562808	-0.15306
1	2.229936	-0.22799	1	-0.37993	2.886231	-4.23773	0.415052
1	0.761676	-0.08419	1	1.088323	1.066242	1.564497	-0.15323
1	2.228976	-0.22790	1	-0.37897	2.883632	-4.22891	0.414189
1	0.762451	-0.08426	1	1.087548	1.066065	1.563822	-0.15316
1	2.229361	-0.22794	1	-0.37936	2.884675	-4.23245	0.414535
1	0.762140	-0.08423	1	1.087859	1.066136	1.564093	-0.15319
1	2.229207	-0.22792	1	-0.37920	2.884257	-4.23103	0.414396
1	0.762265	-0.08425	1	1.087734	1.066108	1.563985	-0.15317
1	2.229269	-0.22793	1	-0.37926	2.884425	-4.23160	0.414452
1	0.762215	-0.08424	1	1.087784	1.066119	1.564028	-0.15318
1	2.229244	-0.22792	1	-0.37924	2.884358	-4.23138	0.414430
1	0.762235	-0.08424	1	1.087764	1.066114	1.564011	-0.15318
1	2.229254	-0.22793	1	-0.37925	2.884385	-4.23147	0.414439
1	0.762227	-0.08424	1	1.087772	1.066116	1.564018	-0.15318
1	2.229250	-0.22793	1	-0.37925	2.884374	-4.23143	0.414435
1	0.762230	-0.08424	1	1.087769	1.066116	1.564015	-0.15318
1	2.229252	-0.22793	1	-0.37925	2.884378	-4.23145	0.414436
1	0.762228	-0.08424	1	1.087771	1.066116	1.564016	-0.15318
1	2.229251	-0.22793	1	-0.37925	2.884376	-4.23144	0.414436

Figure 9-14. A bifurcation in Lin's method of factoring.

Homework Problem. A quartic for which Lin's method diverges is,

$$f(X) = X^4 + 2X^3 + X^2 + X + 1$$

which you should analyze on your spreadsheet. Its solution is given in Figure 9-15. The iterations would probably converge if they were somehow done in "reverse," but lacking this capability, it is often possible to transform the polynomial by inverting its argument—that is, by replacing X with 1/X through-

```
*******      LIN'S METHOD ... RECURSIVELY FACTORING A QUARTIC       *******

REF 8, HILDEBRAND, "INTRO TO NUMERICAL ANALYSIS" McGRAW-HILL, 1956

LONG DIVISION, RESULTS  BELOW ... DESCENDING POWERS / DESCENDING POWERS

  i.e., QUADRATIC is divided TWICE into the quartic giving
        a QUADRATIC remainder, which is used as the NEXT DIVISOR.

        1 = A4      QUARTIC = A4x^4 + A3x^3 + A2x^2 + A1x + A0 = P(x)
        2 = A3      & TRIAL QUADRATIC = A2x^2 + A1x + A0 = Q(x)
        1 = A2      DIVISION GIVES, P(x)/Q(x) = B2x^2 + B1x + ...
        1 = A1      ... + [QUADRATIC REMAINDER]/Q(x)
        1 = A0      NEXT, DIVIDE BY QUADRATIC REMAINDER ... REPEAT!

SUMMARY OF RECURSIONS IS AS FOLLOWS ...
```

~~ DIVISOR ~~~~			~~~~ QUOTIENT ~~~~			~ REMAINDER ~	
x^2	x	constant	x^2	x	constant	x	constant
1	1	1	1	1	-1	1	2
1	0	-1	1	2	2	3	3
1	1.5	0.5	1	0.5	-0.25	1.125	1.125
1	-3	-4	1	5	20	81	81
1	1.05	0.05	1	0.95	-0.0475	1.002375	1.002375
1	-20.0526	-21.0526	1	22.05263	464.2659	9775.019	9775.019
1	1.002153	0.002153	1	0.997846	-0.00214	1.000004	1.000004
1	-464.268	-465.268	1	466.2680	216939.6	1.0E+08	1.0E+08
1	1.000004	0.000004	1	0.999995	-0.00000	1	1
1	-216939.	-216940.	1	216941.6	4.7E+10	1.0E+16	1.0E+16
1	1	2.1E-11	1	1	-2.1E-11	1	1
1	-4.7E+10	-4.7E+10	1	4.7E+10	2.2E+21	1.0E+32	1.0E+32
1	1	4.5E-22	1	1	0	1	1
ERR	ERR	ERR	1	ERR	ERR	ERR	ERR
ERR	ERR	ERR	1	ERR	ERR	ERR	ERR

Figure 9-15. A divergence in Lin's method of factoring.

out. The transformed polynomial with normalized leading coefficient has all the coefficients "reversed." Use your spreadsheet analysis of Lin's method to show that it now converges.

Homework Problem. A quartic polynomial which neither converges nor diverges, and also doesn't bifurcate, but seems to go into essential chaos under Lin's recursion is,

$$f(X) = X^4 + X^3 + X^2 + X + 5$$

Analyze it on 1-2-3 and argue the assertion.

Homework Problem. Using your spreadsheet, can you find any quartics which trifurcate?, quatrofurcate?, etc.

Homework Problem. Develop a spreadsheet for the Newton–Raphson method to extract a root from all preceding quartics.

Homework Problem. Develop a spreadsheet for the modified form of the Newton–Raphson method of extracting a complex root from a quartic and use it to investigate all the preceding polynomials.

9.6.4 Exercise 4

Discuss Integration and Differentiation and show some simple digital approximations which yield recursion formulae. Apply these recursions to a sine-wave and compare the results with known solutions.

Suggested Solution. Differentiation is that process in which an *average* rate of change over an *incremental* distance becomes an *instantaneous* rate of change at a point (in the limit). It is usual to refer to the *derivative* as the *slope* of the curve *at the point*. Similarly, integration is the area under a curve, consisting of the sum of finite rectangles, each approximating a small segment of the curve. In the limit, as the width of these rectangles becomes vanishingly small, the continuum process of integration is defined.

So, differentiation determines the slope of a curve and integration determines the area under a curve. For the most part, simulation of dynamic systems is concerned with the time response that those systems undergo as a result of some specified input stimulus, such as a forcing function (of time) or an initial condition. As a result, the independent variable is almost universally "time," so we elect to adopt this as the variable involved in both the processes of differentiation and integration. In each case, the signal which is to be either differentiated or integrated is called X(t), and the result is called Y(t). While there are many formulae for approximating them digitally (refs. 11–14), the following are two of the simpler ones:

DIFFERENTIATION $Y(t_1) = [X(t_1) - X(t_0)]/[t_1 - t_0]$

INTEGRATION $Y(t_1) = [\frac{1}{2}]*[X(t_1) + X(t_0)]*[t_1 - t_0] + Y(t_0)$

The corresponding recursion formulae are:

DIFFERENTIATION $Y(n + 1) = [X(n + 1) - X(n)]/T$

INTEGRATION $Y(n + 1) = [T/2]*[X(n + 1) + X(n)] + Y(n)$

where T = the time increment between the (n + 1)-st and the nth points in time. It is a constant.

Without regard to certain particulars, integration and differentiation are often considered to be "inverses" of one another, in that performing one after the other will yield the original function. A test of this assertion can be made with a spreadsheet by first differentiating and then integrating a sine wave, according to the recursions given above (or any other such formulae found in the references). We leave it to the student to construct the spreadsheet for 10 samples of a single cycle, and compare it to the GRAPH shown in Figure 9-16. A measure of the accuracy is the RMS-ERROR consisting of the square root of the sum of the squares of the individual errors—that is, the differences between the exact and computed values, given by

RMS-ERROR = SQRT[SUM[(error)^2]]

Homework Problem. Add to the spreadsheet, which you constructed to evaluate the accuracy of the discrete differentiators and integrators, the calculation of the RMS-ERROR. For sampling 10-times each cycle, this error is 14.6% (where 1 cycle is 360-degrees or 2π radians). Then change the sampling rate and tabulate the RMS-ERRORS for each.

Homework Problem. Reverse the procedure by integrating first and then differentiating the sine wave. Compare graphs and errors.

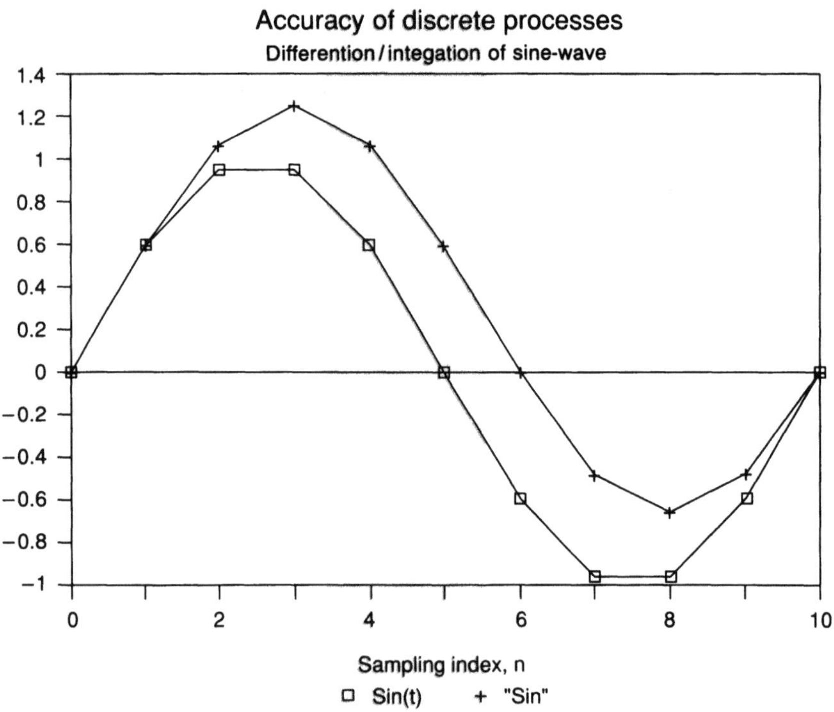

Figure 9-16. Accuracy of discrete differentiation and integration.

Homework Problem. Differentiation is sometimes called a "noisy" process, while integration is called a "smoothing" process. Develop a spreadsheet to verify these assertions by comparing the discrete derivative and integral of a sine wave to the exact cosine wave.

Homework Problem. Digitally differentiate a unit step function, starting the process at some time preceding its leading edge. This investigation may help you understand why differentiation is an unrealizable process. The unit step function is defined as a discontinuity in an otherwise continuous function which has the value 0 (zero) before its leading edge, and 1 (one) thereafter.

9.6.5 Exercise 5

Solve a Linear Ordinary Differential Equation with constant coefficients (whose solution is known by analysis), using a recursion algorithm to approximate integration. This approximate solution can be compared to the exact solution for various values of the "sample time." Typically, we choose the first-order equation,

$$\dot{X} + X = 0, \qquad \text{with initial condition, } X(0) = Xo$$

The value of this equation is that it is representative of the dynamics often encountered in real systems that are simple enough to be linearly modeled, such as the response of a thermometer when dipped into a cold bath, an R-L or R-C circuit excited by a battery, a dc motor's speed response to armature voltage excitation, or the linearized model of a steel ball "falling" through a viscous medium, such as oil.

Suggested Solution. The solution to a differential equation is found by integration (see Homework Problems at end of this section). In this case, the differentials in the equation are treated as algebraic quantities, and the classical solution proceeds by first solving the equation for Xdot, and then the differential dX,

$$\dot{X}(t) - -X(t)$$

where

$$\dot{X} = \frac{dX}{dt}$$

$$dX = -X(t)\,dt$$

and, both sides of the equation are integrated between the appropriate definite limits, giving,

$$X(t) - Xo = -\text{INTGRL}[X(t)*dt], \qquad \text{between the limits, 0 and t}$$

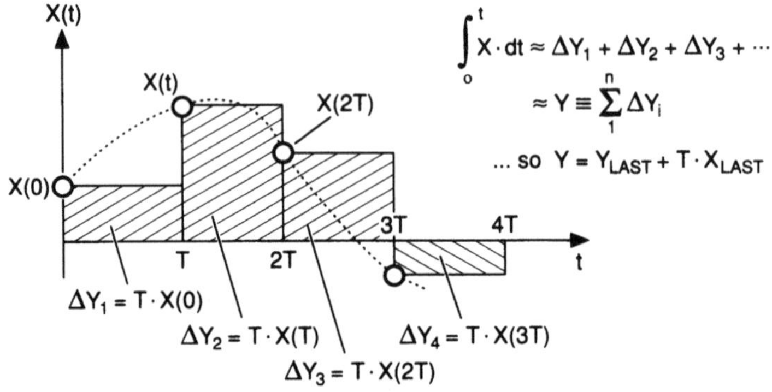

Figure 9-17. A discrete integration algorithm.

in which integration is simulated (modeled) by a running sum, and, although not precisely the same as the continuous process, it closely resembles it by adding the area increments under the curve as in Figure 9-17. While there are many other formulae available for integration, this is the simplest to understand, consisting of equal width (i.e., duration) rectangles of height, X(n*T), where t = n*T, T being the width. Thus, each adds an amount, T*X(n*T) to the preceding sum, producing the recursion formula,

$$\text{INTGRL}[X(t)*T] = Y(t)$$

where Y(next) = Y(last) + T*X(last). The recursion for the entire differential equation is,

$$X(n + 1) = X(n) - T*X(n), \qquad \text{for } n > 0 \text{ starting from } X(0) = Xo$$

which agrees with the differential equation as $T \rightarrow 0$, in

$$[X(n + 1) - X(n)]/T = X(n), \qquad \text{the left side approximating Xdot}$$

Figure 9-18 compares the approximate and exact solutions (you should convince yourself that the fidelity improves for smaller T).

Homework Problem. Instead of simulating the first-order equation with an initial condition, a forcing function can be applied as a function of time. Develop a spreadsheet for the system excited by a step function,

$$\dot{X} + X(t) = U(t)$$

where U(t) = 1 for t > 0 and U(t) = 0 for t < 0, and compare it to the exact result,

$$X(t) = 1 - e^{-t}$$

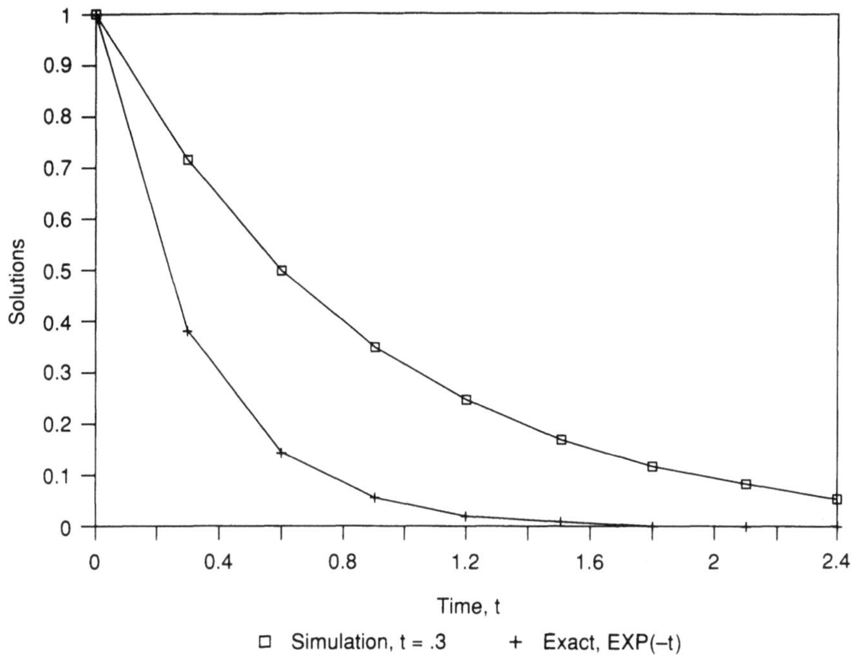

Figure 9-18. Comparing exact and approximate solutions to $\dot{X} + X = 0$.

Homework Problem. Use a spreadsheet to simulate the approximate solution to a linear second-order differential equation and compare it to the exact analytic solution,

$$\ddot{X} + \dot{X} + X = 0$$

Homework Problem. The method of solving the linear ordinary differential equation discussed in Exercise 5 presupposes that it is to be integrated; to do so, it is first "solved" for $\dot{X}$ and then integrated. However, there is seemingly nothing wrong with "solving" instead for X and differentiating. Both procedures are shown in Figure 9-19. Develop a spreadsheet for approximating the solution by differentiating, and stimulate the system with either a step input or from an initial condition. Perhaps this will help you to better understand why differentiation is not a realizable process and why integration is the fundamental theorem of calculus.

9.6.6 Exercise 6

Solve a *nonlinear* ordinary differential equation by an approximate discrete integration algorithm. Just like the previous exercise, there is an analytic solution

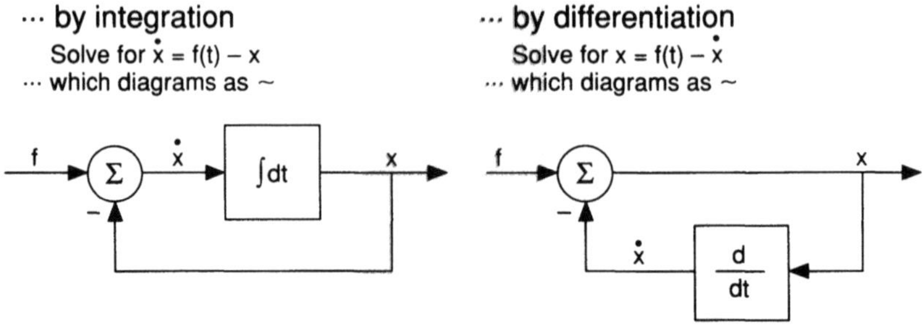

Figure 9-19. Two methods of solving $\dot{x} + x = f(t)$.

with which to compare the result, but it is not so easy to find, nor is it possible in all cases. A respectable approximation can be obtained by "fairing" a curve by eye when the slopes of the solution are identified as in the method of isoclines.

The equation we choose is,

$$\dot{X}^2 = X, \qquad \text{with arbitrary initial condition.}$$

Discussion. That there is an analytic solution to this equation can be verified by recognizing that the variables can be separated, giving,

$$dX/SQRT[X] = \pm dt$$

or

$$\tfrac{2}{3}X^{3/2} = \pm t + c$$

Since the solution involves the square root of X, there can be no real values of time, t, to correspond to negative values of X. Since time is defined as a real variable, we are forced to say that the equation has no solution for negative values of X, which is tantamount to conceding that no solutions exist in the lower half of the solution plane [i.e., a plot of X(t) versus t]. Perhaps a less dramatic and more accurate statement is that the relation is not an equation at all, which can be seen by examining it. Its left side is a square, which must be positive. If it is to be an equation, then the right side must always be positive, thereby denying negative values of X from consideration.

Suggested Solution. Since this is a first-order ordinary differential equation, the procedure is to solve it for its derivative and integrate by a discrete recursion, giving,

$$\pm\dot{X} = \pm\sqrt{X}$$

which tells us two important things about the solution:

1. There is *no solution* in the lower-half plane.
2. There is not one, *but two* solutions.

With these facts, you may proceed to develop the spreadsheet which approximates the solution. To do so, it will be necessary to delineate between the two solutions in the upper half of the plane, because at each point there are two values of the derivative (slope), and your simulation must be able to decide which one is to be followed. Should you be unaware that there are no solutions in the lower-half plane, that fact will become plain when the square root of a negative number is requested. A typical GRAPH of the solutions for various initial conditions is shown in Figure 9-20.

Homework Problem. The problems illustrated in Exercise 6, where there may be more than one solution or there may be no solution in various regions of the solution plane, present potential hazards when simulating any nonlinear equation. Such problems are not encountered in linear equations because there are proofs guaranteeing the existence and uniqueness of their solutions; so we need not be inhibited by the difficulty of more than one value for the derivative,

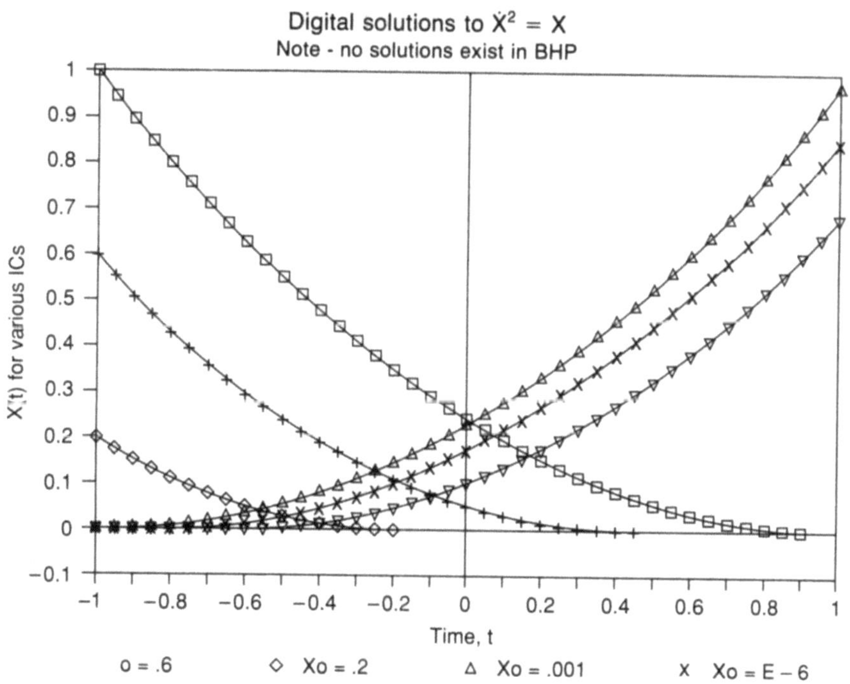

Figure 9-20. Digital approximation to a nonlinear equation, $(\dot{x})^2 = x$.

or other troublesome characteristics. Indeed, the lack of such theorems in the nonlinear world can be the source of frustration (excitement?) for the modeler.

One of the simplest, and yet complete, examples which demonstrates the questions of existence and uniqueness is found in the Clairaut equation as discussed in ref. 15, or many other texts dealing with advanced engineering mathematics,

$$Y(t) = t\dot{Y}(t) + f[\dot{Y}(t)]$$

where f, is arbitrary, except that it must be differentiable.

A particular example of Clairaut's equation, is,

$$f(\dot{Y}) = \dot{Y}^2$$

and Clairaut's equation contains analytic solutions in a region where they are not unique; where they are straight lines which intersect one another; yet there is another region where there are no solutions, corresponding to the assertion that the "equation" is a contradiction. The boundary between these two regions is still another solution, called the singular solution, otherwise unsuspected.

Make a spreadsheet analysis for approximating the solution to either this or any other example of Clairaut's equation, and use it to investigate these peculiarities.

Homework Problem. An interesting nonlinear equation (ref. 16), is,

$$(1 + t^2)^2\dot{Y}^2 = 4Y^3 - 4Y^2$$

with solution, $Y = (1 + t^2)/(Ct + \sqrt{1 - C^2})$.

Investigate with a spreadsheet analysis for various values of the coefficient, C, and show that there is a region of existence bounded by a singular solution,

$$Ys = +1$$

9.6.7 Exercise 7

Investigate the transient behavior of a feedback control system, consisting of a real pole and an integration driven by a saturation as shown in Figure 9-21.

Suggested Solution. Since the input is specifically designated as zero, there is no need to write an equation for the "summer," which leaves only the saturation, the first-order lag, and the integrator needing recursion formulae.

SATURATION:

This recursion is found using the "IF"-statement to determine whether the signal $X = -C$ is inside or outside the saturation region, as

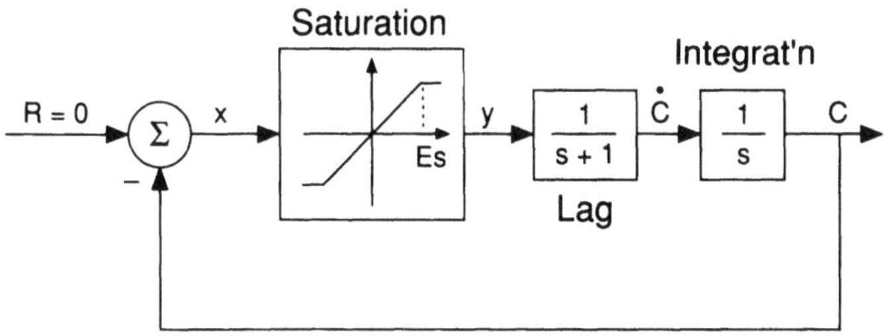

Figure 9-21. Saturating feedback system.

follows, where the index, n, is used to represent the cell into which the formula is deposited,

$$Y(n) = IF(ABS(C(n)) > Es, IF(C(n) > 0, -Es, Es), -C(n))$$

LAG:

The solution to the linear differential equation shown in Exercise 5 can be used here to simulate the first-order lag. You can verify by taking the inverse Laplace transform of its transfer function that this is a first-order linear ordinary differential equation, identical to that examined in Exercise 5.

$$Cdot(n + 1) = Cdot(n) + T*(Y(n) - Cdot(n))$$

INTEGRATION:

The method of summing areas (Exercise 4) is applicable, though you should try other formulae for discrete integration,

$$C(n + 1) = C(n) + T*Cdot(n)$$

The combination of these recursion formulae gives the transient response of the system from various initial conditions as shown in Figure 9-22.

Homework Problem. Another way to approach digital simulations is to insert a sampler and hold immediately upstream of each set of linear dynamics, such as an integrator or a lag. The analysis is carried out with z-transforms, the theory of which is beyond the scope of this tutorial (but is discussed in connection with the theory of discrete simulation in Chapter 8); and, in the case of the first-order lag, a somewhat different recursion formula is,

$$Cdot(n + 1) = @EXP(-T)*Cdot(n) + (1 - @EXP(-T))*Y(n)$$

Use this in your simulation and compare the results.

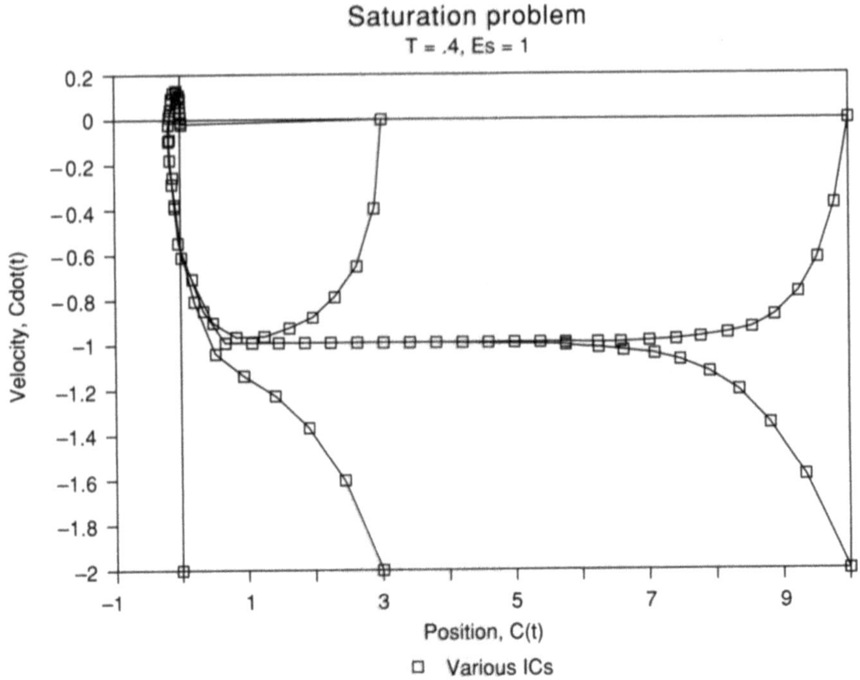

Figure 9-22. "Graph" of transient response in saturating system.

9.7 REFERENCES

1. Anon., 1-2-3 Tutorial, Release 2, LOTUS (1986).
2. Anon., *1-2-3 Reference Manual*, Release 2, LOTUS (1985).
3. Harvey, G., *The Complete Lotus 1-2-3 Handbook*, Release 2.2, SYBEX (1985).
4. Pearson, R., and Campbell, M., *Using EXCEL, IBM Version*, QUE (1988).
5. Anon., ENABLE Spreadsheet Graphics, The Softwear Group (1985).
6. Campbell, M., *Mary Campbell's QUATRO Pro Handbook*, Bantam (1990).
7. Cole, D., *Mastering Symphony*, fourth edition, SYBEX (1986).
8. Hildebrand, F. B., *Introduction to Numerical Analysis*, McGraw-Hill, New York (1956).
9. Karman, T., and Biot, M., *Mathematical Methods in Engineering*, McGraw-Hill, New York (1940).
10. Kolk, W. R., *Modern Flight Dynamics*, Prentice-Hall, Englewood Cliffs, New Jersey (1961).
11. Gerald, C. F., and Wheatley, P. O., *Applied Numerical Analysis*, third edition, Addison-Wesley (1984).
12. Al-Khafaji, A. W., and Tooley, J. R., *Numerical Methods in Engineering Practice*, Holt, Rineholt and Winston, New York (1986).
13. Kelly, L. G., *Handbook of Numerical Methods and Applications*, Addison-Wesley (1967).
14. Froberg, C., *Introduction to Numerical Analysis*, Addison-Wesley (1965).
15. Spiegel, M., *Applied Differential Equations*, third edition, Prentice-Hall, Englewood Cliffs, New Jersey (1981).
16. Ince, E. L., *Ordinary Differential Equations*, Dover, New York (1956).

Chapter 10
An Isobaric Cabin Pressure Control

10.0 INTRODUCTION

Engineering is an endless confrontation with Mother Nature owing to her obstinant denial of linearity. The virtue of Her stubbornness is not obvious in the heat of battle, but consider the alternative. Were Her world faithful to a linear persuasion there would be no engineering, at least as we know it. All problems would be accompanied by their unique solutions, thereby inhibiting creativity. Indeed, there would be little excitement in any aspect of human endeavor under the code of linearity, not unlike discovering the North Pole . . . a hollow victory.

In this chapter we offer a real-life problem, the design of the cabin pressure control for a subsonic commercial airliner. While there is no universal approach to designing a control for a nonlinear system, this example provides an approach which is not atypical. At the outset it must be recognized that failure of the cabin pressure control cannot only damage the airframe, but an abrupt change in pressure can be harmful to the passengers. Accordingly, the aircraft cannot be dispatched with a faulty pressure control, making it a safety of flight item, (unlike the temperature control, which may cause discomfort but no harm). Reliability, then, is the sine qua non, dictating simplicity, and curbing an aspiring engineer's penchant for sophistication.

Earlier in Chapter 3, a suitable model for the cabin dynamics was developed, consisting of a pressure vessel with specified inflow and variable discharge. It was found to be quite nonlinear—due to the gas dynamics—and was linearized around an operating point giving a very simple transfer function, consisting of a gain and a pole,

$$\frac{\Delta P_c}{\Delta A_E}(A) = -\frac{K_c \cdot a}{s + a}$$

The dilemma is the nonlinear dynamics which give rise to huge variations in the gain, K_c, and the bandwidth, a, from one flight condition to another. The quotient of highest to lowest values for the cabin gain exceeds 2500, while the corresponding "swing" on cabin bandwidth approaches 200. Coping with these variations is the job of the controller whose design is our topic.

The usual approach to designing the controller is to determine all the flight conditions which can be expected, and for each, to linearize the cabin dynamics and determine its transfer function. Then, in every case, a linear controller is synthesized to meet the competing requirements of stability and accuracy according to linear design practice. The trick is to find just one such controller which will meet all the requirements at all operating points, a trial-and-error process with no guarantee of success, and indeed, one at which we (the authors) fail. But, this is precisely the challenge of engineering, so in a way, we are beholden to Mother Nature's nonlinear bias, urging us toward a unifying concept. And fittingly, it is nonlinear.

The proof of the pudding is a believable verification that the entire nonlinear system functions satisfactorily throughout its operating range, not just in the neighborhood of each operating point considered in its synthesis. Toward this end a nonlinear simulation of the dynamics is made in which the aircraft can be "flown" over its complete flight spectrum, from takeoff and climb, through cruise and finally into descent and landing. In this way some assurance is obtained that the system performs properly, the burden of proof resting with the fidelity of the simulation.

It should be noted that this example does not make use of all the analytic tools presented in previous chapters. Just as each design problem is unique, so is the choice of tools for its solution.

10.1 BACKGROUND

The cabin pressure control for a commercial airliner is an object lesson in nonlinear design (ref. 1). Because it must accommodate a wide spectrum of operating conditions, the inevitable conflict between stability and accuracy dictates inclusion of a nonlinear dc loop gain. The outflow valve, whose modulation varies the discharge rate, is the means for controlling the cabin pressure. It may be treated as a variable-area orifice through which air flows according to its pressure ratio. This can vary from almost unity when the valve is wide open to 0.53 during cruise when the valve is "choked," and the corresponding effect on its flow coefficient is 60 db.

For this reason, as well as the need for cabin pressure rate limits and differential pressure overrides, the system can easily become complex. And, indeed, it would, were it not for the sobering reality that, unlike other systems such as the cabin temperature control, the pressure control is a flight safety item.

Without it the aircraft cannot be dispatched, providing an immediate economic impact on the carrier. Thus, to be economically viable, the pressure control must be reliable, placing a strong emphasis on its simplicity.

Aside from economic viability, another factor which discourages sophistication is the awesome liability that a malfunction could potentially incur. Extreme pressure fluctuations can seriously damage eardrums and may contribute to cardiac distress. The resulting lawsuits might not only involve the carrier, the airframe manufacturer, and the system vendor, but could realistically place the latter's responsible engineers at risk.

10.2 HOW THE CABIN IS PRESSURIZED

A source of clean high-pressure air is required, which is almost universally found by bleeding the main turbojet engines at one (or more) of their compressor stages. This air is passed through air-conditioning packs and then ducted to the cabin via a flow control valve. Thus, the cabin can be considered to have a relatively constant inflow, $\dot{W}_{IN}$.

The outflow is modulated by a valve usually located in the aft fuselage. Closing it causes pressure in the cabin to rise; therefore, commands directed to reduce the error in cabin pressure become the primary means of control. The transducer sensing pressure is usually located on the flight deck, some 100 feet removed from the outflow valve, and a pneumatic transport delay is incurred, usually denoted as

$$P_{SENSED}/P_{CABIN} = e^{-sT_p}$$

where the sensing delay, $T_p = .09$ is approximated by sonic transmission. A schematic of the cabin pressure control is shown in Figure 10-1.

10.3 BASIC NUMBERS AND CONSTRAINTS

Two quantities are fundamental to the degree of comfort expected by the public. One hundred cubic feet of "living space" and 15 cubic feet per minute of

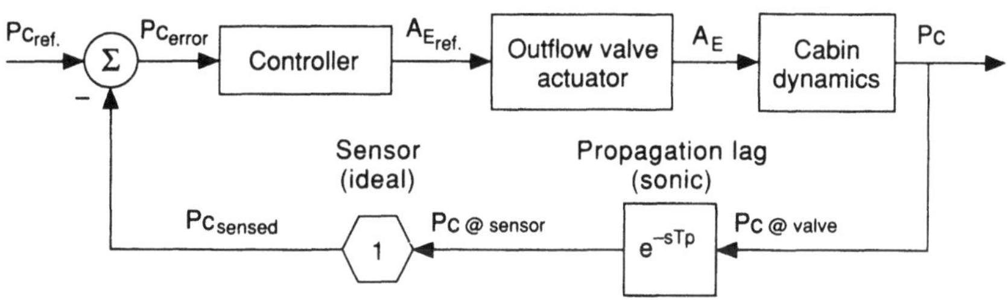

Figure 10-1. Schematic of a cabin pressure control.

ventilating air for each passenger are adequate, if not luxurious, for flights up to 6 or 8 hours.

In our example, we will assume accommodations for 400 people in a subsonic airliner, which translates to a pressurized volume of 40,000 cubic feet and inflow of 6,000 cfm.

The maximum size of the outflow valve is then determined by a safety condition as follows. Awaiting takeoff, the differential pressure must be low enough to allow an exit door to be pulled open manually. An acceptable value is 1 inch of water (.0361 psi), which, when asked to pass the 6,000 cfm, establishes the Effective Maximum Valve Area, 217 square inches (see Section 3.5 and equate inflow to outflow).

The outflow valve actuator is typically an electric servomotor with position feedback, which for simplicity is conservatively modeled by a real pole at 10 radians per second and unit gain.

Additional relevant information pertaining to a subsonic commercial airliner is:

Cruise altitude, 25,000 to 43,000 feet
Max. cabin altitude, 8,000 feet
Cabin pressure accuracy, 100 ft.
Cabin-to-ambient never negative

10.4 CABIN DYNAMICS

The cabin dynamics derive from the ideal gas law and Bernoulli's equation, both of which are nonlinear. The linearized model derived in Chapter 3 is used here, for which the cabin transfer function relating perturbations in cabin pressure and outflow valve area is given as,

Unchoked valve
$$\frac{\Delta P_c}{\Delta A_E}(s) = \frac{-\dfrac{\gamma R\sqrt{T_c}}{V}\cdot k^0 \cdot A_E^0}{s + \dfrac{\gamma R\sqrt{T_c}}{V} k^0 \cdot A_E^0 \cdot \left\{1 + \dfrac{P_c^0}{k^0}\left.\dfrac{\partial k}{\partial P_c}\right|\right\}}$$

where, $k \equiv \kappa \cdot \phi^{1/\gamma} \cdot \sqrt{1 - \phi^{(\gamma-1)/\gamma}}$ Flow Coef.

$$\kappa \equiv \sqrt{\frac{2q}{R}\cdot\frac{\gamma}{\gamma - 1}}$$

$$\phi \equiv P_A/P_c$$

Choked valve
$$\frac{\Delta P_c}{\Delta A_E}(s) = \frac{-\dfrac{\gamma R\sqrt{T_c}}{V}\cdot k \cdot P_c^0}{s + \dfrac{\gamma R\sqrt{T_c}}{V}\cdot k \cdot A_E^0}$$

where,

$$k = \sqrt{\frac{\gamma g}{R} \cdot \left(\frac{2}{\gamma + 1}\right)^{\frac{\gamma+1}{\gamma-1}}} = .532 \frac{\sqrt{R}}{\text{sec}} \text{ flow coeff.}$$

The gain plots of these cabin dynamics are shown in Figure 10-2 for the extreme cabin altitudes of sea level and 8,000 feet. It is from this information that the nature of the controller design problem becomes evident. Over the various flight conditions the dc gain and bandwidth vary over swings of 2,576 and 188, respectively. In some way, the controller design must cope with these wide variations and yet be simple enough to satisfy the needs of reliability.

10.5 DESIGN EVOLUTION

The problem is to find a controller, G_E, which will yield acceptable system performance throughout its entire range of operating points (as established by flight condition). This is described below as a succession of steps starting with the simplest (i.e. proportional control), and increasing in complexity only as necessary.

10.5.1 Proportional Control

Leads to a contradiction between droop and stability.

The loop gain function, $L(s)$, for the system of Figure 10-3 is

$$L(s) = \frac{K_E \cdot K_c \cdot e^{-.09 \cdot s}}{\left(1 + \frac{s}{10}\right) \cdot (1 + T_c \cdot s)}$$

Consider operation at sea level with wide-open valve (i.e., awaiting take-off). The allowable droop error determines the loop position constant, $K_p = K_E K_c$, which in turn specifies the controller gain, K_E, as the other two are known. If the entire accuracy of 100 feet is (naively) allocated to droop alone, the final value theorem sets K_p.

$$\underset{\text{error}}{\Delta P_c}(t = \infty) = \lim_{s \to 0} s \cdot \underset{\text{error}}{\Delta P_c}(s)$$

or,

$$.0524 = \lim_{s \to 0} s \cdot \frac{1}{1 + L(s)} \cdot \frac{\Delta P_{c_{\max}}}{s}$$

where $\Delta P_{c_{\max}}$ is the difference between the extreme cabin altitudes of S.L. and 8000-ft.—i.e., $\Delta P_{c_{\max}} = 3.78$ psi.

$$S_0,\ .0524 = \frac{3.78}{1 + K_p}, \qquad K_p = 71.1\ (37.0\ \text{db})$$

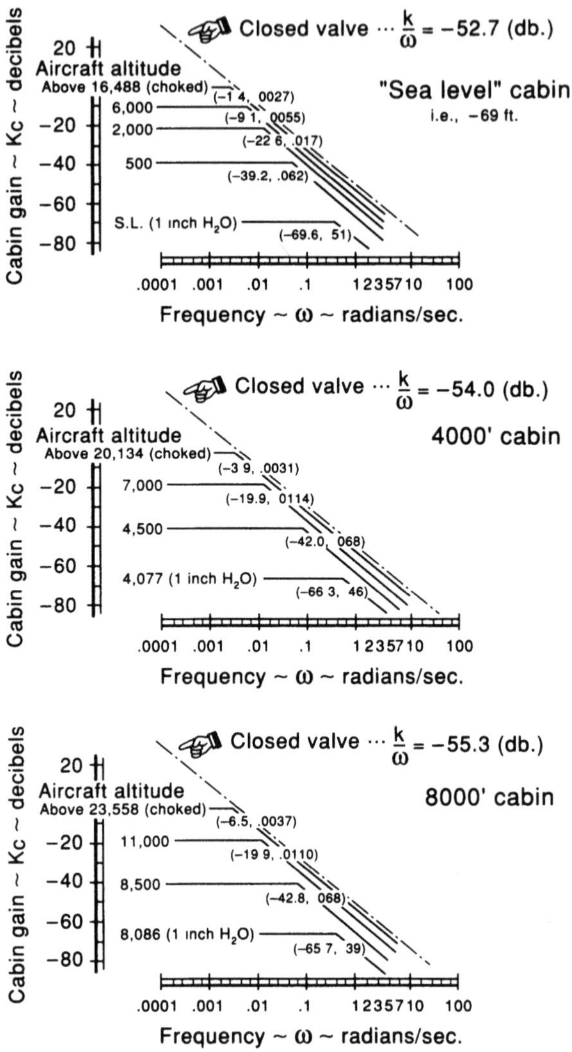

Figure 10-2. Cabin dynamics for three cabin pressures.

Under this condition and K_p, we check the system stability with the gain and phase plots of Figure 10-4. The system is seen to be unstable, having a negative phase margin of some 85°; thus, we uncover the common problem of conflicting accuracy and stability requirements. For a reasonable stability level of, say, 45° phase margin, the position constant, from Figure 10-4, can only be 18.5 db, or 8.4, which is far too low for the accuracy requirement.

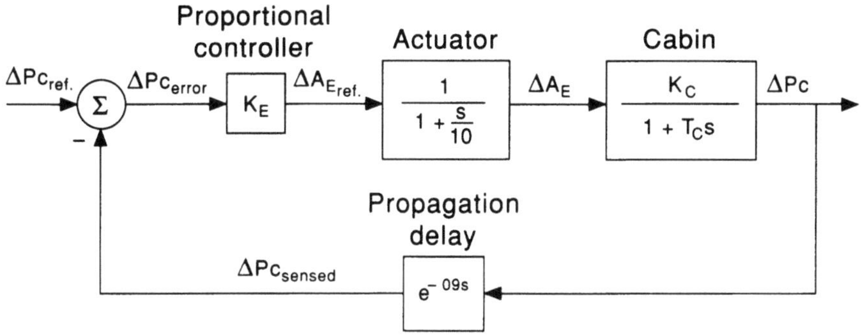

Figure 10-3. Linearized system with proportional control.

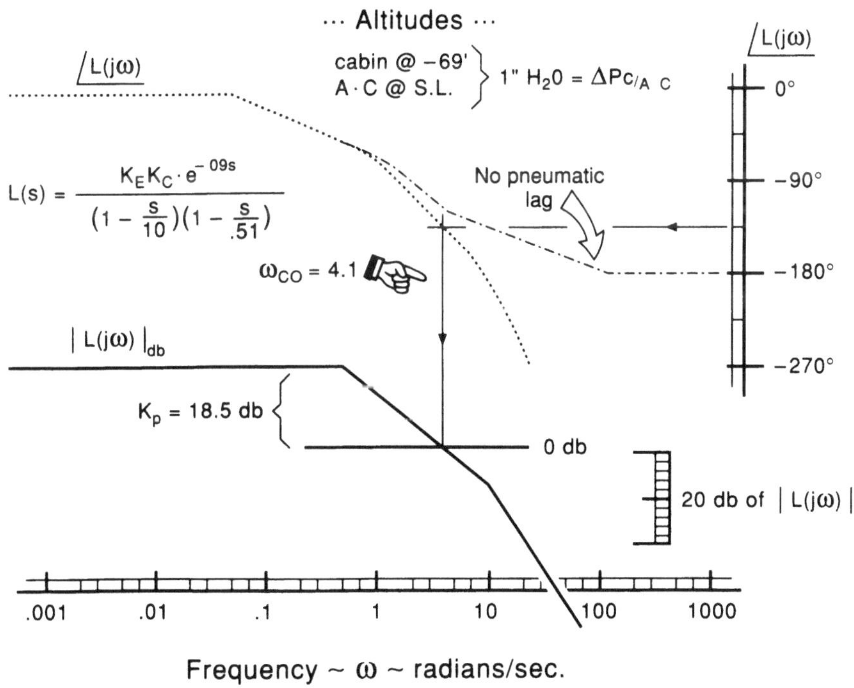

Figure 10-4. Gain and phase . . . proportional control.

We conclude that the proportional controller is not a viable candidate because of its unacceptable droop.

10.5.2 Integral Control

Eliminates droop and stabilizes at sea level (open valve), but unstable at cruise (choked, nearly closed valve). Caveat—Integrator initial condition needs attention.

The droop problem of proportional control is eliminated with integral control, $G_E = K_E/s$, for which the loop gain function is,

$$L(s) = \frac{K_E \cdot K_c \cdot e^{-.09 \cdot s}}{s \cdot \left(1 + \frac{s}{10}\right) \cdot (1 + T_c s)}$$

The gain and phase plots of its frequency response are shown in Figure 10-5, where the velocity constant, $K_v = K_E \cdot K_c$, corresponding to an adequate phase

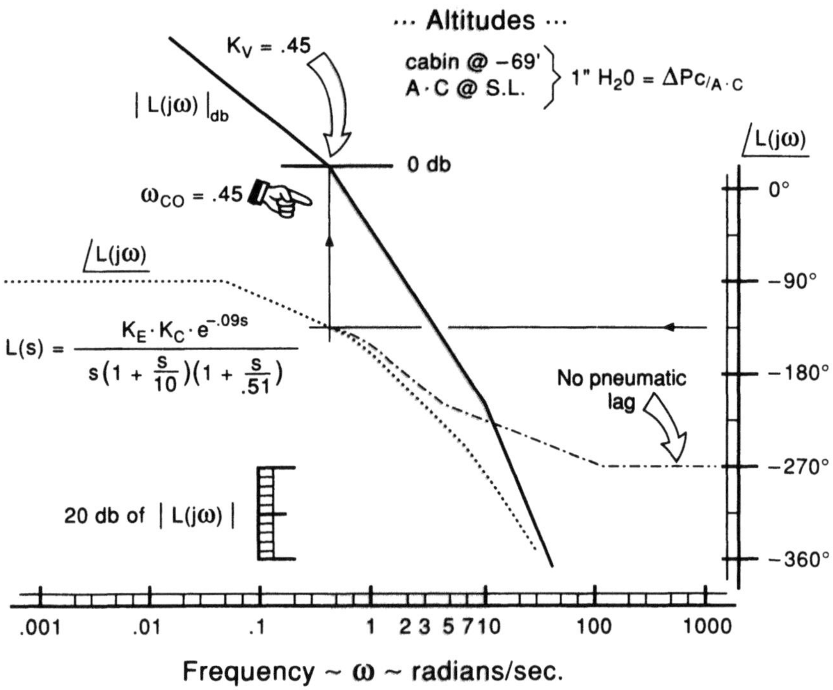

Figure 10-5. Gain and phase . . . integral control.

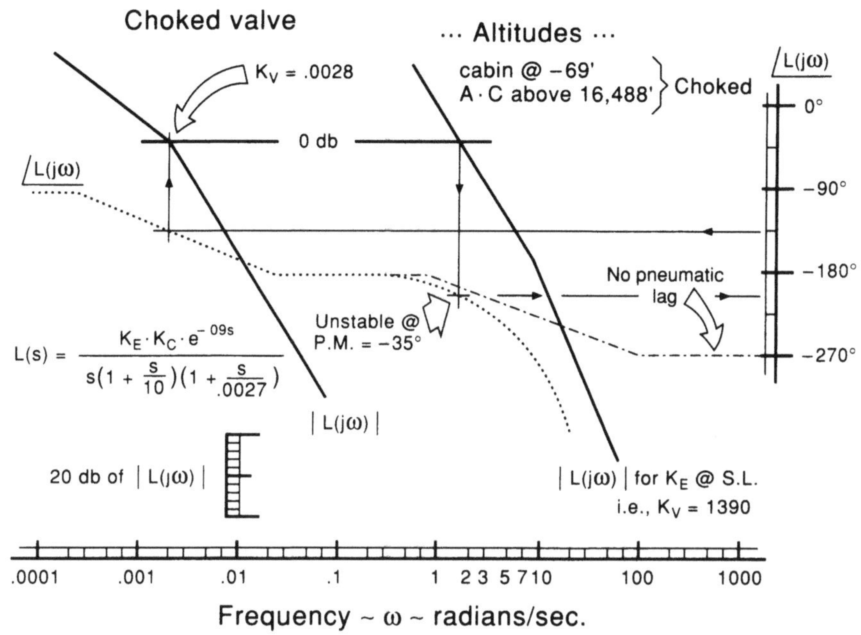

Figure 10-6. Integral control . . . cruise.

margin of 45° is shown to be .45 $\sec^{-1}$ for the sea level, wide-open valve condition. This translates to a controller gain of 1,353.

Having specified the integral controller for the sea level, wide open valve condition, we next check the system performance at cruise, where the valve is choked and operates nearly closed. This information is shown in Figure 10-6 for a sea level cabin and an airplane above 16,500 feet. The system is unacceptable because the cabin gain has increased (from open- to choked-valve) by 68 db, while its bandwidth has been cut 190-fold. Together with the previously established controller gain of 1,353, the system is unstable, having a negative phase margin, −35°, indicated in Figure 10-6. The controller gain must be reduced at this flight condition to .00327 (corresponding to K_v = .0028 $\sec^{-1}$) to provide sufficient stability; also shown in Figure 10-6. And, the system would be overly sluggish due to its crossover frequency of .0028 radians per second.

Before proceeding to the next step, it is worth noting that while integral control is not acceptable, some form of integral reset is ultimately needed to satisfy the droop problem. The attending caveat is the need for preconditioning logic, which drives the integrator to a safe initial condition. This is just as true of a microprocessor-based digital controller as it is of an operational amplifier-based analog controller—for without it, the integrator will start from wherever it last stopped, which may cause an awkward transient in cabin pressure.

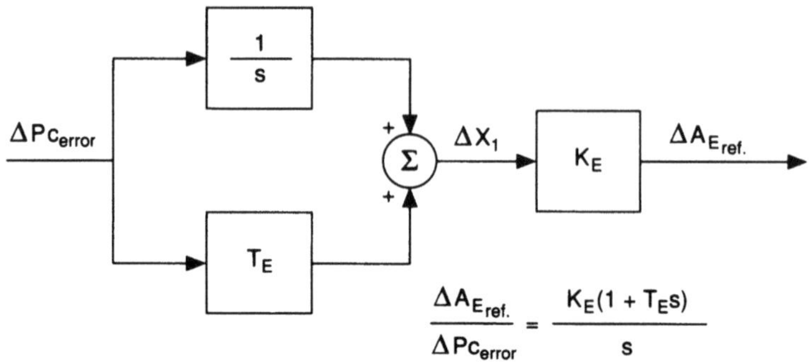

Figure 10-7. Proportional plus—integral controller.

10.5.3 Proportional-Plus-Integral Control

The Compromise that fails.

The additive combination of proportional and integral values of the error provide some phase lead, as shown in Figure 10-7. In order to help the phase margin of the otherwise unstable cruise condition, the phase lead due to the zero at $-1/T_E$ should be placed at a frequency where it will do some good. Referring to Figure 10-6, it is too late at $\omega > 1$, and, in fact, ought to be in the region of .05–.1, which is about as close to "crossover" as possible without being submerged by the pneumatic lag. A typical choice is $T_E = 14$ seconds, placing the zero on the choked cabin pole.

Having made the selection, it remains to examine how well the system performs at various operating points. Figure 10-8 presents the gain and phase plots for the two conditions already discussed (i.e., sea level with a wide open valve and cruise with a choked valve and sea level cabin). If a phase margin for each is set at 45°, the controller gain is $-1{,}743$ and -117, respectively, which are so vastly different that we are forced to abandon the P+I controller.

10.5.4 Add a Nonlinear Gain

Makes proportional-plus-integral work.

Evidently, the P+I controller would work if provision could be made to vary its dc gain, K_E, in some nonlinear way. Exactly how K_E should vary is not immediately obvious, but the data from Figure 10-8 suggest a high gain near open and a low gain near closed. More data should presumably provide a clearer picture of how the gain should vary; accordingly, other flight conditions are examined in Figure 10-9.

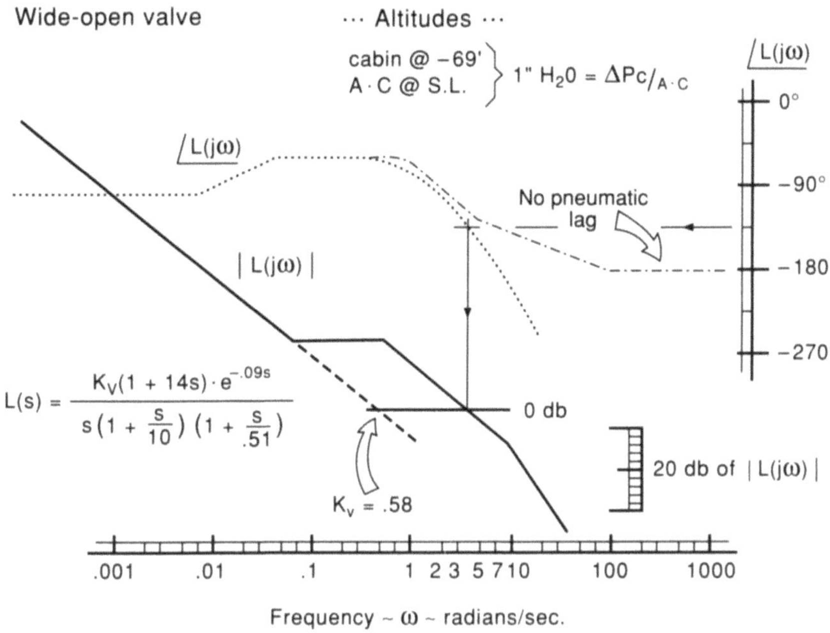

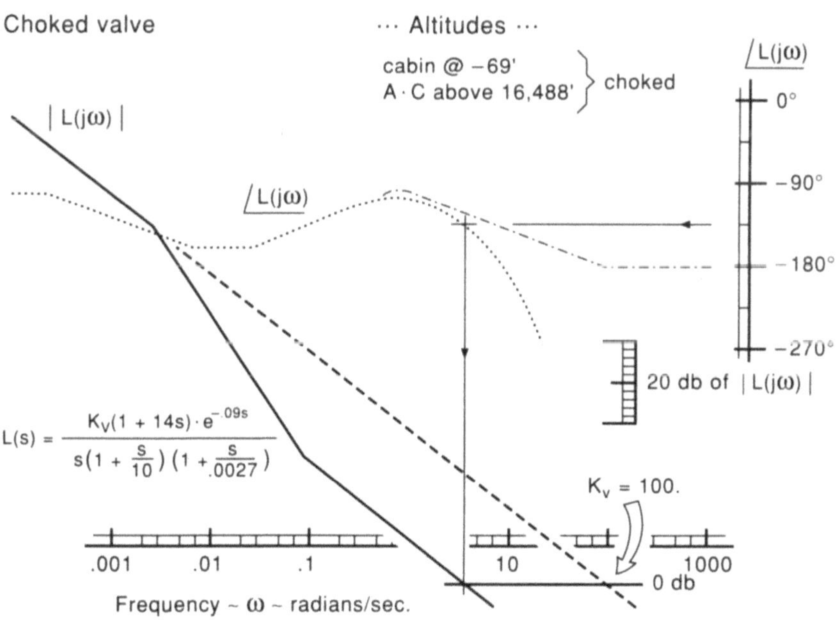

Figure 10-8. Gain and phase . . . proportional plus integral control.

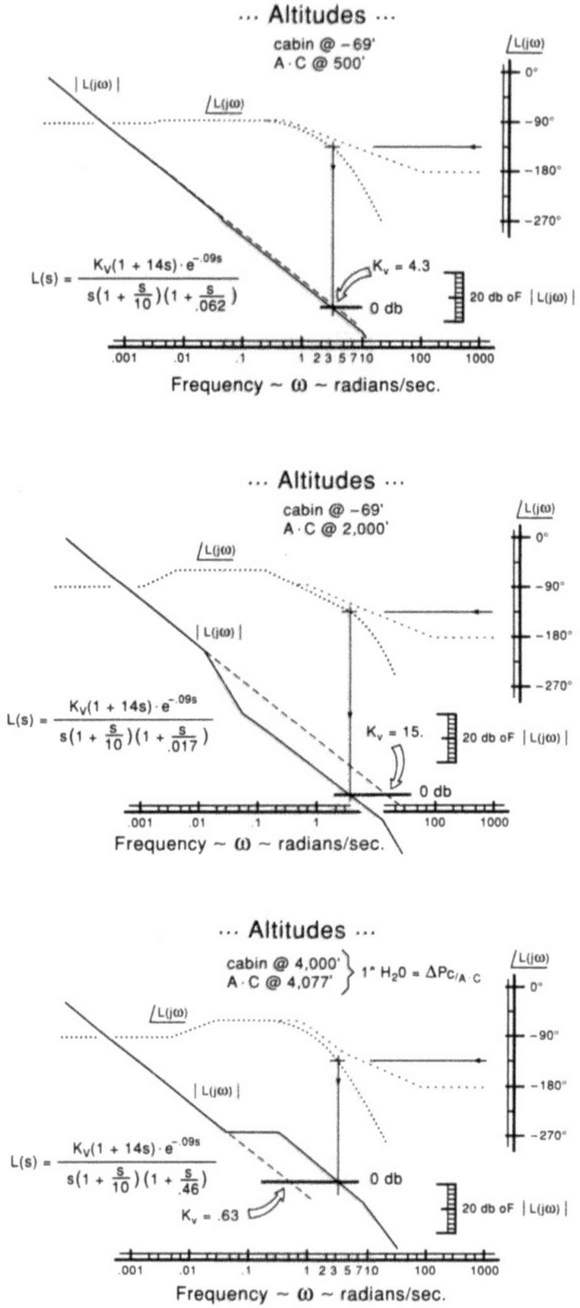

Figure 10-9. Additional flight conditions . . . proportional plus integral control.

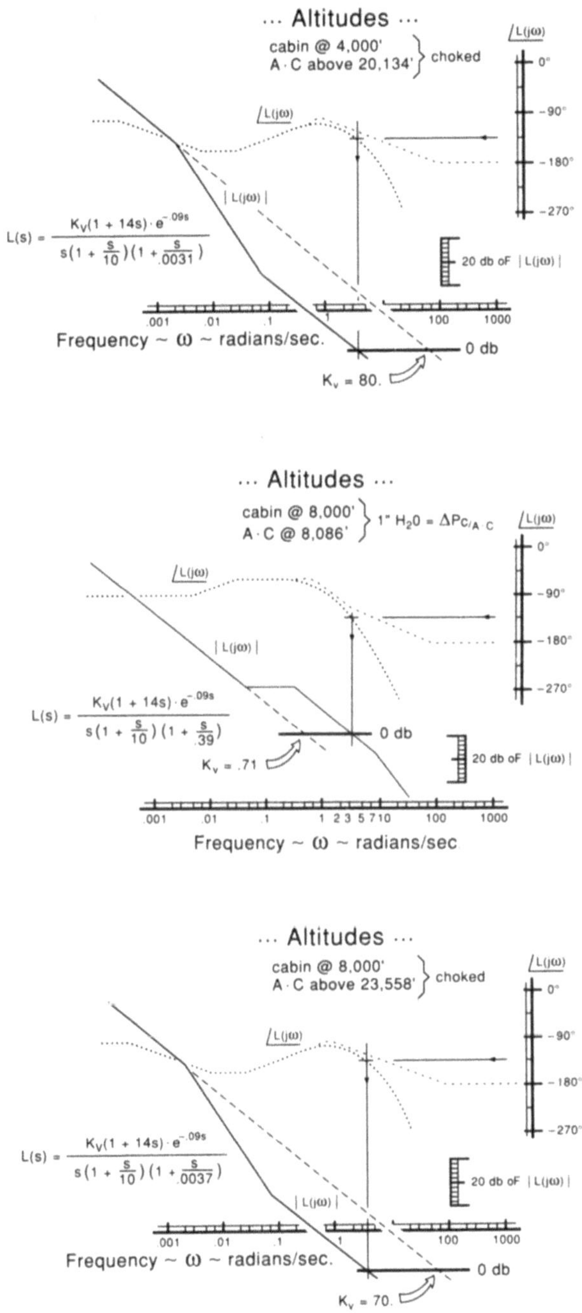

Figure 10-9. *(Continued)*

TABLE 10-1. Controller Gain and Valve Area at Various Flight Conditions[a,b]

Altitudes		Low Frequency Gains			Valve Area	Cross-over Freq.	Correlation
Cabin (ft.)	A.C. (ft.)	Cabin (psi/in^2) $-K_c$	Velocity (@ PM = 45°) K_v	Controller (in^2/psi·sec) $-K_E = K_v/K_c$	A_E^0 (in^2)	ω_{co} (Rad/sec)	$-K_E/A_E^0$ (psi·sec)$^{-1}$
−69	S.L.	.000333	.58	1743.	216.6	4.1	8.05
−69	500	.0110	4.5	391.	76.4	4.3	5.12
−69	2000	.0744	15.	203.	41.7	3.6	4.87
−69	>16488	.857	100.	116.7	17.2	3.8	6.79
4000	4077	.000483	.63	1304.	209.5	4.0	6.22
4000	>20134	.638	80.	125.	19.9	3.6	6.29
8000	8086	.000521	.71	1362.	194.3	4.6	7.01
8000	>23558	.472	70.	148.	23.2	3.5	6.40

[a]Proportional plus integral controller

$$\frac{K_E}{s}(1 + 14 \cdot s)$$

[b]Stability condition: 45° phase margin (PM = 45°).

The data collected from Figures 10-8 and 10-9 are tabulated in Table 10-1, where two interesting correlations become evident. First, the ratio of controller gain to valve area is seen to be nearly constant, which refines our speculation that they are related. Second, we note the rather pleasing fact that crossover occurs at nearly the same frequency for all flight conditions, which means that the system's responsiveness will always be the same. That is, there will be no conditions where it is "sluggish" and others where it will be "quick." This is, in some measure, a cosmetic reward, because we really don't care if its response changes so long as it is acceptable (i.e., has 45° phase margin).

Figure 10-10 shows how the controller gain, K_E, should vary with valve area. A straight line variation is reasonable—typically, $K_E = -6 \cdot A_E$. Since the actuator rapidly converts $A_{E_{REF}}$ into A_E, we can relate K_E to $A_{E_{REF}}$ directly according to the correlation,

$$K_E = -6A_{E_{REF}} \qquad \text{S.T. } K_E \equiv \frac{dx_2}{dx_1}, \qquad A_{E_{REF}} \equiv x_2$$

or,

$$\frac{dx_2}{dx_1} = -6x_2 \Rightarrow A_{E_{REF}} = k \cdot e^{-6x_1}$$

The final step is to establish a reasonable value for the arbitrary constant, k. Recognizing that x_1 is equally likely to be positive or negative, let it be x_1 =

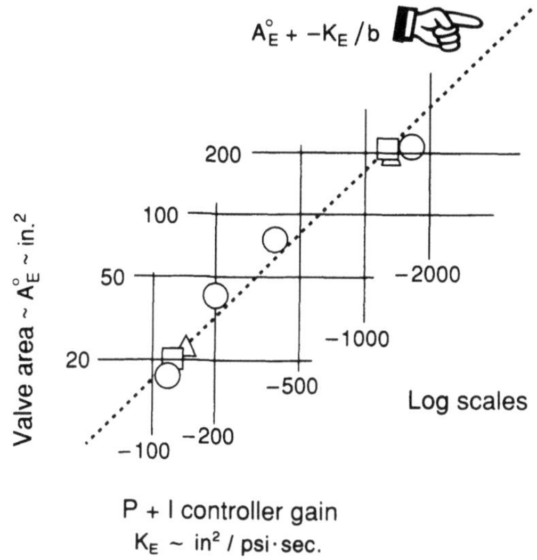

Figure 10-10. Correlation of controller gain with valve area.

0. For this condition, we select some mid-value of $A_{E_{REF}}$, lying between closed ($A_E = 0$) and fully open, 217 in.^{2}, say, 100. Then $k = 100$ and the gain, K_E, becomes the local slope of . . .

$$A_{E_{REF}} = 100 \cdot e^{-6x_1}$$

which is plotted on Figure 10-11.

10.6 A NONLINEAR DIGITAL SIMULATION

Thus far, the cabin pressure system design has made use of the linearized model to evaluate the system's dynamics at representative operating points. This is standard practice because it is the simplest approach, and indeed the linear approximation is something most engineers feel comfortable with. What happens when the perturbations are no longer "small" is a matter of conjecture, and can only be investigated with a nonlinear model. For the most part, this takes us out of the realm of analysis, and we are forced to model the system with either an analog or digital approximation. The former saw considerable use in the 1950s and 1960s, before the digital capabilities were developed. While analog simulation remains today a wonderfully instructive tool, it is no longer fashionable, and only a small band of purists practice it. With the advent of digital simulation packages, notably CSMP, the emphasis has shifted away from

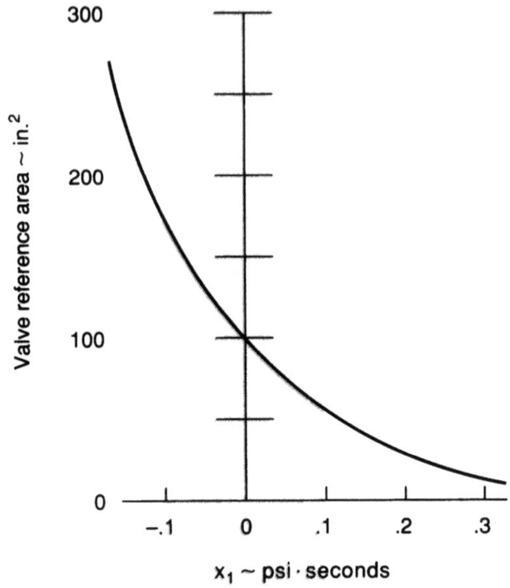

Figure 10-11. Nonlinear gain function.

analog methods, where, today many simple problems can be digitally simulated with spreadsheet packages.

The IBM 1130-version of CSMP (Continuous System Modeling Program) (ref. 2) appeared in the early 1960s, and is available today for many mainframes. Its long-running appeal is probably due to its simplicity, its symbolic resemblance to analog methods, and the fact that, like Coke, its formula has never been made public. In order to verify the system stability, a CSMP simulation has been developed, where some of the more important nonlinearities have been considered. This is shown in Figure 10-12, where the nonlinear nature of the cabin dynamics is placed in evidence—in particular, the calculation of the flow coefficient, k. In addition, other nonlinearities are shown, such as the variable controller gain, the slew rate limit on the outflow valve actuator, and (although not strictly a nonlinearity) the pneumatic lag in sensing cabin pressure.

Other nonlinearities which could be added are the tooth-to-tooth deadband in the actuator gear and the typical nonlinear variation between actuator position and valve area.

The CSMP simulation allows the system evaluation for transients which need not be limited to small perturbations. In particular, Figure 10-13 presents the results of an entire takeoff and climb, which indicates how good the cabin

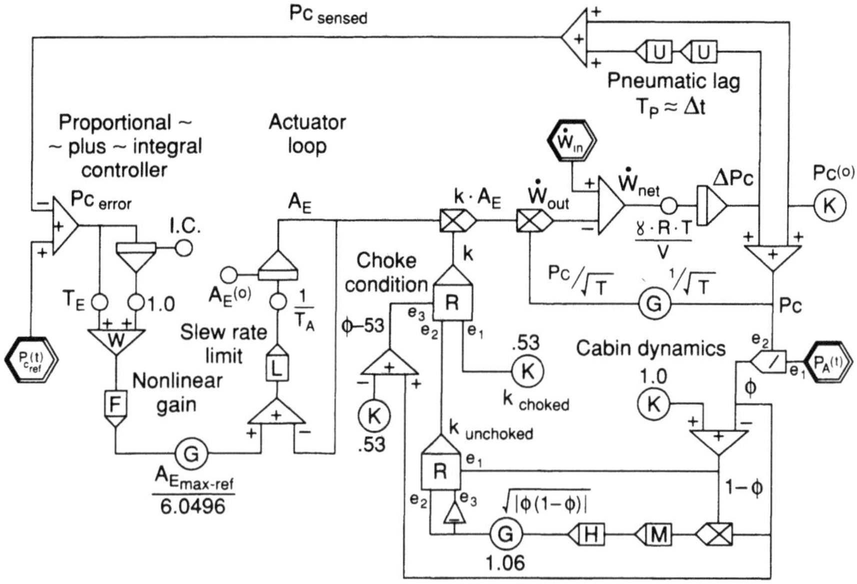

Figure 10-12. Nonlinear CSMP simulation of a cabin pressure control.

pressure regulation is. The condition evaluated asks that the cabin be regulated to sea level while the aircraft rotates at liftoff, causing a slight increase in local ambient pressure at the valve due to ram effects. This propagates into the cabin, forcing the valve to compensate by opening from its initial 217 square inches. The aircraft lifts off and climbs at 3,000 feet per minute to 25,000 feet, at which point the valve is choked. The aircraft simultaneously rotates and lifts off at $t = 0$, at which point there is a ram buildup in local pressure at the valve. At such a low differential pressure, this ram-rise would propagate into the cabin were it not for the control action, which further opens the valve from its normally maximum open position. In this case, the external ram-rise is 100 feet and the control attenuates this to 40 feet in the cabin.

The value of such a simulation is immediately evident as transients due to any and all inputs can be studied. Part of the beauty, however, is our indifference to the fidelity of the various blocks (each of which represents a finite difference equation). Of course, we can, and should, vary the integration interval, but beyond this there is no way to investigate how faithfully the difference equations are representing our dynamics. The easy way out is to ignore the question of simulation fidelity. However, for those who are uncomfortable in the hands of someone else's difference equations, the answer is to develop their own simulation as discussed in Chapters 8 and 9.

A/C Alt (feet)	Time	X 1 (psi · sec) Output 11	AE (in^2) Output 13	PC_E (psi) Output 9	Pc (psi) Output 51
S L	0 000	−0 2574	217.0000	−0.0361	14.7361
	0.100	−0.2610	219.6771	−0 0361	14 7440
	0.200	−0.3757	268.4736	−0 0440	14 7500
	0.300	−0.4638	269.1429	−0 500	14.7526
	0.400	−0.5064	269.4776	−0.0526	14.7544
	0.500	−0.5371	269.6448	−0.0544	14.7556
	0.600	−0.5596	269.7285	−0.0556	14.7564
	0.700	−0.5757	269.7703	−0.0564	14.7568
	0.800	−0.5867	269.7913	−0.0568	14.7569
	0.900	−0.5935	269.8018	−0.0569	14.7567
50	1.000	−0.5965	269.8070	−0.0567	14.7563
	1.100	−0.5964	269.8096	−0.0563	14 7556
	1.200	−0.5934	269.8109	−0.0556	14 7549
	1.300	−0.5880	269.8115	−0.0549	14.7539
	1.400	−0.5802	269.8118	−0.0539	14.7528
	1.500	−0.5704	269.8120	−0.0528	14.7516
	1.600	−0.5588	269.8121	−0.0516	14 7503
	1.700	−0.5454	269 8121	−0.0503	14 7489
	1.800	−0.5305	269.8121	−0.0489	14 7474
	1.900	−0.5142	269.8121	−0.0474	14 7458
100	2.000	−0.4965	269.8121	−0 0458	14.7441
	2.100	−0.4776	269.8121	−0 0441	14.7424
	2 200	−0.4575	269.8121	−0 0424	14 7406
	2.300	−0.4363	269.8121	−0.0406	14.7387
	2.400	−0.4141	269.8121	−0 0387	14.7368
	2.500	−0.3910	269.8121	−0 0368	14.7348
	2.600	−0.3670	269.8121	−0.0348	14 7328
	2.700	−0.3421	269.8121	−0.0328	14 7307
	2.800	−0.3165	269.8121	−0.0307	14.7286
	2.900	−0.2901	269.8121	−0.0286	14.7265
150	3.000	−0.2629	247.4606	−0.0265	14 7245
	3.100	−0.2374	219.3365	−0.0245	14.7232
	3.200	−0.2215	194.4444	−0.0232	14 7226
	3.300	−0.2155	178.0949	−0.0226	14.7225
	3.400	−0.2166	171.2680	−0.0225	14.7226
	3.500	−0.2208	171.4259	−0.0226	14 7227
	3.600	−0.2244	174.7210	−0.0227	14 7227
	3.700	−0.2254	177.6100	−0.0227	14 7223
	3.800	−0.2233	177.8816	−0.0223	14.7219
	3.900	−0.2190	175.1614	−0.0219	14.7214
200	4.000	−0.2140	170.3099	−0.0214	14 7209
	4.100	−0.2095	164.8135	−0.0209	14 7205
	4.200	−0.2064	159.9576	−0.0205	14.7202
	4.300	−0 2046	156.4093	−0.0202	14.7200
	4.400	−0.2037	154.2324	−0.0200	14.7198
	4.500	−0.2030	152.8774	−0.0198	14 7196
	4.600	−0 2020	151.7601	−0.0196	14.7194
	4.700	−0.2004	150.3263	−0.0194	14.7191
	4.800	−0.1984	148.6250	−0.0191	14.7188
	4.900	−0.1963	147.1224	−0 0188	14.7185
250	5.000	−0.1940	145.6033	−0.0185	14.7182

Valve Wide Open, $X_1 < -.3$ (0.200 to 2.800)

Valve Slew Rate, 488 in^2/sec (at 0.100)

Bump in Cabin .0208 psi = −39 7 ft (at 0.800)

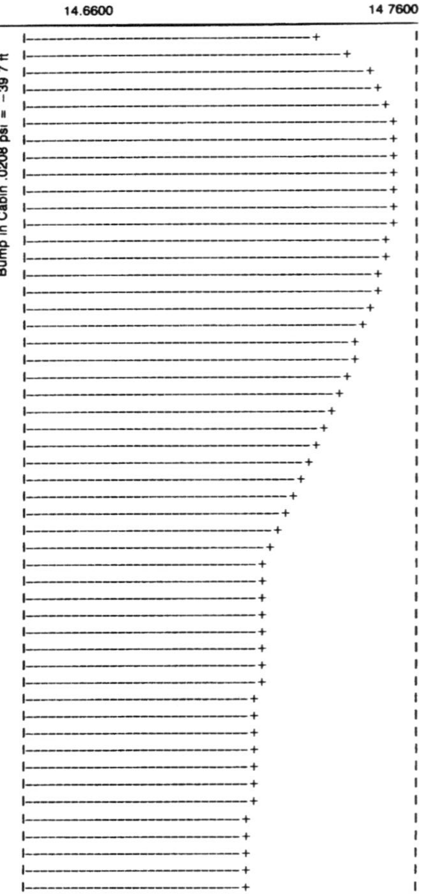

500	10 000	−0 1182		93 9729	−0 0086	14 7084	I------------------------+
	20 000	−0 0510		63 4522	−0.0008	14 7008	I----------------------+
	30.000	−0 0176		51 1387	0 0012	14 6988	I---------------------+
	40 000	0 0017		44 3142	0.0016	14 6984	I---------------------+
	50 000	0.0243		39 7453	0 0019	14 6981	I---------------------+
	60 000	0 0404		36 4912	0 0017	14 6983	I---------------------+
	70.000	0 0526		34.0278	0 0015	14 6985	I---------------------+
	80.000	0 0622		32.1023	0 0012	14 6988	I---------------------+
	90 000	0.0699		30 5370	0 0010	14 6990	I---------------------+
5,000	100 000	0.0763		29 2557	0 0008	14 6992	I---------------------+
	110.000	0.0817		28 1698	0.0007	14 6993	I---------------------+
	120.000	0 0862		27 2534	0 0006	14 6994	I---------------------+
	130.000	0.0901		26 4671	0 0005	14 6995	I----------------------+
	140 000	0 0935		25 7909	0 0004	14 6996	I----------------------+
	150.000	0 0964		25 2059	0 0003	14 6997	I----------------------+
	160.000	0.0989		24 6954	0 0003	14 6997	I----------------------+
	170 000	0 1020		24 2577	0 0003	14 6997	I----------------------+
	180 000	0.1055		23 8700	0 0003	14 6997	I----------------------+
	190 000	0 1086		23 5338	0 0003	14 6997	I----------------------+
10,000	200 000	0 1112		23 2451	0 0003	14 6997	I----------------------+
	210.000	0.1134		22 9976	0 0003	14 6997	I----------------------+
	220 000	0 1153		22 7895	0 0002	14 6998	I----------------------+
	230.000	0.1169		22 6160	0 0002	14 6998	I----------------------+
	240 000	0 1181		22 4776	0 0002	14 6998	I----------------------+
	250 000	0 1191		22 3670	0 0001	14 6999	I----------------------+
	260 000	0 1198		22 2886	0 0001	14 6999	I----------------------+
	270 000	0 1203	Choked Valve	22 2395	0 0001	14 6999	I----------------------+
	280 000	0 1204		22 2171	0 0000	14 7000	I----------------------+
	290.000	0 1205		22 2180	0 0000	14 7000	I----------------------+
15,000	300 000	0 1205		22 2163	0 0000	14 7000	I----------------------+
	310 000	0 1205		22 2172	0 0000	14 7000	I----------------------+
	320.000	0 1204		22 2192	0 0000	14 7000	I----------------------+
	330 000	0 1205		22 2177	0 0000	14 7000	I----------------------+
	340 000	0 1204		22 2193	0 0000	14 7000	I----------------------+
	350.000	0 1204		22 2189	0 0000	14 7000	I----------------------+
	360 000	0 1204		22 2186	0 0000	14 7000	I----------------------+
	370 000	0 1204		22 2185	0 0000	14 7000	I----------------------+
	380 000	0 1204		22 2185	0 0000	14 7000	I----------------------+
	390 000	0 1205		22 2184	0 0000	14 7000	I----------------------+
20,000	400 000	0 1205		22 2162	0 0000	14 7000	I----------------------+
	410 000	0 1205		22 2184	0 0000	14 7000	I----------------------+
	420 000	0 1205		22 2184	0 0000	14 7000	I----------------------+
	430 000	0 1205		22 2184	0 0000	14 7000	I----------------------+
	440 000	0 1205		22 2184	0 0000	14 7000	I----------------------+
	450 000	0 1205		22 2184	0 0000	14 7000	I----------------------+
	460 000	0 1205		22 2184	0 0000	14 7000	I----------------------+
	470 000	0 1205		22 2184	0 0000	14 7000	I----------------------+
	480.000	0 1205		22 2184	0 0000	14 7000	I----------------------+
	490 000	0 1205		22 2184	0 0000	14 7000	I----------------------+
25,000	500,000	0 1205		22 2184	0 0000	14 7000	I----------------------+

Figure 10-13. Takeoff-rotation and initial climb . . . CSMP simulation.

10.7 LIMITING CABIN PRESSURE RATE

Anyone who has ridden a many-storied elevator or driven a long uphill stretch is familiar with a pressure sensation in the ears. It is "cleared" by swallowing, particularly in riding downward. The effect is caused by the shape of the eustachian tube, which acts as a check valve, trapping air on the inner side of the eardrum. Differential pressure across the eardrum causes the discomfort, and it is generally recognized that increasing cabin pressure is more uncomfortable than decreasing. Studies of the comfort level (ref. 3) indicate that cabin pressure rates should be kept below equivalent sea-level tape-line rates of 300 feet-per-minute ascent and 200 descent. Aircraft climb and descent rates are easily 10 times; thus, provision for a rate limit must be made in order to guarantee passenger comfort. In our case so far, control has been an isobaric regulation aimed at maintaining a constant cabin pressure. There is, however, nothing to prevent the reference pressure from being changed abruptly at a rate exceeding the comfort level. A limit on this rate could be imposed if the rate signal were available, but generally it is not because the flight engineer hasn't the time to regulate cabin pressure by means of a rate-setting. With other things to do, he simply wants to select a destination cabin altitude and let the system do the rest, within constrained rate limits.

One solution is indicated in Figure 10-14, where a feedback loop with an integrator is incorporated between P_{CSET} and P_{CREF}. The loop error becomes the rate signal which can be limited before being integrated.

10.8 INITIALIZING INTEGRATORS

Integrators have been found necessary at two places in the cabin pressure control: one in the isobaric loop to eliminate droop, and the other in the reference loop to provide a rate signal which can be limited for passenger comfort. They both need to be properly initialized in order to protect against the possibility of an errant transient when the system is activated. Every integrator, whether it be a mathematical function, a recursion formula, a motor, or based on an operational amplifier, must start integrating from an initial condition. It is like a

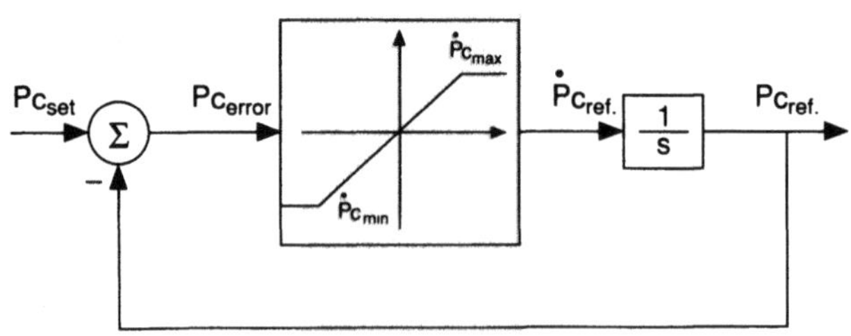

Figure 10-14. A method of rate-limiting cabin pressure.

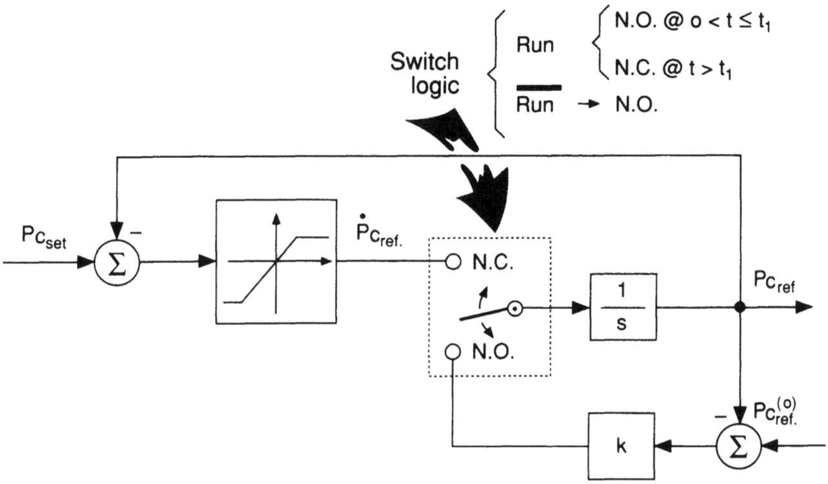

Figure 10-15. Preconditioning logic.

counter which increments at a rate determined by the size of its input signal. But a counter must be told where to start if it is to be accurate, so a voting machine must be set to zero as otherwise it gives a bad tally.

Another example is steering a car. The driver notices its position, compares it against her desired position (say, in a traffic lane), and the error suggests motion of the front wheels via the steering wheel. This results in a sideward velocity which integrates in time to a change in position (left or right). Consider then, a parked car, the angle of the front wheels being the initial condition. If they are inadvertently cocked when the clutch is engaged, an undesirable transient can happen until the wheels castor or command is taken over them. By habit, learned through experience, drivers pre-position the front wheels in order to avert a calamity.

A method of avoiding the startup transient is shown in the logic of Figure 10-15. Prior to activating the system, the switch is set to "normally open." From activation at $t = 0$ until a selected time, t_1, the integrator output is driven to its desired initial condition, $P_{CREF}(0)$. After t_1, the system is switched to "normally closed" and the pressure control is activated. This method has been used to initialize integrators of analog computers when run in a repetitive mode (ref. 4), making use of the integrator to initialize its own output.

10.9 PROBLEMS

1. The approach taken in this chapter is to place the nonlinearity in the controller gain. Another possibility is to place it in the feedback path around

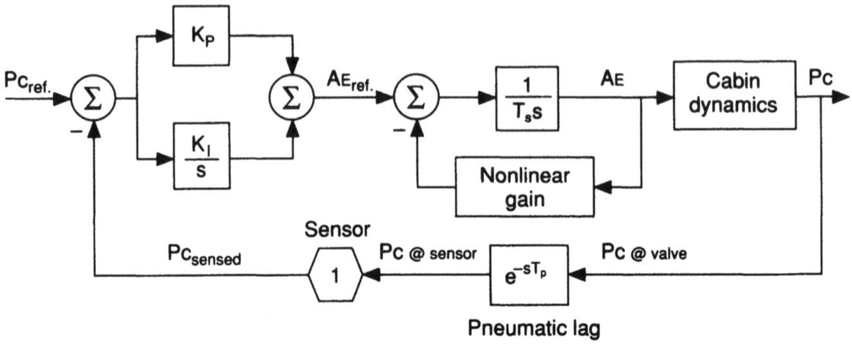

Figure 10-16. Proposed nonlinear actuator feedback.

the actuator. Redesign the system using this approach, as shown below in Figure 10-16. What does this do to the actuator dynamics? Is it a reasonable design approach?

2. In order to show that the choice of lead time constant in the P+I controller is not critical, repeat the calculations of Sections 10.4.3 and 10.4.4, using $T_E = 70$ seconds. Find the relation between valve area and desired controller gain, K_E, for 45° phase margin.
3. Show that the results of Problem 2 and those of Table 10-1 lead to the conjecture, $T_E \cdot K_E/A_E^0 =$ constant.
4. Design an override for the cabin pressure control to prevent negative differential pressure between cabin and ambient.
5. In the rate-limiting network suggested by Figure 10-14, a limit is placed on the integrator's output, Figure 10-17, in an attempt to prevent startup transients. Discuss the potential difficulty which it may cause.
6. Design an override for the cabin pressure control to prevent exceeding some maximum differential pressure between cabin and ambient.
7. Increasing pressure is not only more uncomfortable than decreasing pressure, but can damage the eardrum should the pressure abruptly change by 2 or 3 psi. A runaway closed valve during taxi could cause such a change. Is there anything which can be added to the system to guard against such an event?

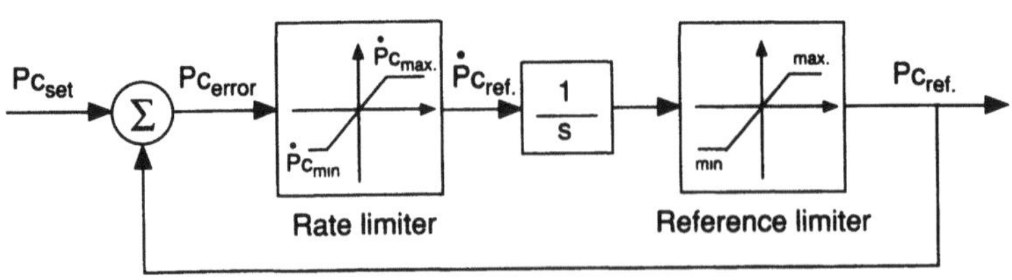

10.10 REFERENCES AND RELATED LITERATURE

10.10.1 References

1. Kolk, W. Richard, Cabin pressure control—An exercise in nonlinear design, First Annual Advanced Control Conference, Purdue University (1974).
2. Anonymous, 1130 Continuous system modelling program (1130-CX-13X), H20-0281-1, White Plains, New York (1968).
3. Mallonee, Richard C., Human reaction to short duration pressure changes in aircraft cabins, The Boeing Company D6-17453, Seattle, Washington (1966).
4. Anonymous, *MINIAC Reference Handbook*, Electronic Associates, West Long Branch, New Jersey (1971).

10.9.2 Related Literature

1. Anonymous, Part 25 airworthiness standards: Transport category airplanes. Federal Aviation Agency, Ca 1969.
2. McDonnel-Douglas Corp., Eardrum differential pressure during transient conditions. 1-Page chart, prepared & distributed at Douglas-Santa Monica (California) Circa 1969

Index

MIX
Papier aus verantwortungsvollen Quellen
Paper from responsible sources
FSC® C105338

If you have any concerns about our products,
you can contact us on
ProductSafety@springernature.com

In case Publisher is established outside the EU,
the EU authorized representative is:
Springer Nature Customer Service Center GmbH
Europaplatz 3, 69115 Heidelberg, Germany

Printed by Libri Plureos GmbH
in Hamburg, Germany